IDENTITY IN EDUCATION

The Future of Minority Studies

A timely series that represents the most innovative work being done in the broad field defined as "minority studies." Drawing on the intellectual and political vision of the Future of Minority Studies (FMS) Research Project, this book series will publish studies of the lives, experiences, and cultures of "minority" groups—broadly defined to include all those whose access to social and cultural institutions is limited primarily because of their social identities.

For more information about the Future of Minority Studies (FMS) International Research Project, visit www.fmsproject.cornell.edu

Series Editors:

Linda Martín Alcoff, Hunter College, CUNY
Michael Hames-García, University of Oregon
Satya P. Mohanty, Cornell University
Paula M. L. Moya, Stanford University
Tobin Siebers, University of Michigan

Identity Politics Reconsidered
edited by Linda Martín Alcoff, Michael Hames-García, Satya P. Mohanty, and Paula M. L. Moya

Ambiguity and Sexuality: A Theory of Sexual Identity
by William S. Wilkerson

Identity in Education
edited by Susan Sánchez-Casal and Amie A. Macdonald

IDENTITY IN EDUCATION

Edited by
Susan Sánchez-Casal and
Amie A. Macdonald

palgrave
macmillan

IDENTITY IN EDUCATION
Copyright © Susan Sánchez-Casal and Amie A. Macdonald, 2009.

All rights reserved.

First published in 2009 by
PALGRAVE MACMILLAN®
in the United States—a division of St. Martin's Press LLC,
175 Fifth Avenue, New York, NY 10010.

Where this book is distributed in the UK, Europe and the rest of the world, this is by Palgrave Macmillan, a division of Macmillan Publishers Limited, registered in England, company number 785998, of Houndmills, Basingstoke, Hampshire RG21 6XS.

Palgrave Macmillan is the global academic imprint of the above companies and has companies and representatives throughout the world.

Palgrave® and Macmillan® are registered trademarks in the United States, the United Kingdom, Europe and other countries.

ISBN: 978–0–230–60916–7 (hardcover)
ISBN: 978–0–230–60917–4 (paperback)

Library of Congress Cataloging-in-Publication Data is available from the Library of Congress.

A catalogue record of the book is available from the British Library.

Design by Newgen Imaging Systems (P) Ltd., Chennai, India.

First edition: June 2009

10 9 8 7 6 5 4 3 2 1

Printed in the United States of America.

We dedicate this book to Satya P. Mohanty, cofounder of the Future of Minority Studies (FMS) Research Project, and critical reader and supporter of our work for many years. We like the word "beautiful" to describe your enthusiasm for FMS, your commitment to its ideals and objectives, and your nurturing of the people who participate in it. We thank you.

Contents

List of Tables   ix

Acknowledgments   xi

Notes on Contributors   xiii

Introduction   1
Amie A. Macdonald and Susan Sánchez-Casal

Part 1: Critical Access and Progressive Education

1. Identity, Realist Pedagogy, and Racial Democracy in Higher Education   9
   Susan Sánchez-Casal and Amie A. Macdonald

2. What's Identity Got to Do with It? Mobilizing Identities in the Multicultural Classroom   45
   Paula M. L. Moya

3. Fostering Cross-Racial Mentoring: White Faculty and African American Students at Harvard College   65
   Richard J. Reddick

Part 2: Curriculum and Identity

4. Which America Is Ours? Martí's "Truth" and the Foundations of "American Literature"   103
   Michael Hames-García

5. The Mis-education of Mixed Race   131
   Michele Elam

6. Ethnic Studies Requirements and the Predominantly White Classroom   151
   Kay Yandell

7. Historicizing Difference in *The English Patient*: Teaching Kip Alongside His Sources   171
   Paulo Lemos Horta

Part 3: Realist Pedagogical Strategies

8. Teaching Disclosure: Overcoming the Invisibility of Whiteness in the American Indian Studies Classroom   191
   Sean Kicummah Teuton

9. Religious Identities and *Communities of Meaning* in the Realist Classroom   211
   William S. Wilkerson

10. Postethnic America? A Multicultural Training Camp for Americanists and Future EFL Teachers   225
    Barbara Buchenau, Carola Hecke, Paula M. L. Moya, and J. Nicole Shelton

11. The Uses of Error: Toward a Realist Methodology of Student Evaluation   251
    John J. Su

Index   273

Tables

3.1	Mentoring Functions as described by Kram (1988)	69
3.2	Participant Sample, Areas of Study, and Rank	72
3.3	Student Demographics for Harvard College, 2006 and U.S. Undergraduate Students, 2004	73
3.4	Faculty Demographics for Harvard, 2006 and U.S. Faculty at Four-Year Institutions, 2004	73

Acknowledgments

We would like to thank our colleagues and friends in the Future of Minority Studies (FMS) Research Project for constantly inspiring and challenging us as we develop our scholarship on racial democracy and social justice in higher education. For their support and collaborative spirit, we especially want to thank all of the FMS Coordinating Team members.

We want to give special thanks to Linda Martín Alcoff, Tobin Siebers, Satya Mohanty, Michael Hames-García and John Su for staying close to this work as we completed the book, and for providing us with invaluable comments on the chapters and on the concept for this anthology.

We want to acknowledge and express thanks to Dr. Jane Bowers, Provost, John Jay College/CUNY, for her encouragement and support for this project. Thank you to the Professional Staff Congress of the City University of New York (PSC-CUNY) for contributing funds necessary to completion of the manuscript.

We thank Julia Cohen, Kristy Lilas, Samantha Hasey, and our entire editorial team at Palgrave Macmillan, for all of their careful attention to our manuscript. We would also like to thank the production people at Newgen Imaging, especially Maran Elancheran, who made the process of copyediting the manuscript quite lovely.

We thank the Harvard Civil Rights Project, and especially Dr. Andrew Grant-Thomas (now Deputy Director of the Kirwan Institute for the Study of Race and Ethnicity), for giving us the opportunity to present the first version of our work on racial democracy at the *Color Lines Conference: Segregation and Integration in America's Present and Future*, Harvard University, in August 2003.

We thank our families and friends, and especially our husbands, for encouraging us and for keeping the home fires burning as we envisioned and completed this important project.

Our most special thanks go to each of our contributors, who outline so elegantly **their** productive theories about how educators in higher education can work with identity in progressive ways. We thank you for your eloquent and inspiring essays, for the ideas, questions, insights, and problems that you raise in each of your chapters. We thank you for your patience with us during the years that we edited and brought this volume to publication. Thanks to you, this volume promises to make a significant contribution to ongoing conversations and movements aimed at providing *critical access* to higher education for minority students across the nation and the world.

And most of all, we want to acknowledge and express gratitude to all of our students, past and present, especially those who have suffered the effects of educational inequity as they have pursued their college degrees. Your intelligence, courage, resourcefulness, and unsinkable spirit have inspired and motivated us to theorize and work for democratizing changes in what remains a very imperfect educational system. We thank you. And we'll keep pushing forward, in your name.

Notes on Contributors

Barbara Bucheneau is Assistant Professor of American Studies at Göttingen University, Germany. She was a Researcher in the Göttingen Collaborative Research Center, The Internationality of National Literatures, from 1997–2000, and a Visiting Scholar at Stanford University in 2004–2005. She received her Ph.D. for work on the difficult positioning of early U.S. American fiction in a predominantly Atlantic literary culture, and is currently completing a book on racial and ethnic typecasting in literature about the Haudenosaunee/Iroquois. Her primary interest is in the politics and practices of literature.

Michele Elam is Martin Luther King Jr. Centennial Professor and Director of African & African American Studies at Stanford University. She is the author of *Race, Work & Desire in American Literature, 1860s–1930s* (Cambridge University Press, 2003) and *Mixtries: Mixed Race in the New Millennium* (Stanford University Press, forthcoming), and is currently writing a book on post-race and post-apartheid performance in the U.S. and South Africa with her husband, Professor Harry J. Elam, Jr.

Michael Hames-García currently chairs the Department of Ethnic Studies at the University of Oregon, where he teaches courses on prisons, African American and U.S. Latina and Latino literatures, and theories of gender, race, and sexuality. At the time he wrote "Which America Is Ours?" he taught in the English Department at Binghamton University, State University of New York. He is the author of *Fugitive Thought: Prison Movements, Race, and the Meaning of Justice* (University of Minnesota Press, 2004) and co-editor of *Reclaiming Identity: Realist Theory and the Predicament of Postmodernism* (University of California Press, 2000) and *Identity Politics Reconsidered* (Palgrave Macmillan, 2006). He is completing another book, tentatively titled *Identity Complex: Gender, Race, and Sexuality from Oz to Abu Ghraib* and a co-edited collection, tentatively titled *Gay Latino Studies: A Critical Reader*.

Carola Hecke lectures in the field of Teaching Methodology at Georg-August-Universität Göttingen, Germany. She studied English and Spanish Languages and Cultures for Teaching Profession and Art History at Georg-August-Universität Göttingen, Germany. During this time she received a two-semester scholarship for the University of California Berkeley where she took Art History classes. She graduated from Georg-August-Universität Göttingen in 2005. Currently she is working on her Ph.D. thesis on the use of images to teach literature in foreign language classes.

Paulo Lemos Horta is Assistant Professor of World Literature at Simon Fraser University. He is the first faculty member of the program, which he designed. He is the author of literary translations and articles on translation, cosmopolitanism, and multiculturalism.

Amie A. Macdonald is Associate Professor of Philosophy at John Jay College of Criminal Justice/City University of New York. She is the co-author and co-editor, with Susan Sánchez-Casal, of *Twenty-First Century Feminist Classrooms: Pedagogies of Identity and Difference* (Palgrave Macmillan, 2002). Her teaching and research are focused on racial democracy, racial justice, and equity in education, with a focus on urban public higher education and implementing policy changes that increase access to graduate and professional schools for first-generation college students. She is a member of the Future of Minority Studies (FMS) Coordinating Team.

Paula M. L. Moya is Associate Professor and Vice-Chair of the Department of English at Stanford University. Her publications include several essays on Chicana feminism, feminist theory, multicultural education, and the epistemic significance of minority perspectives. She is the author of *Learning from Experience: Minority Identities, Multicultural Struggles* (University of California Press, 2002) and co-editor of *Reclaiming Identity: Realist Theory and the Predicament of Postmodernism* (University of California Press, 2000), and *Identity Politics Reconsidered* (Palgrave Macmillan, 2006). Professor Moya served as Director of the Undergraduate Program of the Center for Comparative Studies in Race and Ethnicity (CCSRE), and Chair of the Comparative Studies in Race and Ethnicity (CSRE) major at Stanford University from 2002 to 2005, and was a Stanford Fellow from 2003 to 2005. In addition, she is a founding organizer and coordinating team member of the FMS research project.

Richard J. Reddick is Assistant Professor of higher education and coordinator of the M.Ed. program in College and University Student Personnel Administration at The University of Texas at Austin. Dr. Reddick's current research agenda is focused on the effects of mentoring on the college aspirations of graduates of high poverty, high minority population high schools. Dr. Reddick is an affiliated faculty member with the Center for African and African American Studies and teaches in the Plan II Liberal Arts Honors Program, both at the University of Texas at Austin.

Susan Sánchez-Casal is an independent educator and scholar who resides in Madrid, Spain, and was former Associate Professor and Chair of Hispanic studies at Hamilton College in Clinton, NY. She has published essays on Latin American and Latin@literature, and is the co-author and co-editor, with Amie A. Macdonald, of *Twenty-first-Century Feminist Classrooms: Pedagogies of Identity and Difference* (Palgrave Macmillan, 2002). Professor Sánchez-Casal's current research focusses on realist pedagogy and educational policy, specially on increasing critical access and racial democracy for minority students in higher education. She is a member of the FMS coordinating Team.

J. Nicole Shelton is Associate Professor of Psychology at Princeton University. She earned her B. A. in Psychology from the College of William and Mary in 1993 and her Ph. D. in Psychology from the University of Virginia in 1998. She was a postdoctoral fellow at the University of Michigan from 1998 to 2000. Her primary research focuses on how Whites and ethnic minorities navigate issues of prejudice in interracial interactions. She is also interested in the consequences of confronting perpetrators of prejudice.

John J. Su is Associate Professor of Contemporary Anglophone Literature at Marquette University. He is the author of *Ethics and Nostalgia in the Contemporary Novel* (Cambridge University Press, 2005).

Sean Kicummah Teuton is Associate Professor of English and American Indian Studies at the University of Wisconsin-Madison. He is the author of *Red Land, Red Power: Grounding Knowledge in the American Indian Novel* (Duke University Press, 2008).

William Wilkerson is Associate Professor of Philosophy at the University of Alabama in Huntsville. His research and teaching interests are in the fields of Continental Philosophy and gender/sexuality issues. He is the co-editor (with Jeffrey Paris) of *New Critical Theory: Essays on Liberation* (Rowman & Littlefield, 2002) and the author of *Ambiguity and Sexuality: A Theory of Sexual Identity* (Palgrave Macmillan, 2007).

Kay Yandell is Assistant Professor of English at the University of Wisconsin-Madison. Her current book project is tentatively titled *Transcendental Telegraphy: Nineteenth-Century Telecommunication in American Literature*.

Introduction

Amie A. Macdonald and Susan Sánchez-Casal

I

Our work is centrally concerned with racial democracy in higher education. We believe that educational opportunity, and specifically the opportunity to earn a college degree, is one of the most significant factors in the continuing effort to democratize our society and our world. The current state of social inequality is reflected quite accurately by the disparate measures of educational attainment seen in different racial groups; these alarming statistics come into sharp relief when we measure what minority students actually accomplish in the educational sphere compared to what they have attempted to accomplish—by educational attainment we mean how much education people have actually acquired, what type of degrees they have actually earned, and so on. A close examination of educational attainment in the United States is therefore one way of getting a sense of whether different demographic groups are closing the so-called achievement gap between rich and poor, male and female, immigrant and native-born citizen, racially priveleged and racially disadvantaged, and so on. It is also an indicator of where educational policy-makers and academic institutions need to focus their efforts and work together toward the creation of racial democracy in education.

Historically, the disparity between white and non-white access to education has been and continues to be one of the most significant features of the racial social structure in the United States, as well as one of the primary causes of racial inequality in education and in society at large. While earning a baccalaureate degree is by no means a guarantee of future economic or social stability, statistics show that there are consistent and dramatic differences in the earning power of people who have graduated from college and those who have not. For example, the median income in 2005 of black men who graduated from high school but never went to college is $21,241. Compare this to the median income in 2005 of black men who have earned a college degree: $43,496. For black women in the same statistical groups the difference is equally striking: $15,768 for those who graduated high school but never attended college versus $40,784 for those black women who earned a college degree. We can also see the ways that, at least in terms of economic status, a college education can mitigate some of the long-range economic disparities that result from structural racism.[1] The U.S. Census Bureau reports show that the median income of white women college graduates working full time in 2005 was $43,100 and that the median income of black women college graduates working full time in 2005 was $45,273.

Of course educational attainment not only largely determines economic outcomes, but also other critical indices of social equality and inequality—of opportunity—that determine people's life chances at the same time that they measure the extent to which we are living in a democracy. Differences in access to and quality of health care, housing, transportation, employment, police protection, and financial products such as consumer loans and mortgages, create enormous and unjust disparities in the lives of people in the United States. Differences in citizenship status and the racist practices of law enforcement generate further opportunities for some as they produce injustices for others. There is no doubt that the undemocratic distribution of all educational resources, and especially of postsecondary education, contributes significantly to the striking inequalities in basic measures of social stability between white people and people of color.

But people of color not only face the staggering obstacle of lack of access to higher education. Those who do manage to attain access are confronted by the expression of racial bias at all levels of institutional life, in particular in the form of exclusive Eurocentric curricula, lack of equal access to educational resources, and overt and implicit racism in the classroom and on campus. We don't mean to imply that there are no students of color flourishing on college and university campuses across the United States, because we know that this is not true. In fact, what we like to do is call attention to the fact that a condition of these students' success is the ability to negotiate and survive racial bias at all levels of the institution. Therefore we argue that while equal "access" to higher education is crucial, this first level of access must be accompanied by a structural revamping of the institution at all levels, in order to establish what we call *critical access*. Critical access aims to create equality in educational opportunities and conditions for minority students by pressing for not only equal access in admissions, but also for the racially democratic redistribution and rearticulation of policies, resources, curricula and pedagogies. The assumption that grounds the theory of critical access is that increased numbers of minority students in higher education alone will not change the fact that students of color continue to confront structural racism in the academy. In order to work toward racial democracy, colleges and universities must be aggressive in seeking new sources of funding for structural diversity initiatives; more importantly, they must be willing to support critical access for minority students by reallocating existing monies in order to achieve the following objectives:

1. Diversify curricula across disciplines to represent accurately the multiracial foundation of human knowledge.
2. Racially diversify tenured and tenure-line faculty through aggressive recruitment policies, Target of Opportunity lines, and the establishment of incentive programs to reward departments that hire minority faculty, etcetera.
3. Racially diversify the student body by increasing support for national and institution-specific affirmative action admissions policies; reallocate current admissions funding to support the recruitment of and to increase the "yield" of all racial minorities, with special attention to Native Americans, Alaska Natives, Latin@s, African Americans, and immigrant and economically marginalized Asian Americans.
4. Develop university/college-specific financial aid programs that redress the national trend of dramatic declines in state and federal grant programs (e.g., Pell

Grant), and the concurrent trend to substitute these grants with government and private loan programs.
5. Create development and retention programs for both faculty and students of color.
6. Work toward parity in graduation rates between white and minority students; work toward parity in tenure and promotion rates between white and minority faculty.
7. Decolonize teaching methods and classroom dynamics to acknowledge both the epistemic significance and the social vulnerability of the subjugated knowledges that minority students produce in the classroom.
8. Democratize the distribution of student authority in the classroom.
9. Institute non-racist, attentive and responsive academic advising and mentoring.
10. Train professors to have equally high expectations of students of color and white students.
11. Guarantee to minority students equality of choice in field of study.
12. Assure students of color equal access to non-academic educational resources that give students economic and social mobility after graduation, such as, career services, internships, campus employment interviews, travel opportunities, research assistantships, alumni career liaisons, and so on.
13. Implement hospitable, secure, and racially supportive residential options.
14. Provide fair and non-racist protection by campus security personnel.
15. Implement Universal Design in order to guarantee students of all abilities equality of access to all educational and social resources, as well as to all physical structures that students access on and off campus (i.e., study abroad programs).
16. Establish universal, affordable, and no-cost student health insurance policies; implement a management system that ensures racial equality in treatment by health-care providers and establishes a grievance process for students who report mistreatment on the basis of race, class, gender, sexual orientation, ability, religion, and nationality, and all pertinent social categories.

In closing this section, we would like to emphasize that we approach our work on broadening democracy in higher education very seriously but also with this awareness: access to higher education is a crucial component of the struggle to create equality among different social groups and individuals, yes, but education is by no means the only necessary area for social action. We therefore see our work on higher education as a part of a much larger progressive movement to theorize the nature, causes and effects of social inequality, and to conceptualize and implement concrete policies aimed at creating structural equality and opportunity for disenfranchised minority[2] demographic groups.

II

The editors and authors of *Identity in Education* attempt to redress the racially undemocratic status of higher education in the United States by advocating for the objectives of critical access listed above. Operating within a realist framework, the contributors (all of whom are minority scholars) consider ways to productively engage identity in the classroom and in the educational system. As realists, all authors in the volume hold the theoretical position that identities are both real and constructed, and that identities are always epistemically salient. Thus the essays gathered here seek to (1) discuss the political and epistemic salience of social identity in society and in academia (2) explain how established assumptions, practices and policies in higher education contribute to social inequality and intellectual deficiency, and (3) offer analyses of how educators can mobilize identity productively and progressively and contribute to efforts to democratize conditions in the classroom and on campus. While the pedagogical topics undertaken by the contributors to *Identity in Education* are wide-ranging, each chapter-length essay endeavors to analyze and critique what we have defined as "realist pedagogy:" a teaching practice that activates identities in the classroom and works toward the wide-ranging objectives of social justice in higher education and in our world. Thus the book argues—from diverse disciplinary and educational contexts—that mobilizing identities in the classroom is a necessary part of progressive (antiracist, feminist, anticolonial) educators' efforts to transform knowledge-making and to work toward a more just and democratic educational system and society.

Part 1: Critical Access and Progressive Education

In our opening essay "Identity, Realist Pedagogy, and Racial Democracy in Higher Education," we focus on the structural goal of racial democracy in higher education, and explore various ways that activating a realist conception of racial identity in institutional strategies and in the classroom will contribute to establishing critical access for minority students to higher education. By identifying the many ways that students of color are disadvantaged—in relation to their white counterparts—in their pursuit of undergraduate education, we begin to articulate a theory of critical access for minority students. Thus we advocate for a realist pedagogy that has concrete policy implications, and can challenge asymmetrical power relations in the classroom, the curriculum, and across campus (in housing, financial aid, health services, and so on).

In her influential essay "What's Identity Got to Do with It? Mobilizing Identities in the Multicultural Classroom," Paula M. L. Moya elaborates the realist account of identities as epistemic resources that educators can mobilize in order to draw out the diverse knowledge that students produce. Contending that educators can dismiss the epistemic power of identity about as easily as we can dismiss the physical power of gravity, Moya makes a convincing case for the democratizing pedagogical role of identity to empower students in the classroom to produce more objective knowledge about reality.

Richard J. Reddick advances a new framework for faculty-student mentoring, which he terms *critical theory of difference*, in his essay, "Fostering Cross-Racial Mentoring: White Faculty and African American Students at Harvard College." He demonstrates that mentoring relationships between white faculty and African American undergraduate students at predominantly white institutions (PWIs) such as Harvard College can create greater access for these students to necessary educational resources that are often denied to them because of their minority status. His conclusions are significant, and offer a contextual solution to the problem of overburdening faculty of color at PWIs with mentoring responsibilities for minority students. Since African American faculty are often a hyper-minority at PWIs, Reddick's case for cross-racial mentoring is a crucial step in providing critical access to students of color.

PART 2: CURRICULUM AND IDENTITY

Section 2 opens with Michael Hames-García's groundbreaking essay, "Which America Is Ours? Marti's 'Truth' and the Foundations of 'American Literature.'" Arguing that the North American literary canon continues to frame the inclusion of minority texts within a cultural relativism that robs it of epistemic power, Hames-García advocates for the inclusion of ethnic/racial narratives that offer students the chance to engage deeply with perspectives that contradict white master narratives. Hames-García concludes by advancing the realist notion that only by including minority texts that present information and perspectives on the conflictive and violent racial history of the United States will students be able to consider and debate varying narratives of Americanness.

In her powerful essay "The Mis-education of Mixed Race" Michele Elam situates the emerging "mixed race" literary genre, and mixed-race social identities, in a historical framework. Elam shows how the category itself is often mistakenly taken to be progressive and asserts that the classroom is a primary source of current (mis)understanding of "mixed race." She argues convincingly for a realist pedagogy that is driven by a more politically complex account of "mixed race identity," one based on the often exploitative circumstances surrounding race mixing in American history.

Kay Yandell, in "Ethnic Studies Requirements and the Predominantly White Classroom," writes from the challenging, and often unsettling, experience of teaching the Ethnic Studies Requirement at predominantly white institutions. On the basis of her teaching, Yandell makes six pedagogical recommendations, each of which is based on a realist approach to identity and knowledge. Her critique of what she calls "Identity Blindness"—the view that it is socially unacceptable to talk about inequality—will be of particular interest to anyone teaching or learning about cultural diversity.

In "Historicizing Difference in *The English Patient*: Teaching Kip Alongside His Sources," Paulo Lemos Horta explains how his students gained greater epistemic objectivity about the representation of minority experience in the *The English Patient* when he asked them to research Ondaatje's acknowledged historical sources for the Sikh character, Kip. Approaching identity from a realist (as opposed to an essentialist) standpoint, Lemos Horta empowers his students to see themselves as members of *communities of meaning* (Sánchez-Casal and Macdonald, 2002) and, allows them to engage the epistemic value of their own identities and the minority identities presented in the text.

PART 3: REALIST PEDAGOGICAL STRATEGIES

Proceeding from a realist perspective on how identity shapes our values, Sean Kicummah Teuton, in "Teaching Disclosure: Overcoming the Invisibility of Whiteness in the American Indian Studies Classroom," theorizes the role of undisclosed identity politics in the American Indian Studies classroom, analyzing how the invisible power of whiteness can silence students of color. He argues that it is possible to mediate white identity in a way that can allow students to see the differences between white skin and white dominance, thereby inviting them to exchange white dominance for politically progressive white identities.

In his timely essay "Religious Identities and *Communities of Meaning* in the Realist Classroom," William Wilkerson describes how to engage problematic religious identities constructively by employing *communities of meaning* (Sánchez-Casal and Macdonald, 2002) to generate collective thinking about ethics across religious lines. Wilkerson goes on to explain how his work with these smaller epistemic communities in the classroom allowed him to elicit from students a more complex account of ethical reasoning, and a strong set of questions about the Christian bible as a historical (rather than an infallible) text.

Theorizing from the nexus of literary/cultural studies and foreign language teaching (specifically, teaching the English language and American culture to German students), the authors of "Postethnic America? A Multicultural Training Camp for Americanists and Future EFL Teachers" advocate for a multicultural emphasis on racial/ethnic difference rather than a raceless postethnic ideal. From diverse disciplinary perspectives Barbara Buchenau, Carola Hecke, Paula M. L. Moya, and J. Nicole Shelton argue convincingly that minority identities can be mobilized for both progressive and regressive social change, and that accurate teaching about American literature and society must take the epistemic function of cross-cultural identities into account.

John J. Su, in "The Uses of Error: Toward a Realist Methodology of Student Evaluation," provides an innovative method for faculty to democratize the classroom and contribute to critical access for marginalized students by implementing a realist method for evaluating student performance. Encouraging faculty to move past the more narrow process of "grading," Su argues from a realist standpoint that we would benefit our students by approaching "error" as a useful and necessary part of developing more objective accounts of reality.

Notes

1. See the second edition of Melvin Oliver and Thomas Shapiro's *Black Wealth/White Wealth: A New Perspective on Racial Inequality* (New York: Routledge, 2006) for an excellent discussion and analysis of enduring economic inequality—based on an accounting of wealth, not income—between black and white families.
2. We are following the articulation of "minority identity" as it is understood by participants in the Future of Minority Studies (FMS) Research Project. That is, we take the term "minority" to describe not simply numerical minorities in our society, but demographic groups whose access to social power, to cultural capital (in addition, in some cases to actual capital), and to institutions (such as education, health, justice, government, etc.) is constrained by their identity. So on our view, *minority identity* is a term that encompasses people who are socially or politically marginalized on the basis of race, gender, class, sexual orientation, ability/disability, religion, and/or national origin, and whose marginalization is typically historical.

1
CRITICAL ACCESS AND PROGRESSIVE EDUCATION

1

IDENTITY, REALIST PEDAGOGY, AND RACIAL DEMOCRACY IN HIGHER EDUCATION

Susan Sánchez-Casal and Amie A. Macdonald

> ...a truly multiperspectival, multicultural education is a necessary component of a just and democratic society
>
> Paula M. L. Moya, *Learning from Experience: Minority Identities, Multicultural Struggles*

INTRODUCTION

The fundamental assumption grounding our work is that college and university campuses today continue to be racially undemocratic. The philosopher Charles Mills refers to race-based inequality as the "racial contract" (Mills, 1997), a global historical social contract that distinguishes between white persons and non-white racialized sub-persons, and guarantees unequal and undemocratic processes of representation and distribution. Following Mills, our main objective in this chapter is to document and analyze how this "racial contract" currently reproduces racially undemocratic conditions in higher education. By racially undemocratic we refer to the ways that an unspoken white "racial project" (Omi and Winant, 1994) informs student and faculty demographics, admissions policies, governing committees, institutional curricula, teaching methods, residential living options, career planning services, and general educational resources. This racial privileging creates a two-class system of citizenship on campus by making it easier for white students to get into college (thereby making it harder for students of color to gain admission), giving white students greater social and educational freedom during college (thereby limiting these—and other—freedoms for students of color), maintaining curricula and pedagogies that privilege white students (thereby marginalizing students of color in the classroom), and creating greater economic and political mobility for white students upon graduation from college (thereby contributing to broader social conditions of racial inequality). Because our work is centrally concerned with racial democracy in higher education, and in education at large, we believe that it is crucial to identify and name the fundamental reality of racism so that we can begin to discern how to reconstruct the academy according to racially democratic principles, strategies and practices. We would like to clarify that the insistence on the reality of institutional racism is meant not as an indictment against white faculty, students, administrators, staff and others, but as an invitation to dialogue on racial justice across racial lines. Thus the intent of our chapter is constructive, and we hope that it will contribute meaningfully to concerted efforts toward building racial equality into colleges and universities in the United States.

REDEFINING DEMOCRATIC ACCESS TO EDUCATION

Racial democracy begins with access. The struggle for equal access to education in the United States begins with black resistance to enforced illiteracy during slavery. Black struggle for literacy and citizenship—both individual and collective—also played a major role in the momentous African American civil rights victory in *Brown v. the Board of Education* (1954). The triumphant move to repeal the "separate but equal" precedent established under *Plessy v. Ferguson* paved the way for the eventual desegregation of schools in the United States.[1] Equal access to education was also named as a central component of racial democracy by the black, Chican@, and Puerto Rican civil rights movements of the 1960s and early 1970s, each calling for increasing minority representation among faculty, diversified curricula (that would include and centralize minority histories, literatures, and art), and democratized financial aid and admissions policies that would further desegregate colleges and universities. Sadly, these initiatives to democratize education have been fiercely opposed from a variety of political standpoints, a conflict that crystallized in the 2003 legal battle over the future of race-sensitive affirmative action in admissions policies at the University of Michigan. Thwarting right-wing and neoconservative attempts to repeal the legality of race-conscious admissions policies established in *Bakke* (1978), the Supreme Court upheld the constitutionality of affirmative action admissions at the University of Michigan (*Grutter v. Bollinger*, 2003). In the *Grutter* opinion, the Supreme Court resituates in the twenty-first century the fundamental rationale of *Brown*, that "education is the very foundation of good citizenship" and therefore that "the diffusion of knowledge and opportunity through public institutions of higher education must be accessible to all individuals regardless of race or ethnicity" (*Brown v. Board of Educ.*, 347 U.S. 483, 493 [1954], cf. "Reaffirming Diversity: A Legal Analysis of the University of Michigan Affirmative Action Cases" 6). Stating that "the path to leadership" must remain open to all races and ethnicities, the court identifies racial integration in higher education as a fundamental government objective *that has yet to be realized*. Since the *Civil Rights Act* of 1964 most colleges and universities have made progress in enrolling and graduating students of color and in hiring and tenuring faculty of color;[2] however, the context of the *Grutter* decision shows that racial movements that call for equal access to higher education continue to be vigorously contested. Further evidence of the contestation of the meaning of race in higher education is the fact that 50 years after *Brown*, the Supreme Court found it necessary to once again reassert the constitutionality of admissions actions aimed at promoting racial equality in access to higher education.

Despite the progress made in improving educational access for racial minority groups, the current inequality in enrollment and educational attainment in higher education continues to be socially devastating.³ The scenario becomes even gloomier when you consider that the dropout rates for those students of color who do make it to college are much higher than for white students. Even at Ivy League institutions⁴—where the graduation rate for black students is significantly better than the 39% national average—the difference in graduation rates between whites and blacks ranges from three to fourteen percent. Nationally, the average difference in graduation rates between white and black students is 20%.⁵ According to the U.S. Census Bureau statistics from the year 2000, every racial group experiences a significant gap between the number of students who begin college and the number who actually attain a bachelor's degree. However, the rate of attrition for every non-white racial group is two to three times what it is for white students.⁶ Latin@ students⁷ lag behind African American students in national college enrollments and have the highest dropout rates of any minority group.⁸ Only 12.7% of the Latin@ population over age 25 is in possession of a bachelor's degree.⁹ Native American students share this dismal reality—while nearly 71% of these students over age 25 have graduated from high school, and 41.7% begin college, only 11.5% have completed a bachelor's degree.¹⁰

The disparity between white and non-white access to education is one of the most significant features of the racial social structure in the United States, and one of the primary causes of racial inequality in society at large. The undemocratic distribution of all educational resources,¹¹ and especially of postsecondary education, contributes significantly to the striking inequalities in basic measures of social stability between whites and people of color—and here we are referring to median net worth¹², median family income, access to health care and insurance, infant mortality rates, rates of infection and treatment for HIV/AIDS, rates of incarceration, labor status (i.e., professional versus service and manual), immigration status, and so on. There can be no doubt that equitable access to education for minority groups would help to create a more egalitarian distribution of social resources and political representation—democratizing changes that would improve the lives and life chances of Native Americans, African Americans, Latin@s, Asian Americans,¹³ and other peoples of color in the United States.

We cannot overemphasize the importance of equal access to education for the realization of racial democracy. Increasing the numbers of minority students in higher education must be a primary objective, and should be pursued aggressively and systematically. Yet this constitutes only the first level of "access." Why do we say this? Our concern is that greater minority enrollment alone will not automatically transform racist structural issues that create serious obstacles for students of color at the same time that they impoverish intellectual inquiry for all students. While it's true that achieving a critical mass of students of color on university campuses will create the right conditions for democratizing changes,¹⁴ it's also true that numbers alone will not transform classroom dynamics and teaching methods, reshape the curriculum, restructure the content and valuing of academic discourses, reevaluate the legitimacy of knowledges and truth claims, build the intellectual esteem of students of color, or redistribute students' authority to speak about what our shared social world is actually like. We are concerned that integrationist policies aimed at establishing a more democratic representation of people of color are limited in scope and vision because they leave untheorized the ways that white supremacy and white separatism continue to operate even in diverse classrooms and institutions.[15] For even in "majority-minority" classrooms, we are still faced with the predominance and presumed superiority of the Eurocentric curriculum,[16] and the marginalization of the histories and intellectual production of peoples of color in the United States and of entire geographic regions and nations of color across the globe. Majority-minority classrooms are not automatically equipped with decolonized pedagogies and antiracist teachers who understand that racial identity operates in the classroom, offering differing weights of legitimacy to students' arguments—on the basis of race—, shaping students' ideas, perspectives, and judgments about what the world is and what it should be. Moreover, students of color are commonly ghettoized in majority-minority schools, and because these institutions have the least economic resources,[17] democratic access to education is severely curtailed; insufficient funding makes it impossible for colleges and universities to provide adequate career services, academic and personal advising, tutoring and remediation, internships, full-time non-adjunct faculty, faculty office hours, computer facilities, appropriate library and research assistance, and so on.[18] Thus the problem with the simple integrationist approach—the inversion of the white-student-to-student-of-color ratio—lies in the fact that it alone will not dismantle the racial hegemony that characterizes U.S. colleges and universities, nor will it transform teaching methods to accommodate the social and intellectual needs of diverse student constituencies.

Consequently, while we strongly advocate for efforts geared toward establishing equal access to higher education across racial lines, we advance the transformative model of *critical access* to redress the historical legacies of racism, segregation, and white privilege in education. Conceptually, our model of *critical access* couples a racially democratic politics of representation with the systematic redistribution and/or rearticulation of policies, resources, curricula and pedagogies. *Critical access* means creating academic worlds and campus communities that are responsive to the pervasiveness of white privilege, and that transform all areas of educational life in ways that acknowledge, support, and develop the intellect and full humanity of students of color. Establishing *critical access* requires that we work toward these structural goals: (1) diversified curricula grounded in the multicultural and multiracial foundation of human knowledge; (2) antiracist pedagogies that acknowledge how racial identity "performs" in the classroom and that take seriously both the epistemic significance and the social vulnerability of the knowledge that minority students produce in the classroom; (3) professors who have equally high academic expectations for students of color and white students; (4) equality of choice in field of study; (5) non-racist, attentive and responsive academic and personal advising; (6) equal protection by security personnel;[19] (7) hospitable, secure, and non-racist residential options; (8) a variety of additional non-academic educational resources that give students economic and social mobility after graduation—that is, career services, internships, campus employment interviews, travel/study abroad opportunities, research assistantships, and so on.

The structural changes proposed in our *critical access* model would insure that minority students who gain access to higher education will not be tokenly accommodated within a racial order that continues to privilege the educational and social needs of white students over those of students of color. Integrationist movements that lack this understanding often support policy decisions that negatively affect the social and intellectual lives of minority students on campus. For example, institutions regularly assume an "integrationist" standpoint to justify the elimination or modification of supportive infrastructure that includes race-based housing initiatives,[20] "special interest" disciplines (minority, women's, gay/lesbian, disability studies), student organizations, and identity-based innovation in curricula and pedagogy. These policy decisions are misguided, since they eliminate important parts of a democratizing infrastructure that supports the intellectual and social flourishing of minority students. We don't mean to suggest that all minority students will elect to study in Minority Studies departments, live in race-based housing, or join race-focused student solidarity groups; many, in fact, will not. But we are suggesting that minority students will have a greater chance of academic success and social stability in an educational system that acknowledges how racial hierarchies operate on college and university campuses, and that provides intellectual, cultural, social and residential options for students who elect to use these resources to bolster them in racially adverse conditions.[21] No matter how strongly integrationists believe that the best way to promote a color-blind campus is to eliminate spaces where minority students tend to "cluster" and thereby promote interaction and relationships among white and non-white students, in practice the dismantling of race-based infrastructure does little or nothing to change the conditions of racial segregation that exist on college and university campuses today, conditions that are not imposed by minority students. On the ground, a predominantly white college or university campus with no race-based social options usually translates into minimal white student contact with isolated minority students who are constantly obliged to interact in a disproportionately white environment.

While "integration" models assume that both white students and students of color will benefit from "colorblind" policies, they actually privilege the social and intellectual needs of white students: one of the strongest integrationist arguments against race-based housing options, for example, is that white students need to learn about students of color, and that they can only do so by having regular contact with them. In this scenario, underrepresented minority students are construed as "cultural ambassadors" who provide white students with valuable knowledge about racial and cultural "others."[22] The advantage for white students is clear. But how are the social and intellectual needs of students of color theorized in the colorblind policies of the academy, and how do these students stand to benefit?[23] Moreover, the insertion of isolated students of color into a sea of whiteness does not automatically invite white students to question their investment in and practice of racial superiority, nor does it make them aware that they occupy—willingly or not—a racially dominant social position. And ultimately, these unequal interactions routinely take place in markedly adverse racial conditions that reproduce the social vulnerability of students of color and the corresponding dominance of white students.

We suggest that aiming for racial democracy means keeping the enduring consequences of historical racism and colonization front and center as we consider ways to reform institutional systems and curricular strategies. Thus the goal of a racially integrated environment must be developed within a comprehensive and transformative framework of structural change. First and foremost, prevailing institutional curricula must become racially inclusive (and thereby epistemically accurate), something that would allow students of color *critical access* to the histories and the cultural and intellectual production of peoples of color;[24] doing so would also give all students an accurate understanding of the interracial and multicultural foundations of human knowledge. We agree with Paula Moya who argues that educators need to take responsibility for teaching culturally non-dominant students the codes and rules of whiteness (which Lisa Delpit calls the "culture of power") and simultaneously work to change unequal structures of power in curricula that heroicize white intellectual culture while denigrating cultures and epistemologies that are non-white.[25] Thus, because we are aiming for *critical access*, representational reform cannot be unhinged from systematic efforts to transform deeper institutional power relations that exist in established curricula, and thereby, to dismantle the "racial contract" at all institutional levels.

THE EPISTEMIC EFFECTS OF WHITENESS IN THE CLASSROOM

We believe that institutional policies aimed at correcting the de facto exclusion of people of color from the academy—policies that determine student admissions and development, curriculum, pedagogy, residential life,[26] faculty hiring and research, and so on—are crucial but insufficient responses to institutionalized racism; we go further to add that these policies fall short of racially democratizing college and university campuses and of providing *critical access* for students of color. In this section, we will examine how the epistemic and social effects of whiteness materialize in the classroom, where students are stratified by the same unequal and undemocratic social relations that structure society in general. As antiracist feminist teachers, we are particularly concerned about how whiteness in the classroom competes against the interests of minority students, and impedes the imperatives of objective, democratic, multicultural knowledge-making.

The great majority of classrooms on college campuses today are governed by an unarticulated white male standpoint that privileges Eurocentric curricula and unjustly reinscribes the intellectual superiority of white histories and epistemologies.[27] Students of all racial identities, but especially students of color, are cognitively harmed by this racial hegemony, since it reconfigures historically racist notions of black, Latin@, Native American, and Asian American inferiority and marginality in a contemporary context.[28] Furthermore, the dominance of white epistemologies results in impoverished and inaccurate accounts of human agency and intellectual production in the United States and in the world, something that falsely imbues white students with a sense of natural superiority just as it robs students of color of a sense of cultural and historical value and legitimacy. The failure to undertake this kind of curricular reform at the undergraduate level bolsters tacit racist presumptions of the superiority of white and European intellect and ways of reading the world, and thus negatively impacts the learning community for students of all racial identities. For example, introductory philosophy courses are typically focused exclusively on the so-called canonical texts of Plato, Descartes, Hume, John Stuart Mill, and so on. At best, introductory textbooks and syllabi may include the standard "boxed" selections on feminist ethics or on Martin Luther King's position on civil disobedience. In no way, however, has there been a broad scale restructuring of introductory or upper level undergraduate philosophy curricula that legitimizes and contextualizes the philosophical production and histories of non-white, non-European peoples.[29] In fact, nearly all of the leading philosophy doctoral programs in the nation remain committed to Eurocentric curricula that institutionalize white male hegemony in the field of philosophy; moreover, the Eurocentric masculinist dominance in the field is manifest in the dearth of "respected" publishing venues for race-conscious, feminist, non-Western, multicultural, and alternative philosophical projects and perspectives.

English departments and American Studies Programs are similarly dominated by the white male canon, and by a parallel tendency to marginalize intellectually those minority narratives that do make it into the curriculum. For example, by devoting very little page-space to non-white authors, texts, and critical perspectives, American literature anthologies reinforce the idea that literariness is a prime feature of whiteness, and an anomalous feature of non-whiteness, thus implicitly inviting students to see black, Latin@, and Native American writers as "exceptions" to the general rule of non-literariness (and non-literacy) of peoples of color in the United States. At the same time, this curricular distortion insures that hegemonic, non-critical constructions of "Americanness" (read: white Americanness) will not be substantively challenged by the critical perspectives of "America's" racialized others.[30] This racial stilting of the American literature and American Studies curriculum, as Michael Hames-García argues, reproduces nationalist myths about the democratic, egalitarian origins of the white-European United States, rather than allowing students to examine critically our history as a racially balkanized nation that has always privileged whiteness. In his analysis of American literature anthologies Hames-García concludes that minority writers such as African American Ida B. Wells-Barnett and Cuban José Martí are excluded from canonical discussions of the abolitionist movement because "they are not 'pro-American...'" [and] "[t]hey present an unflattering view of the United States, suggesting that its true nature is violent and racist, and that there is no simple way to correct this by appealing to its own noble legacies (Hames-García, 2003, 39; reprinted in chapter 4 of this volume). Thus the white Eurocentric curricula that inform the intellectual content of college and university classrooms not only fail to represent non-dominant perspectives on social reality (that is, from non-dominant racial, class, gender, sexual identities), they also—without explicitly making the case—generate official histories of white exceptionality. This tacit assumption obscures and tokenizes the intellectual contributions of people of color, denying minority students access to what John Wills calls a "usable past" (Wills, 1996, 365–389).

We want to acknowledge that many individual professors in "traditional" classrooms and disciplines make sincere attempts to decenter the predominance of whiteness in their syllabi, and in classroom dynamics. Unfortunately, many of these attempts to diversify and democratize curricula and pedagogy are limited to what James Banks describes as the "additive approach" to multicultural inquiry (cf. Moya, 2002, 146). While the additive approach "incorporates key concepts and themes related to recognizable minority racial groups into an existing curriculum" (146), it fails to transform the goals, structure and perspectives of the curriculum at large. By presenting minority themes as "being unique to the groups in question and marginal to the history of the dominant population" (146),[31] the additive approach to curricular reform tokenizes minority narratives because it places them outside of history, and thus fails to racially democratize the curriculum or challenge the white-centered knowledges it (re)produces. In addition, many of the well-intentioned attempts to diversify curricula are largely idiosyncratic since they are generated by individuals who tend to operate without institutional guidance and support, often in the face of great opposition at various institutional levels; moreover, if these faculty members leave the institution where these progressive changes are enacted, the changes are often immediately or eventually dismantled.[32]

ACTIVATING IDENTITY IN THE REALIST CLASSROOM

The classroom is a primary locus of white privilege and racist power relations in academia. Today's college and university classrooms reveal the persistence of racism:[33] in the selection of course readings, in the intellectual exclusions of traditional academic disciplines, in the political standpoints of professors, in the assumed intellectual authority of white students and faculty, and in the interactions among students and between faculty and students.[34] All of these institutional spheres are infused with white ways of knowing that privilege white subjects—whether the "subjects" are disciplines, areas within disciplines, people, ideas, values, social norms, communities, and so on. Consequently we argue that our teaching and curricular strategies must directly engage the intellectual, epistemic, and ideological effects of racial identity.

Our theory of realist pedagogy[35] acknowledges and centralizes the *epistemic function of social identity*—and of racial identity in particular—by foregrounding the identifiable but mediated relationship that exists between *what we know* (epistemic access) and *who we are* (our social identities). As Satya Mohanty explains:

> Whether we inherit an identity—masculinity, being black—or we actively choose one on the basis of our radical political predilections—radical lesbianism, black nationalism, socialism—our identities are ways of making sense of our experiences. Identities are theoretical constructions that enable us to read the world in specific ways. It is in this sense that they are valuable, and their epistemic status should be taken very seriously. (1997, 216)

In developing realist pedagogy, we therefore build on the realist theory of identity[36] that emphasizes the indeterminate but significant connections among experience, identity, and knowledge. Realist pedagogy doesn't "get" students to bring their identities into the classroom, but rather acknowledges that they already do: students inevitably access their experiences as members of distinct social groups when they evaluate the truth claims of texts, students, teachers, when they analyze literature and art, when they generate theories about racially coded disparities in economic, educational, housing, employment, and health-care opportunities, and so on. The beauty of the realist classroom is that it can work to promote among all students the ability to evaluate how their social identities refer outward to causal features of the world (such as colonialism, racism, classism, sexism, homophobia and homohatred, ableism, etc.). It can help to explain and understand how one's social location determines to a great extent the types of experiences one will have, and that the analysis of these experiences is what yields each of our "truths" about the world we share. Realism thereby gives us the pedagogical advantage of activating the explanatory power of identity—which is crucial to antiracist, feminist teaching and learning—in a way that allows us to engage the experiences of oppression in a non-essentialist framework.[37] This means that we can solicit the experiential knowledge of both white and minority students, but at the same time, we can retain a pedagogical stance that allows us to examine how those knowledges are conditioned and mediated by ideologies that yield either *more* or *less* accurate truths. It also means that we can promote in white students the notion that identities are historical and thus mutable, a concept that allows them to re-evaluate their beliefs from a new perspective that includes racially diverse sources of knowledge. This inclusive re-evaluation allows white students to proliferate the meanings of whiteness to include racially progressive analyses, and inevitably opens the door, for some white students, to antiracist identity and practice. Therefore, our pedagogical focus on identity simultaneously engages both what Satya Mohanty refers to above as "inherited" identities (that is, our race, sex, religion, etc.) and "political" identities (various political commitments that may actually modify our inherited identities over time),[38] a democratizing practice that attends to the epistemic and ethical needs of all students in the classroom.

Significantly, because realist pedagogy provides a way for us to acknowledge the mediated link between social identity and epistemology, realist teachers can legitimize the explanatory power of identity without making the untenable claim that all identity-based knowledge is automatically accurate and reliable or inaccurate and unreliable, a stance that can avoid polarizing the classroom along racial lines (that is, minority students are not falsely identified as the knowers of absolute truths, not even about race and race relations, and white students are not completely disqualified as the knowers of only false "knowledges of domination"). Thus realist pedagogy "allows for an acknowledgement of how the social categories of race, class, gender, and sexuality function in individual lives, without reducing individuals to those social determinants"(Moya, 2002, 38). This is crucial in a classroom that privileges oppositional narratives and contradicts exclusive white narratives, because we can simultaneously legitimize the epistemic advantage of black and Native American identities, for example, without foreclosing the possibility that white students may also produce accurate analyses of racism because they have developed political identities as antiracists, and therefore as intellectual and political allies of African Americans and Native Americans. Thus, realist pedagogy is democratizing for all students because while it establishes the contingency of knowledge on the theories we hold about the world (and thus on our social identities), it rejects the essentialist notion that there is a fixed relationship between who one is and what one can know. In fact, realist pedagogy relies upon teachers and students to engage in a constant process of acknowledging, assessing, and reinterpreting the experienced-based theories that we advocate and defend.

REALIST PEDAGOGY AND *CRITICAL ACCESS* FOR STUDENTS OF COLOR

By centralizing the epistemic function of identity, realist pedagogy redistributes intellectual authority in the classroom in a variety of ways that together help to create *critical access* for students of color. More specifically, the realist focus on identity allows us to engage one of the prime consequences of racism and white dominance in the classroom—the unequal epistemic positioning between white students and students of color.[39] Typical classroom dynamics include the presumption that white students are speaking *objectively* and with the authority of the "[W]estern rationalistic tradition" (Searle), while students of color are routinely deemed to be speaking *subjectively* and from an inferior cultural standpoint.[40] Framing classroom dynamics with the awareness that *every student in the class is speaking subjectively—that is, from a social identity*—contradicts the presumed "objectivity" of white knowers, and restructures the unequal intellectual ground upon which racially diverse students engage with each other, their professor, and the subject matter under study. Emphasizing the fact that all knowledges are positioned, including those of white Americans, presents a formidable challenge to those white students who may feel inclined to dismiss the critical insights of students of color—as well as the discourse of minority scholars—as emotive, self-interested, irrational, angry, biased, and untrue. Thus, the focus on identity is democratizing, because it supports the at-risk intellectual authority of minority students, and of the subjugated histories and analyses that are central to their intellectual and political empowerment.

The realist classroom further redistributes intellectual authority by privileging the scholarly and artistic production of Native Americans, Latin@s, blacks, and Asian Americans as epistemically imperative, and therefore *more* significant to the construction of objective truths than overrepresented knowledges, precisely because these minority perspectives represent histories, cultural elaborations, and theories of our shared social reality that have been omitted from larger discussions over what is true about the world. Thus the realist classroom assumes that the intellectual legacies of racially oppressed peoples, as well as the histories of institutional racist violence and resistance, must be prioritized in order to know human history more accurately, and to legitimize and valorize the life experiences and intellectual inheritance of students of color in the classroom.

For example, Native American accounts of the genocide of native peoples perpetrated by the U.S. military over a period of four hundred years provide an invaluable contribution to the construction of accurate knowledge about the institutional history of white racial violence in the United States. Furthermore, the inclusion of these narratives also demonstrates how dominant academic discourses suppress histories of state violence against peoples of color. But most significantly, a prioritized focus on Native American intellectual history would reveal the extent to which the foundational philosophical principles of the American nation are actually predicated on Native American theories of government and justice. The *Declaration of Independence* and the *Bill of Rights* appeal to the notion that the authority of government is based always and only upon the consent of the governed—a philosophical principle whose origin is commonly attributed to Anglo-European thought in general, and to Rousseau's social contract theory in particular. However, the notion that government gains its authority from the consent of the governed was already fully operative in the political structure of the Five Nations of the Iroquois, a political advance that the U.S. colonial government admired and attempted to emulate. Indeed, Lt. Gov. Cadwaller Colden of New York wrote in 1727 about the governmental structure of the Five Nations that "Their authority is only the Esteem of the People, and ceases the Moment that Esteem is lost. Here we see the natural Origin of all Power and Authority among a free People" (cf. Loewen, 1995, 110). In fact, as James Loewen argues, "most Indian societies north of Mexico were much more democratic than Spain, France, or even England in the seventeenth and eighteenth centuries" (110). He goes further to assert that "Native American ideas may be partly responsible for our democratic institutions. We have seen how Native ideas of liberty, fraternity and equality found their way to Europe to influence social philosophers...[who] then influenced Americans such as Franklin, Jefferson, and Madison" (Loewen, 1995, 111). This fact was corroborated in the 1775 Continental Congress declaration that stated "the six nations are a wise people, let us harken to their counsel and teach our children to follow it" (111).

Yet, in spite of the incontrovertible evidence of the advanced philosophical and political status of Native societies, American history textbooks, professors, and curricula not only exclude the history of the Iroquois Confederacy's contributions to the burgeoning colonial democracy, but in so doing reify racist and intellectually flawed accounts of white European/American superiority. By prioritizing subjugated Native American discourses in discussions of U.S. political history, the realist classroom helps to write Native American students into the historical past and present by legitimizing their intellectual and experiential narratives, and by reclaiming the suppressed Native American influence on what is ubiquitously constructed as the "white," "Anglo," and "European" origins of American democracy. This democratized curriculum interrupts the dichotomy of presenting Native Americans as either passive victims of white atrocities that have allegedly long since ceased to occur, or as exalted, but ahistorical, and nearly supernatural beings whose relationship to nature and "special" knowledges is irrelevant to contemporary political, ethical, and scientific issues.

The realist classroom thereby helps to dislodge Native American students and other students of color from the dehumanizing position of being seen by their white counterparts (students and professors) alternately as ethnographic informants who have unmitigated access only to "their" exotic truths, or as people outside of history altogether and without any repositories of knowledge.[41] At the same time, the realist insistence on the epistemic relevance of social identity obliges white students to reconsider their own relationship to history and to learn about themselves and about the history of white dominance from students of color. This strategy repositions minority students more democratically in the curriculum and in the classroom, both as learners whose intellectual legacies are privileged in the construction of truths, and as critical thinkers who develop the tools to critique the ideologies, exclusions, and intellectual deficiencies of white-dominated discourses.[42] Thus in this example, the realist classroom creates *critical access* for minority students by opening an analytical space for all students to conceive of American colonial history in a way that acknowledges the multicultural foundations of democracy and the epistemic weight of generally unknown minority discourses and histories.[43]

Furthermore, by taking seriously the epistemic relevance of identity, realist pedagogy works to make students aware of how their identities (e.g., race, gender, and class) provide a specific lens through which to read the world; thus it challenges students to think about the fallibility of truth, and about how they are distinctly implicated in the process of creating knowledge. Significantly, realist pedagogy helps us to avoid problematic versions of epistemic privilege that mistakenly posit an automatic and self-evident relationship between the experience of oppression and an unmediated access to truth about the world. Consequently, the realist classroom assumes that racially oppressed students—in spite of their experiences with racism—may produce inaccurate knowledge about racism and their relationship to it. For example, students of color who hold neo-conservative views may attribute the widespread poverty among Native Americans, African Americans, and Latin@s to a lack of social competence among individuals in these racial groups, rather than to the racist effects of capitalism and the enduring economic consequences of colonization. Alternatively, progressive minority students may evaluate the social status of people of color from within a context that acknowledges and critiques the racist structure of capitalist economic opportunity. These opposing "truths" held by minority students are based, at least partly, on experiential knowledge that emanates from a *shared* social location of racial oppression; clearly then, these different epistemic outcomes are determined not solely by experience and identity, but by the theoretical biases that these students employ to mediate their experience-based knowledges, that is, by the theories they hold about themselves and others, and about how our social world is structured.

A realist approach to teaching further democratizes racial power dynamics by accentuating the moral agency of minority students as they are called upon to adjudicate meaning, value, and truth as full citizens of the academic community. By using the critical perspectives of minority theorists to construct a more authentic view of human history, students of color are prompted to recognize that their perspectives on social reality are crucial to the collective project of knowing. It is important to emphasize here that the realist classroom seeks to offer students of color *critical access*, and therefore *equal rights*—not special rights—by acknowledging the necessity of including and prioritizing their histories and mediated perspectives in the quest to produce more reliable and accurate knowledge. By focusing on the epistemic, political, and ontological power of *all* of the racial identities in the classroom—in a context that gives credence to a diverse range of experience-based knowledges and that promotes constant examination and revision of our "truths"—we empower all students to evaluate and create knowledge from a racially democratic position.

Engaging White Racism in the Realist Classroom

A powerful obstacle to the realization of racial democracy in higher education and to *critical access* for students of color is the persistence of white racism in the curriculum and in classroom interactions. White students obviously differ dramatically from one another in gender, sexuality, class, religion, political standpoint, and so on, and therefore do not display unified epistemologies of whiteness. However, all white students have been, to a greater or lesser degree, miseducated about the realties of racial politics, racist histories, and racist epistemologies. Wherever white students fall on a spectrum that can range from vehement and unapologetic racism, to a variable degree of race-consciousness about the consequences of white peoples' unearned advantages within an economic and political system that subordinates peoples of color, most white students are deeply disturbed by analyses of racism;[44] many white students are outwardly hostile and aggressive to students, teachers, and/or texts that articulate the incidence, causes, and outcomes of racism.

We would like to suggest that realist pedagogy offers concrete and reliable strategies for engaging the educational manifestations of white racism in a productive and progressive manner. Because realism highlights the epistemic and non-essentialist aspects of racial identity, antiracist realist pedagogy is not preoccupied with the project of transmitting to white students the progressive "truth" about racism. Instead, realist pedagogy aims to theorize certain identities (in this case, white racial identities) for the purpose of, first, understanding the connection between identity and knowledge; and second, for the purpose of reevaluating what people have taken to be "true" and thereby reconstructing more democratic, more objective, and more accurate accounts of (again, in this case), race, racism, and racial justice. Realist pedagogy thus helps to create an epistemological laboratory that invites students to revise their theories about the world by focusing specifically on the mediated connection between identity and knowledge-making. This approach makes it possible for students to be more open to radical and progressive curricula, because it works to expose the experiential and theoretical biases that yield received belief; thus realist pedagogy stands a better chance of loosening students' grip on pernicious ideological knowledges. Since white students are already predisposed to reject new knowledges that contradict the mainstream curricula, one of the greatest pedagogical challenges of teaching from an antiracist perspective centers on how to get these students to "try-on" subjugated knowledges before rejecting them out-of-hand.

For instance, our theoretical focus on the epistemic component of racial identity creates opportunities for white students to chart out the ways in which whiteness is embedded in a variety of seemingly race-neutral social concepts such as choice, opportunity, and achievement. In turn, white students are positioned to theorize how white racial identity is causally connected to certain sets of their own experiences (in many cases, unearned advantages in education, housing, employment, justice, health care, banking, immigration, public safety, etc.) and to the ways that these experiences determine their assertions about the world (i.e., that racial inequality is the result of the extraordinary individual initiative of whites and the lack of individual determination among peoples of color). The following example will illustrate the process: in a discussion about the racist politics of welfare reform, a white middle-class student unabashedly critiques black women on public assistance as "welfare queens" who "live off of white middle class America," and who "make money by having a lot of kids and not working." This student may eventually revise her racist theory of social reality in the face of new information and analyses—offered in course readings—about how the racist, sexist structure of the U.S. capitalist economy creates widespread poverty among millions of Americans, among women in general, and especially among women and children of color.[45] But the question of whether or not she revises her theory about poor black women will depend on her willingness to challenge received belief about "welfare queens" commonly generated by people with white racial identities, and her concurrent willingness to consider the possibility that her own racial identity is implicated in this production of false knowledge.

Thus if this white woman student begins to develop an analysis of how the condition of *whiteness* largely predisposes her to certain social situations—including white segregation, a tradition of anti-black racism and prejudice in the family, a particular relationship to racist images of black women in the media, economic privilege—and distinct unearned advantages—functioning public schools, college educated parents, societal approbation, access to health care, unbiased police protection, superior and safer housing options, and so on—she may, during the course of a semester, begin to reevaluate her life experiences and social status in relation to progressive analyses of racism and white privilege. This reevaluation may eventually lead her to "try-on" a new theory about what it means to have a *white* racial identity, and therefore to a new awareness of the social consequences of the identity *white*—which in this case includes social privelege and a corresponding tendency to reproduce racist stereotypes about poor black women.

In this context, the student learns something crucial about the social world—that the U.S. economy reproduces racism and sexism, that the welfare system is not universally exploited by poor black women, that young and old white women comprise the majority of welfare recipients, and that the politics of poverty in this country are not just racialized, but are also gendered. But she also learns about the relationship between being a white middle-class woman (i.e., *who* she is) and her tendency to produce false knowledge about women of color and poor people (i.e., *what* she knows), a realist knowing-strategy that she can now take forward into other areas of knowledge-making. Subsequently, she may begin to see as suspect other received racist theories about people of color, which would enable her to seek more reliable knowledge about what the world is like for people of different social identities, especially for those who do not share her racial or class locations. She may then be able to produce more accurate knowledge not only about women of color, and about the epistemic and ideological effects of whiteness, but also about the racial state in general and her relationship to it.

The example above suggests that a realist classroom can democratize knowledge-making and power between students and teacher, and among students, by legitimizing and engaging all social identities in the classroom in the quest for collective, theory-mediated knowledge. This means that while we privilege students of color as the most significant *knowers* in discussions about racism—in part because their analyses have typically been excluded from accounts of the "truth" about race and racism—we regard the truth claims from students of all racial backgrounds as valuable sources of knowledge about the world, whether or not they are politically progressive or epistemically reliable.[46] Students—like the young white woman mentioned above—who make racist, sexist, classist and/or homohating assertions should not be undemocratically silenced or dismissed in a realist classroom, precisely because their distorted narratives—as harrowing as they can be—tell us something important about the social world. What we have to do in these cases is to support students in identifying the interpretive error in these analyses—by exposing the distorted perspective of the world that underlies it—and to reveal these errors as indicative of the unequal social effects of identity. Paula Moya confirms the epistemic value of undertheorized identities when she states:

> Identities based on ideological mystifications can, when examined closely, be important sources of knowledge about the world. To the extent that identities do not work well as explanations of the social world—to the extent that they refer imperfectly to aspects of the society from which they emerge—they can help reveal the fundamental contradiction and mystification within which members of that society live. (2002, 114–115)

When all social identities and experiential narratives are actively and critically engaged in a classroom, then teachers and students are empowered to scrutinize and challenge the ideological mystifications implicit in any cultural standpoint. This process is important not only in terms of interracial classroom dynamics, but also in intraracial contexts, because it opens a space for the scrutiny of intragroup dominance and social differences. In other words, the democratizing imperatives and practices of the realist classroom support the epistemic and social needs of both students of color and white students who find themselves subordinated within their own racial groups on the basis of gender, class, color, sexual orientation, ability, and so on.

Activating Teachers' Identities in the Realist Classroom

As argued above, since realist pedagogy is not based on the premise that students currently produce knowledge in a theory-independent and subject-independent manner, our goal is not to "get" students to bring their experiences and identities into play in knowledge production. Our goal is to shine a light on the connection between identity and knowledge that we see playing itself out in our classrooms everyday. For example, when white students assert their belief in the American meritocracy they are inadvertently exposing the negative epistemic consequences of their privileged racial identities. The fact that they are not aware of this means just that: they are not aware that the opinions they hold emanate from their particular social identity and the blindspots it creates. Their lack of awareness does not, however, negate the link between identity and knowledge-making in the classroom; it simply makes the connection invisible and inaccessible. Similarly, minority students who readily acknowledge the racist underpinnings of white intellectual history (including the myth of the meritocracy) do not automatically see this awareness as resulting from the epistemic link between the consequences of their racial identities and what they know. They simply find these things easy to believe, given their lived experiences. Ironically, in some cases it is first-generation college students of color who defend racist and classist accounts of poverty and social disenfranchisement. In response to these students too, a pedagogy that shows the theory-mediated relationship between who people are and what we know can empower students to analyze their own internalization of oppression, and the extent to which they are seeing the world through a racist and/or classist lens. In this way everyone in the classroom can access a positive epistemic framework through which to read and analyze course materials and our individual lived experiences.

Analogously, it is not our "goal" to encourage our faculty colleagues to bring their identities into their classrooms; just as students enter the learning environment with a set of social identities that are connected to the ways they see and think about reality, so do professors. The teacher in the realist classroom can therefore situate her own social and racial identity in discussions about histories of domination, in order to place herself intellectually and politically in relation to those histories; she may then discuss candidly with students her own challenges as a situated knower who holds varying degrees of privileged and underprivileged identity positions. For both white and minority teachers, this process is significant, since it models for students a realist knowing-strategy that they can emulate; but it is particularly crucial for professors who are racially privileged, since they must be careful not to position themselves as unsituated "objective" authorities. Indeed, the realist classroom empowers white teachers to be antiracist by explicitly bringing their experience of privilege to the forefront of classroom discussions. When teachers activate their identities in this way they contribute to the development of both *critical access* for minority students and positive racial identities for white students.

For example, in response to white student resistance to the idea that whites continue to benefit from the history and persistence of racism in the United States, a white teacher can use her own awareness of the effects of white race privilege as a way to dislodge students from this cognitive impasse. In addition to constructing diversified curricula that expose the theoretical errors in exclusive white epistemologies, the teacher can prompt white students to engage the lives of racial others, and to take on the interests and struggles of peoples of color as they define what our world is and what it should be. In the example given above about a white student's racist assertions about "welfare queens," a white teacher can effectively engage this student by acknowledging that she too (the teacher) has inherited racist beliefs about the unproductiveness and social inferiority of people of color, and by organizing a class discussion in which common racist narratives are named and progressively dismantled. Thus the white teacher may assume epistemic and social responsibility for her whiteness, rather than leaving the class (of both white and minority students) to assume that her current status as an intellectual ally to racially oppressed peoples is something she came by "naturally," without struggle, self-reflection, ambivalence, and risk.

Faculty of color in the realist classroom are supported by the focus on the epistemic relevance of identity, particularly since the antiracist curriculum repositions minority teachers as intellectual "insiders" whose life experiences and identities may grant them epistemic advantage. In addition, the minority teacher has the opportunity to play a pivotal role in developing *critical access* for minority students, since her presence in the classroom tends to reinforce the idea among students of color that they "belong" in institutions of higher learning. Even more significantly, minority faculty can model realist knowing strategies in the classroom by articulating the theoretically mediated relationship between identity and knowledge. For example, as we have discussed earlier, students of color sometimes defend racist analyses of the economic disparities between whites and non-whites. In response, Native American, African American, Latin@, and Asian faculty can critique racist analyses from the standpoint of a racially oppressed identity, thereby exemplifying the significant but indeterminate relationship between one's racial identity and one's perspectives on racial reality. For example, African American faculty can dispute essentialist accounts of the relationship between identity and knowledge by critiquing the racist belief that black people are "lazy" or "economically irresponsible" (these views are often held by black students and other students of color). Furthermore, faculty of color can push minority students to examine how their assumptions about racial politics, opportunity, achievement, and "choice" may be influenced by internalized oppression. Realist pedagogy thereby gives faculty of color concrete methods for showing students of all racial backgrounds the theory mediated relationship between racial identity and knowledge production—enabling teachers to activate the explanatory power of identity without reverting to essentializing accounts of the epistemic component of non-white racial identity.

We do not mean to suggest that the realist classroom creates instant utopia for minority teachers. In fact, since realist pedagogy turns the tables on white intellectual privilege, the minority teacher in predominantly white colleges and universities may find herself confronting a hotbed of white student hostility, and on the other hand, the expectation, on the part of minority students, that she show her allegiance to students of color by putting white students on the defensive. Given the tendencies of white students and of mainstream academic departments to undervalue the intellectual production of minority scholars and professors, white resistance can be particularly threatening to junior minority colleagues who rely on student evaluations to get reappointed and tenured. Conversely, if the minority professor does not respond to minority students' expectations that she demonstrate open disdain for white students, this will also put her in a precarious position vis-a-vis student evaluations. This can be particularly true for light-skinned faculty of color, or for feminist or queer faculty of color who may be perceived by minority students as thwarting the conventional values and social norms of their racial group. The challenges for minority teachers are formidable and constant, and realist pedagogy will not dissolve or dilute them. However, these risks notwithstanding, the realist commitment to articulate and examine the epistemic and social consequences of identity facilitates an open engagement with these issues, and ensures that these racial dynamics—alongside the course curriculum—be given serious consideration by both teachers and students as part of the knowledge-making process in the classroom.[47]

COMMUNITIES OF MEANING

Realist pedagogy assumes that there is a collective aspect to knowledge-making that clusters students into intellectual, identity-based affinity groups.[48] We call these epistemic affinity groups *communities of meaning*.[49] *Communities of meaning* are ways of thinking about how a common social location and a series of identity-based experiences can lead a group of students (or any group of people who share a social location) to arrive at theory-mediated objective truths about the world we live in. A common example of a *community of meaning* at a predominantly white private liberal arts college would be white middle- to upper-class students who come up with the same ideas about racism—"it's a thing of the past; the racial playing field is now even" or about welfare— "the welfare lines are populated mostly by black and Latin@ people who are too lazy to improve their lot in life;" or about affirmative action—"it's reverse discrimination against white people and most black people who benefit from it are unqualified."[50] Another example of a *community of meaning* in the white liberal arts classroom would be two or three students of color who come up with the same ideas about racism—"it exists everywhere and I experience it on a daily basis;" or about white privilege—"white people have more opportunity and are treated better than people of color;" or about affirmative action—"it's necessary and justified, but it isn't doing enough to change racial inequality." In direct contrast, public, urban, majority-minority, nonresidential, universities and colleges differ dramatically from the private predominantly white institutions (PWIs) in almost every respect. These large city universities encompass a wide range of *communities of meaning* that include: on the one hand, new immigrants to the United States of all racial backgrounds who are reluctant to challenge any form of authority, be it the police, government, teachers, etc, and who are skeptical about the existence of racism—"the complaints about racism from African Americans are unjustified;" and alternatively, young black and Latin@ students who readily acknowledge the existence and history of racism, but are simultaneously committed to "boot-strap" theories of individual social success and failure, and who therefore resist the systematic analysis of racism—"my parents had to fight for everything they have, and any black or Latin@ person with enough initiative can do the same."

Communities of meaning are formed anytime a group of students generates common perspectives about the world from similar social locations—perspectives that can be either more or less accurate, thus *communities of meaning* have no intrinsic subversive character. What is subversive about structuring the classroom according to an awareness of *communities of meaning* is that this conceptual focus helps students become aware that people who share the same social identity tend to base their beliefs on shared experiences, and situated—not idiosyncratic—accounts of the social world. Here is an example to illustrate our theory. As reasoned above, socially privileged white students commonly proceed from the notion that American society is a meritocracy and not in need of racial reform, whereas underprivileged students of color commonly believe that our society is unjust on a number of levels and in dire need of racial reform.[51] Asian American, black, and Latina women students may also see that in relation to them, the men in their communities receive unearned privileges in the family and in certain racialized social contexts, while their male counterparts tend to see this inequality as either the "natural order of things" or as part of a gender division that "protects" women against racist sexism; some men of color do not see the gap at all. In these two scenarios, white students and men of color—by virtue of the social hierarchies of race and gender—hold and exercise more epistemic authority than those who oppose their views. Since hegemonic thinking weighs so heavily in the classroom, students who speak in opposition to dominant views (i.e., students who theorize from a position of inter- or intragroup difference) often feel disempowered and reluctant to develop their arguments openly. *Communities of meaning* provide *critical access* for non-dominant students (in the first scenario non-dominant students would be all students of color, and in the second, women of color) by shifting epistemic authority to them. *Communities of meaning* support students in exposing and critiquing underlying assumptions (theories) about the world that exclude subjugated perspectives, and in opposing hegemonic knowledges; in this way *communities of meaning* equip students with potentially subversive epistemic tools as they highlight not only the situated character of knowledge-making, but the inherently collective process of determining the truth.

In the classroom we observe that students tend to produce knowledge in tandem with others who share their experiences and interests—people who are "like them;" however, in the same way that students are unaware of the link between identity and knowledge, they are usually not aware that they are thinking collectively in the classroom. Thus, as in the case of identity in the classroom, the realist conception of *communities of meaning* does not "invent" collective thinking but *activates* it, cultivating among students an awareness of their participation in epistemic collectives that produce knowledge, meaning, and moral judgment in the classroom. In so doing, the realist classroom prompts students to visualize how and where they stand with other people—and on the basis of which social formations—when judging what is right and wrong about the world, what is more and less valuable, and ultimately what needs to be changed. Racially defined *communities of meaning* thus work to empower the intellectual production of students of color by engaging them actively in a communal struggle for truth and justice, and opening a space for them to produce collective knowledge not only about *what the world is*, but *what it should be*. The moral aspect of these intellectual affinity groups supports students of color as they work collectively—based on an awareness of identity-based experiences, knowledges, and interests—to establish normative claims about our shared social world; so in addition to creating more reliable and inclusive knowledge about how our world is structured, *communities of meaning* can simultaneously promote political coalition aimed at constructing a racially democratic future. When students are trained to ask questions collectively about how their identities situate them in unequal social relations, they stand a much greater chance of extending this knowledge outside the classroom, in activist groups that struggle for social change—whether it be to change academic structures that are racially biased, or oppressive structures in society at large. Thus *communities of meaning* function as epistemic, moral, and political affinity groups that empower students of color to think collectively about how to transform our unjust society.

Communities of meaning democratize the relationships between dominant and oppressed communities because they take seriously the subjugated claims issuing from minority groups. But in addition to this, *communities of meaning* engage the cultural and social heterogeneity of racially defined groups by assuming that unequal intragroup differences exist and should be made explicit. Thus *communities of meaning* work to provide *critical access* to minority students who occupy multiply disadvantaged social categories and who may be marginalized not only outside of their racial group but also within it. More specifically, *communities of meaning* are grounded in the inclusive antiracist proposals of minority scholars and activists such as the Combahee River Collective, Angela

Davis, Audre Lorde, Leslie Feinberg Cherríe Moraga, Gloria Anzaldúa, Bernice Johnson Reagon, Chandra Talpade Mohanty, Jacqui Alexander, Melanie Kaye/Kantrowitz, María Lugones, Michael Hames-García, and others.[52] Because these theorists take seriously the social consequences of the multiplicity of social locations and identities in communities of color—largely because of how their own social identities are multiply subjugated—they urge racially defined communities to be responsive to the multiple oppressions suffered by members of racial groups who occupy more than one subjugated social location, that is, women, gay/lesbian/bisexual/transgendered people, biracial people, and poor people, to name a few. Realist pedagogy shares with these minority scholars and activists the idea that the struggle for racial democracy must include these diverse knowledges and interests in order to understand more fully how the apparatuses of oppression are interrelated—that is, how racism is gendered, for example—, and thus to clarify, strengthen, and make more democratic identity-based movements that oppose institutional racism. Realist pedagogy considers that the inclusion of intragroup difference in articulating racially defined knowledges is a fundamental moral principle of racial democracy. Michael R. Hames-García emphasizes this point when he writes:

> Emancipatory struggle can only be successful when straight people of color and white lesbians and gay men come to see the interests of queer people of color as *their own*. They must come to expand their sense of what their own interests are and who their own people are. Coalitions must cease to be coalitions of people with "different" interests, and the fragmentation within them must be healed…
>
> In other words, fighting racism and homophobia must be seen as a primary interest of all feminists and fighting sexism and homophobia must be seen as a primary interest of all people of color. (2000, 121).

By highlighting the epistemic, moral, and political relevance of intragroup differences, this pluralist moral epistemology strives to generate the most inclusive and reliable context for the articulation of a racially defined group's experiences, social status, needs, interests, and knowledges. In the classroom, realist *communities of meaning* assume—in line with the genealogy of black, Chican@, and third world antiracist theorists named above—that the cultivation of understanding and solidarity along the lines of intragroup heterogeneity will yield both the most accurate accounts of social reality, and the most effective and democratic political movements for racial justice.

CONCLUSION: *COMMUNITIES OF MEANING* AND INTERCULTURAL COOPERATION

We would like to conclude with an example that shows what *communities of meaning* can look like on the ground, and how *communities of meaning* can operate as projects of moral inquiry that challenge power relations in the classroom. The collective nature of this inquiry is significant for all students, but plays a particularly important role for minority students whose multiple social locations often place them in the position of double or triple jeopardy. In Susan's experience teaching Junot Díaz's collection of short stories *Drown* and Amie's experience teaching Marlon Riggs' films *Tongues Untied* and *Black Is/Black Ain't*, we have both struggled with students' delegitimizing of the experiences of queer people of color. Many heterosexual students of color in our classes do not consider GLBT people of color as "members" of their racial communities. In Susan's classes, racially diverse groups of men have declared the sexual identity and sexual freedom of men of color to be unworthy topics for class discussion ("since what we should be talking about is racism"). In Amie's classes, the overwhelming majority of students (divided equally among blacks, Latin@s, and whites, with small numbers of Asians) have openly condemned gay men of color—and all GLBT people for that matter—as "disgusting" and morally bankrupt. Although the classroom racial climate in our teaching experiences is entirely different, student responses in both these racial contexts dehumanize queer people of color, delegitimize their struggles for full citizenship inside and outside their communities, and disenfranchise students in the class who consider that racial equality should include the political and moral value of sexual freedom.

In spite of enormous odds against them, students who align themselves with GLBT people can confront these dehumanizing gestures by enacting realist knowing strategies. In Susan's class, a gay black student and his allies (a diverse group in terms of race and gender) were able to launch a serious political and moral challenge to homophobic and homohating ideas, institutions, and social and cultural norms by questioning their heterosexist underpinnings. In Amie's class, a small contingent of Latina and black women were able to initiate support for Riggs' declaration that sexual autonomy is a civil rights issue by theorizing the ways in which black and Latin@ communities are harmed and weakened by racist homohatred. In so doing, each group of students activated their knowledge making from a *community of meaning*. The *communities of meaning* that value sexual freedom and affirm gay and lesbian human rights are able to draw upon a diverse array of identity-based knowledges (queer, straight, black, Latin@, white, male, female) to construct coherent arguments about why homohatred is politically pernicious and morally wrong; therefore the students who form these *communities* stand on firmer epistemic ground, since their ideas are explicitly framed by the moral and political theory that all people have the right to express themselves sexually, and that the struggle for the sexual freedom of people of color should be a significant aspect of social movements for racial democracy. In effect then, the difference between students' silencing or encouraging the discussion about the relationships between racism and homohatred is no longer a mere "difference of opinion"; it is a moral and political conflict in which subversive *communities of meaning* gain an epistemically reliable standpoint from which to argue for and establish their truths.

From a realist standpoint, the black gay student in Susan's class, by virtue of his experience of both homohatred and racism, has a different, richer theoretical commitment than do the Latin@ and white students who dismiss *Drown*'s analysis of gay men of color and Latin@ masculinity as "missing the point." Therefore, the knowledge he produces is "better," in the sense that it includes the experiences of non-straight people of color in deciding what is true about racism and about the world. Similarly, the straight-identified women of color in Amie's class, and their epistemic allies, mobilize their own experiences of being marginalized on the basis of sex within communities of color, and build their support of Riggs' analysis on those raced and gendered experiences. On both epistemic and moral grounds, students can determine that the experience-based knowledge that emerges from each of these *communities of meaning* is more democratic and thus more valuable than the collectively produced dismissals of Díaz and Riggs. First, the epistemic reason: gay men of color and all women of color are in marginalized positions within communities of color, gay and feminist perspectives on the world are usually silenced by abusive and discriminatory power; therefore the experiential knowledges that they advance are subjugated knowledges whose absence impoverishes our attempts to know about Latin@s and African Americans and also to know the unique effects of racism on gay Latin@ and African American men. Latin@ and African American experiences (and all identity-based experience, for that matter) are not monolithic but multiple, composed of many different ways of experiencing race, gender, class, religion, sexuality, national origin, language, and so on; even the common experience of racism is not the same for Latin@ and African American people of different social locations. Thus, if we do not admit the experiential and theoretical knowledge of gay African Americans and Latin@s, we see less and know less about ourselves and the world. Second, the moral reason: the suppression of these subjugated knowledges is morally reprehensible because it is based upon the devaluing and denial of gay humanity and citizenship, and therefore serves as a rationale for the violation of human rights.[53]

We understand that the commitment to a heterogeneous articulation of identity-based knowledges within racially defined groups might be seen as a disunifying influence on communities of students (and faculty) of color. Clearly, cultivating the awareness and articulation of difference and disagreement among white students is less controversial and more transparently liberatory, since it tends to disrupt unifying narratives of white dominance; and indeed, highlighting the intragroup differences among students of color may be seen as suspect, since it can be interpreted as a disempowering force among students who already struggle to overcome adverse racial conditions in higher education. We know that our proposal may be seen as controversial in this regard, and we take this consideration very seriously.[54] We want to clarify that our goals in this project are the promotion of accurate, democratic and ethical student-centered knowledge about history, race, economy, literature, philosophy, art, music, and so on—which for us means knowledge that values equally the diverse and multiple spectrum of experiences, social locations, identities, and interests of all members of racially defined groups—but also greater solidarity and coalition both within oppressed student groups and across social differences. We believe, along with Satya Mohanty, that a diversity of cultural perspectives (in this case a proliferation of the existence and epistemic consequences of intracultural differences) ensures that more of the relevant information about the social world will be available to students as they produce knowledge together, and that questions about what the world is like will be more robust and reliable, since they "will be shaped and honed by a reasonable array of competing theoretical perspectives"(1997, 241). We also believe, along with Audre Lorde, that if we are to transform the formidable and interrelated structures of racism, we must "devise ways to use each other's difference to enrich our visions and our joint struggles" (1984, 122). Using each other's differences to measure the validity and error of identity-based truth claims, students of color in racially defined *communities of meaning* can enter into critical dialogue about how to productively and justly engage the social heterogeneity implicit in the struggle for racial freedom. Thus *communities of meaning* are predicated on the idea that the formation of racial democracy is contingent on the promotion of epistemic and political cooperation across social differences, both in interracial and intraracial contexts. This assumption is consciously grounded in the moral premise that the lives and interests of all people of color—including those who are further marginalized within their own racial groups as a result of related systems of oppression and discrimination—must be equally considered as we organize and define struggles to reshape our world democratically within a robust vision of racial justice.

In addition, because *communities of meaning* activate the awareness that people of all racial identities have *multiple* zones of identification that intersect with those of other individuals who may or may not share their racial identity,[55] they open a space for cross-cultural and cross-racial coalition by prompting students to identify intellectual and political allies across social and racial differences. While race is a causal social formation that produces common experiences for people who share a racial identity, race also tends to be a social category that casts a homogenizing light on members of racial groups, eclipsing other differences that might be less immediately accessible. Once students of color recognize and begin to theorize the relevance of the multiplicity of their own knowledge-making, they become increasingly aware of the epistemic consequences of the shifting identity standpoints and experience-based knowledge claims of those with whom they are learning. In turn, students of all racial identities are empowered to assess and create knowledge in a collective of others who—in spite of other social differences—share some important experiences, and thus share at least some significant intellectual and political commitments. These epistemic contact zones can further prompt students within a *community of meaning* to scrutinize more vigorously the quality of knowledge they do produce, since the expression of intragroup differences and the formation of cross-cultural affinity groups acts as a constant source of challenge to what is often perceived as uniform thinking within identity-based social groups. Facilitating cross-racial coalition therefore requires that we encourage the expression of social difference among students whose heterogeneity may not be apparent, in order to promote the proliferation of diverse *communities of meaning* that traverse social categories. We believe that both racial and intragroup subjugated differences are politically, socially, morally, and above all, epistemically relevant. Thus cultivating realist *communities of meaning* can help to create *critical access* for students of color by utilizing the constructive/collective power of student difference to build racial democracy into our classrooms.

NOTES

We first presented portions of this chapter at the Colorlines Conference, *Segregation and Integration in America's Present and Future,* sponsored by the Civil Rights Project at Harvard University, August 2003, and at the Future of Minority Studies National Conference held at the University of Wisconsin, Madison, November 2004. We subsequently presented other portions of this chapter at the The Future of Minority Studies National Research Project Conference, *Reading Identities: Literature, Pedagogy and Social Thought,* University of Wisconsin, Madison, October 2003, and at the *The Global Ethnic Literatures Seminar,* sponsored by The Program in Comparative Literature at the University of Michigan at Ann Arbor, December 2004.

We are grateful to the many people who have given us thoughtful, challenging feedback on this chapter, and would especially like to acknowledge Tobin Siebers, Satya P. Mohanty, Linda Martín Alcoff, Paula M. L. Moya, and John J. Su.

1. The Supreme Court unanimously agreed that separate educational facilities were "inherently unequal," that they had deprived black children of equal educational opportunities, and that they violated the Fourteenth Amendment. Nevertheless, moves to integrate schools and universities were violently contested at local and state levels. See *The Shape of the River: Long Term Consequences of Considering Race in College and University Admissions,* William G. Bowen and Derek Bok, for a discussion about the limited effects of Brown, that is, massive resistance at the federal, state, and local levels of government to implement and enforce the Supreme Court ruling.
2. See the *Journal of Blacks in Higher Education* report on "Ranking America's Leading Liberal Arts Colleges on Their Success in Integrating African-Americans," July 2003. However, in terms of faculty representation, these increases are deceiving. The *Chronicle of Higher Education* notes that, "[t]aken together, African-American, Hispanic, and American Indian scholars represent only 8% of the full-time faculty nationwide. And while 5% of professors are African American, about half of them work at historically black institutions. The proportion of black faculty members at predominantly white universities—2.3%—is virtually the same as it was 20 years ago" (Wilson, 2002, A10).
3. We are relying in this chapter primarily on the Census 2000 that provides the most comprehensive data and analysis in the area of educational attainment. In the ten years between each national census the federal government undertakes two measures of data regarding the population of the United States and this data is analyzed by the National Center for Education Statistics (NCES). These are the Current Population Survey (CPS) and American Community Survey (ACS). We have included in this essay the data from the most recent (2007) CPS and ACS on percent of people over age 25 who have completed four or more years of college. However, the CPS and ACS data we are using here differ from that available in Census 2000 in at least two important areas that are relevant for our analysis: first, the CPS and ACS data do not include racial or ethnic categories for Native Americans or Native Hawaiians; and second, the CPS and ACS data stipulate the percent of people, by racial group, over age 25 who have completed four or more years of college, but do not stipulate that those individuals have attained the B. A. degree. http://www.census.gov/population/cen2000/phc-t1/tab01.xls;http://nces.ed.gov/pubs2002/digest2001/tables/dt187.asp; http://www.census.gov/prod/2003pubs/c2kbr-24.pdf—this is the report issued in August 2003, a Census 2000 Brief, titled "Educational Attainment 2000." All URLs last accessed in September 2008.
4. All Ivies except for University of Pennsylvania, where the discrepancy between white and black graduation rates is 20%.
5. See http://www.brownalumnimagazine.com/january/february_2000/the_race-report_card.html. Last accessed September 2008.
6. The six major race categories used in Census 2000 are white, black or African American, American Indian and Alaska Native, Asian, Native Hawaiian and Other Pacific Islander, and some other race. According to the Census 2000 Brief "Educational Attainment" 54.1% of whites over 25 have some college education with 26.1% having attained a bachelor's degree. The numbers for the remaining racial groups are as follows: 42.5% black or African American have some college whereas only 14.3% have a bachelor's degree; 41.7% of American Indian or Alaska Native have some college and only 11.5% have the bachelor's degree; 44.6% of Native Hawaiian and other Pacific Islander have some college and only 13.8% have attained the B.A.; 30.3% of Hispanic or Latin@ students have some college

and only 10.4% succeed in attaining the B.A.; Asian students have the highest rates of educational attainment with 64.6% having some college education and 44.1% attaining the B.A. All statistics are from the U.S. Census Brief, Educational Attainment 2000, Table 2 "Educational Attainment of the Population 25 Years and Over by Age, Sex, Race, and Hispanic or Latin@ Origin: 2000" (Issued by the U.S. Census Bureau, August 2003).

7. Incidentally, given the massive presence of undocumented Latin@ students living in the United States, and the obvious underreporting of actual numbers of Latin@s in the U.S. population, the educational attainment of Latin@s is likely far worse than even these dismal numbers from the U.S. Census suggest.

8. See http://www.census.gov/population/cen2000/phc-t1/tab01.xls; http://nces.ed.gov/quicktables/Detail.asp?Key=901; http://nces.ed.gov/quicktables/Detail.asp?Key=711; http://nces.ed.gov/pubs2002/digest2001/tables/dt187.asp. Last accessed September 2008.

9. U.S. Census Bureau. http://www.census.gov/population/www/socdemo/educ-attn.html, Table A-2, "Percent of People 25 Years and Over Who Have Completed High School or College By Race, Hispanic Origin, and Sex: Selected Years 1940–2007." Compare these statistics with the percent of people 25 years and over in 2007 who have completed four years of college or more: for white non-hispanics, the number is 31.8%, for blacks the number is 18.5%, for Asians, the number is 52.1%. The 2007 report (which is based on the Current Population Survey and American Community Survey) does not include data for Native Americans.

10. National Public Radio, Morning Edition, August 12, 2003, "NPR: Tribal College Confronts Funding Woes," http://www.npr.org/templates/story/story.php?storyId=1393124.

11. By "educational resources" we mean schools, scholarships, prep and training programs for standardized tests, access to public libraries, etc.

12. See the *Journal of Blacks in Higher Education* comparison of Median Net Worth (dollar value of all assets including home, auto, securities, and money in banks) for black families versus white families. On the basis of Census 2000 data and Pew Hispanic Trust analysis of this data, the median Black family net worth is $5,598, as compared to a median net worth for white families that is nearly 15 times greater at $88,651.

13. We recognize that there are stark differences in the basic measures of social stability among these various groups of color in the United States. Moreover, we are well aware of the popularized and often racist identification of Asians and Asian Americans as so-called "model minorities." Even though many statistical analyses conclude that Asian Americans dramatically outpace other minority groups in terms of educational attainment, median incomes, percentage of population living in poverty, and so on, we maintain that there are great numbers of Asians and Asian Americans whose lives and life chances are negatively affected by racism. Analyses that work with the category "Asian American" typically do not include large numbers of Asian and Asian American people currently living and working in the United States, e.g., undocumented immigrants from China—both those who have recently arrived and those who have been here for decades—as well as undocumented immigrants from Tibet, Pakistan, Indonesia, Cambodia, the Philippines, Burma, Malaysia, and Vietnam. Demographic figures on undocumented immigrants are notoriously difficult to ascertain; however, at the very least we must acknowledge that current statistics on basic measures of social stability among Asian Americans would change dramatically if they included undocumented Asians and Asian Americans, many of whom have low educational attainment, and live in extreme poverty. Political asylum seekers are at a special disadvantage, often forced to conceal their actual identities in the asylum process (and therefore unable to access their educational attainment from other countries). For these reasons we believe that Asians and Asian Americans should be included in the process of articulating the goals of racial democracy for minority students.

14. We want to emphasize that increasing the numbers of people of color in higher education is vital to racialized political struggle because individuals who share minority identities tend to group together to fight for what they need—and in the case of the struggle for racial democracy, a critical mass of students of color on campus highlights

that what these students need from the college or university is often denied to them on the basis of race. For example, at a predominantly white institution, when a number of Latin@ and Black students discover in talking with one another that they have all been steered away from taking Introductory Chemistry (a course that must be taken in a specific sequence in order to prepare students for the MCAT and application to medical school), they begin to suspect that they have been denied access to a medical school preparatory track on the basis of race. Faculty explanations for this sort of advising are unconvincing: they typically include the suspect rationale that the students do not have the necessary high school science and math preparation and/or that "college is so hard for these students that they need to be protected in their course selection from an overly challenging program."

15. Of freshman college students nationwide 50% report that they often have a serious conversation with a student of a different race or ethnic group. Unfortunately, the likelihood of these interracial conversations drops slightly during the four years of college, with only 49% of all senior college students nationwide reporting that they often have a serious conversation with a student of a different race or ethnic group (National Survey of Student Engagement), *Journal of Blacks in Higher Education*, 39 (Spring 2003). We interpret these figures to mean that classroom and residential experiences on college and university campuses do not automatically create interracial contact and relationship.

16. For a rigorous and clarifying discussion about the hidden and explicit forms of white hegemony in the classroom, see Margaret Hunter, "Decentering the White and Male Standpoints in Race and Ethnicity Courses," in *Twenty-first-Century Feminist Classrooms: Pedagogies of Identity and Difference*, eds. Amie A. Macdonald and Susan Sánchez-Casal (NY: Palgrave Macmillan, 2002, 251–280). See also James W. Loewen, *Lies My Teacher Told Me: Everything Your American History Textbook Got Wrong* (New York: Touchstone/Simon and Schuster, 1995).

17. We are not assuming that rich schools ensure students of color the best kind of education. For example, CUNY—the public university whose mission is to educate the children of the people of the city of New York—makes it possible for students who also have children, full-time jobs, and family responsibilities to earn an advanced degree. Some Native American students attending Haskell Indian Nations University in Lawrence, Kansas, report that the all-Indian student body is far more culturally hospitable and therefore conducive to learning and achievement. However, contemporary students at Haskell have to face the reality of severe underfunding from the Bureau of Indian Affairs (the BIA is currently allotting only $3,800 of the budgeted $6,000 per student), in addition to grappling with Haskell's history as a residential "school" where Native American children were forcibly detained after being taken from their homes against their parents' will. We therefore believe that whether students of color are attending PWIs or "majority-minority" institutions (with the possible exception of the elite historically black colleges such as Spelman College, Morehouse College, Wilberforce University, Howard University, Fisk University etc.), they are in a less favorable educational environment than their white peers.

18. The City University of New York—CUNY—is a prime example of this sort of academic ghettoization.

19. The *Journal of Blacks in Higher Education* released an August 2003 report entitled "Ranking America's Leading Liberal Arts Colleges on Their Success in Integrating African Americans." This report does not include the experiences of Latin@s, Asians, or other peoples of color. The authors emphasize that there are "other factors that go into the overall racial climate at a given college which cannot be measured by the standard indices of institutional racial integration. These include attitudes of faculty toward black students, patterns of residential segregation on campus, attitudes of white students toward racial minorities, and particularly, the seriousness and frequency of campus incidents of racial animosity or violence" ("Ranking America's Leading Liberal Arts Colleges on Their Success in Integrating African-Americans," 9). We believe that these findings apply to other students of color as well.

20. See especially the cases of Harvard University and Hamilton College.

21. See especially the work of Claude Steele on the academic effects of stereotype threat for racial minority students in college classrooms. "Race and the Schooling of Black Americans," *The Atlantic Monthly* (April 1992): 68–78.
22. This is a common argument offered to defeat proposals for race-based housing or other racially defined spaces proposed by minority groups. See especially Elizabeth Anderson, "Integration, Affirmative Action, and Strict Scrutiny," *NYU Law Review*, 77 (2002): 1195–1271.
23. Admittedly, there are some benefits for students of color. As Lisa Delpit points out , students of color can benefit by learning about the cultural codes of the "culture of power;" they can also learn how to negotiate the racially stilted, white dominant environment of academia, something that undoubtedly helps them to negotiate the larger, racially undemocratic world we live in. However, given the dramatic ratio of white students to non-white students in PWIs, students of color can be assured contact with whiteness whether race-based social and academic spaces exist or not.
24. For an excellent discussion, see Lisa Delpit, "The Silenced Dialogue: Power and Pedagogy in Educating Other People's Children," *Harvard Educational Review*, 58.3 (1988): 280–298.
25. Delpit suggests that one of the ways to do this is by creating institutional and curricular strategies in consultation with people of color.
26. See Amie A. Macdonald, "Racial Authenticity and White Separatism" for a defense of the existence of race-based program housing, in *Reclaiming Identity: Realist Theory and the Predicament of Postmodernism*, eds. Paula M. L. Moya and Michael Hames-García (Berkeley: University of California Press, 2000), 205–225. Although we are not discussing race-based housing in this chapter, our position on critical access and the creation of racial democracy in education includes a defense of race-based program housing.
27. See especially Paula M. L. Moya, *Learning from Experience, Minority Identities, Multicultural Struggles* (Berkeley: University of California Press, 2002), and Loewen, *Lies My Teacher Told Me*. See also Paula Rothenberg, *White Privilege: Essential Readings on the Other Side of Racism* (New York: Worth, 2002) and *Invisible Privilege: A Memoir about Race, Class, and Gender* (Lawrence: University Press of Kansas, 2000); George Lipsitz "The Possessive Investment in Whiteness: Racialized Social Democracy and the 'White' Problem in American Studies," *American Quarterly*, 47.3 (September 1995): 369–387; Gary Orfield, *Expanding Opportunity in Higher Education: Leveraging Promise* (Albany, NY: SUNY Press, 2006); Peggy McIntosh, "White Privilege and Male Privilege: A Personal Account of Coming to See Correspondences through Work in Women's Studies," in *Critical White Studies: Looking behind the Mirror*, eds. Richard Delgado and Jean Stefancic (Philadelphia: Temple University Press, 1997, 291–299).
28. White students are clearly harmed by this type of racism in the classroom as well, since it reinforces their state of radical unknowing about U.S. history and world history. However, the "harm" to white students is far less damaging than the harm done to students of color, since whites are not barred from accessing educational resources on the basis of race, nor are white students generally given the message that their cultural inheritance is inferior or insignificant.
29. Margaret Hunter makes this case especially well in "Decentering the White and Male Standpoints in Race and Ethnicity Courses," in *Twenty-first-Century Feminist Classrooms: Pedagogies of Identity and Difference*.
30. For an excellent discussion of the epistemic exclusions inherent in celebratory constructions of American "character," see Michael Hames-García "Which America Is Ours?: Martí's 'Truth' and the Foundations of 'American Literature,'" Modern Fiction Studies 49.1 (2003): 18–53, and reprinted in this volume, chapter 4.
31. For a rigorous and clarifying discussion of the history and status of the integration of ethnic content in educational curricula, see Moya, "Learning How to Learn from Others: Realist Proposals for Multicultural Education," in *Learning from Experience: Minority Identities, Multicultural Struggles*.

32. Here we are referring to resistance at all levels of the institution: from students, faculty, colleagues, alumni, departments, administrators, and trustees.
33. Racism is clearly not the only discriminatory practice present in higher education classrooms. While we are focused in this essay primarily on the reality of racism in the classroom, we are aware of and have argued elsewhere to demonstrate the presence of sexism, homophobia, nationalism, classism, abelism, religious discrimination, and other forms of social discrimination in the classroom.
34. We want to acknowledge the tremendous effort and progress that numerous people (students, faculty, staff and administrators) and collectives have made in transforming areas of curriculum and pedagogy within the traditional disciplines, as well as in continuing to democratize the content of "insurgent" disciplines. By "insurgent" disciplines, we mean disciplines that explicitly critique structures of domination, such as Women's Studies, African American Studies, Latin@ Studies, Queer Studies, Native American Studies, Disability Studies, Chican@Studies, Ethnic Studies, and so on.
35. Sánchez-Casal and Macdonald formulate the theory of realist pedagogy in the introductory essay to our co-edited anthology *Twenty-first-Century Feminist Classroom: Pedagogies of Identity and Difference* "Feminist Reflections on the Pedagogical Relevance of Identity" (Palgrave Macmillan, 2002), 1–28.
36. In 1993 Satya P. Mohanty introduced the theory of realism in his groundbreaking essay "The Epistemic Status of Cultural Identity: On Beloved and the Post-colonial Condition," reprinted in *Reclaiming Identity: Realist Theory and the Predicament of Postmodernism*, 29–66. Since then several important books have been published elaborating the realist theory of identity. See S. P. Mohanty, *Literary Theory and the Claims of History: Postmodernism, Objectivity, Multicultural Politic* (Ithaca, NY: Cornell University Press, 1997); Paula M. L. Moya and Michael Hames-García, *Reclaiming Identity: Realist Theory and the Predicament of Postmodernism*; Moya, *Learning from Experience*; Michael Hames-Gárcia, *Fugitive Thought: Prison Movements, Race, and the Meaning of Justice* (Minneapolis: University of Minnesota Press, 2004); Linda Martín Alcoff, *Visible Identities: Race, Gender and the Self*; (New York: Palgrave Macmillan, 2006); Sean Kicummah Teuton, *Red Land, Red Power: Grounding Knowledge in the American Indian Novel* (Durham, NC: Duke University Press, 2008). See especially Moya, *Learning from Experience*, 37–45, for an outline of the six claims of the realist theory of identity. Realist theories of identity are also the subject of an ongoing multi-disciplinary research project (The Future of Minority Studies Research Project, www.fmsproject.cornell.edu).
37. Essentialism suggests that individuals or groups have a pre-social, unchanging and immutable essence; essentialist theories tend to see one social category (race or gender for example) as overly determining of the social identity of an individual or group, and tends to posit the existence of an unmediated relationship between experience and "truth."
38. We are well aware that the distinction between "inherited" and "political" identities is not hard and fast. There are of course many important controversies about whether certain identities are inherited or political (i.e., some combination of socially constructed and chosen): for example, sexuality, religion, and Jewish identity to name a few. Moreover, many feminist theorists have argued that masculinity and femininity are themselves only "inherited" in a partial way, and that to a large extent, are socially constructed by a sexist system that privileges the rights and experiences of men over those of women. At the same time, it is indisputable that there is a conceptually meaningful and politically important distinction to be made between inherited and political identities.
39. This unequal racial positioning is also often the case between white faculty and faculty of color, and between the texts/histories of whiteness versus those of and by people of color. See also Loewen, *Lies My Teacher Told Me*; Howard Zinn, *A People's History of the United States: 1492–Present* (New York: Perennial Press, 2003) and Haunani Kay-Trask,

From a Native Daughter: Colonialism and Sovereignty in Hawai'i (Honolulu: University of Hawai'i Press, 1999).

40. See Susan Sánchez-Casal's "Unleashing the Demons of History: White Resistance in the U.S. Latin@ Studies Classroom," in *Twenty-first-Century Feminist Classrooms: Pedagogies of Identity and Difference*, 59–85. Sánchez-Casal argues that "decontextualized multicultural narratives in the educational system miseducate students about the historical struggles of peoples of color in the United States, to such an extent that the content and focus of antiracist courses in the academy are often construed as 'propaganda.'"

41. See especially Edward Said's *Orientalism* (New York: Random House, 1979) and Edwin N. Wilmsen's *Land Filled with Flies: A Political Economy of the Kalahari* (Chicago: University of Chicago Press, 1989), for further discussion of how peoples of color are viewed as without history by white and Western cultural imperialists.

42. The realist classroom also prepares minority students to critique subjugated discourses that advance problematical constructions of social reality, i.e., sexist and/or homophobic theories of Black or Latin@ liberation. We will discuss this complexity further in the section on *communities of meaning*. In addition, teaching disenfranchised students the strategies and power of critical thinking also improves their ability to access educational resources outside the classroom that are so important to social success (i.e., fair advising, student development, career and internship planning services, hospitable residential and social options, and so on). We of course recognize that institutions have to make structural changes to the way that they offer all of these additional educational resources and that this transformation is necessary and independent of any action taken on the part of students of color.

43. Students of all racial backgrounds are often astounded upon learning basic facts about American history (the way that Hawai'i became a state is an excellent example), and the tone in a classroom shifts markedly when students are given the opportunity to reflect on the cultural meaning of having been completely wrong or entirely uninformed about basic historical truths regarding U.S. history, and global history for that matter.

44. For a compelling and accessible discussion on the psychological effects on white students of discussing race and racism in the classroom, see Beverly Tatum, "Talking about Race, Learning about Racism: An Application of Racial Identity Development Theory in the Classroom," *Harvard Educational Review* 62.1 (1992): 1–24.

45. What most students do not know is that the majority of people on welfare are white women, young and old.

46. Epistemic authority necessarily shifts continually in a classroom, depending upon the subject and area of discussion. The realist critique of essentialism, however, requires that we reject more classic notions of epistemic privilege and the notion that experience of oppression necessarily leads to the best and most accurate knowledge about that experience. Nevertheless, in discussions of homohating, GLBT (gay, lesbian, bisexual, transgendered) identities are granted epistemic authority; in discussions of American nationalism, non-Americans; in discussions of sexism, women, etc.

47. Although it is beyond the scope of this essay, we are well aware that many progressive and radical professors of all racial identities often experience demoralizing encounters with students of all backgrounds who defend racist, woman-hating, classist, homohating, nationalist, and/or xenophobic worldviews. It is our contention that realist pedagogy can begin to offer educators a useful set of strategies for engaging these encounters productively, based on a coherent theoretical perspective that defends both curriculum and methodology.

48. For a psychosocial analysis of the benefits of and motivation for occasional self-segregation of racial minority groups, and for the ways that this collective self-segregation empowers antiracist community building, see Beverly Daniel Tatum's *Why Are All the Black Kids Sitting Together in the Cafeteria? And Other Conversations about Race: A Psychologist Explains the Development of Racial Identity* (New York: Basic Books, 2003).

49. See Sánchez-Casal and Macdonald, "Feminist Reflections on the Pedagogical Relevance of Identity," in *Twenty-first-Century Feminist Classrooms: Pedagogies of Identity and*

Difference. For a discussion of *communities of meaning* in the context of race-based program housing, see Amie A. Macdonald "Racial Authenticity and White Separatism: The Future of Racial Program Housing on College Campuses."

50. We want to clarify that we are using "black" in this example precisely because most white students identify welfare recipients in this way. We are not suggesting that *communities of meaning* are valuable only in classrooms that study welfare, affirmative action, or the structures of the racial state; we use these examples to show how knowledges such as these inform students' acceptance or rejection of antiracist theories of society.

51. For a variety of reasons, the discussion of which are beyond the scope of this work, socially marginalized students sometimes share the view that American society is fair and that those who don't "make it" aren't trying hard enough. For a discussion of the epistemic complexities underlying this phenomenon, see Sánchez-Casal, "Unleashing the Demons of History" and Macdonald "Feminist Pedagogy and the Appeal to Epistemic Privilege."

52. The Combahee River Collective, "A Black Feminist Statement," in *Capitalist Patriarchy and the Case for Socialist Feminism*, ed. Zillah R. Eisenstein (New York: Monthly Review Press, 1979), 362–372; Angela Davis, *Women, Race, and Class* (New York: Vintage Books, 1981); Audre Lorde, *Sister Outsider* (Freedom, CA: Crossing Press, 1984); Cherríe Moraga, *Loving in the War Years: lo que nunca pasó por sus labios* (Boston: South End Press, 1983); Gloria Anzaldúa, *Borderlands/la frontera The New Mestiza* (San Francisco: Spinster/Aunt Lute, 1987); Moraga and Anzaldúa, *This Bridge Called My Back: Writings by Radical Women of Color* (New York: Kitchen Table: Women of Color Press, 1983); Bernice Johnson Reagon, "Coalition Politics: Turning the Century" in *Home Girls: A Black Feminist Anthology*, ed. Barbara Smith (New York: Kitchen Table/Women of Color Press, 1983), 356–368; Chandra Talpade Mohanty, *Feminism Without Borders: Decolonizing Theory, Practicing Solidarity* (Durham, NC: Duke University Press, 2003); M. Jacqui Alexander and Chandra Talpade Mohanty, *Feminist Genealogies, Colonial Legacies, Democratic Futures* (New York: Routledge, 1997); Melanie Kaye/Kantrowitz, *The Issue Is Power: Essays on Women, Jews, Violence and Resistance* (San Francisco, CA: Aunt Lute, 1992) and *The Color of Jews: Racial Politics and Radical Diasporism* (Bloomington: Indiana University Press, 2007); María Lugones, *Pilgrimages/Peregrinajes: Theorizing Coalition Against Multiple Oppression* (New York: Rowman and Littlefield, 2003); Michael Hames-García "Who Are Our Own People? Challenges for a Theory of Social Identity," in *Reclaiming Identity: Realist Theory and the Predicament of Postmodernism*, 102–129.

53. We believe that the realist classroom supports students in creating communities of meaning that value equally queer and straight people, and that prioritize the discussion of sexual freedom as a justice issue. Significantly, the example we offer shows how communities of meaning that contest racist homohatred do not rely for their existence on the presence of a significant number of "out" queer students of color. The Black gay student in Susan's class was able to elicit epistemic support for his critique from a variety of students who are not black and/or gay. Amie's students—heterosexual women of color—were able to develop a critique that many epistemic essentialists might see as only possible if there were a significant number of out gay people in the classroom (including many out gay people of color). But because realism takes seriously the epistemic function of identity—as it asserts that social identities can lead to either more or less accurate truths—, progressive *communities of meaning* do not depend for their existence on the presence (and willingness) of people who "look like" each other (i.e., share the same social identities as) to consider and determine what is true.

54. Along with other antiracist scholars we contend that surfacing differences among Latin@, black, and Native American students can lead to knowledge about subjugated intragroup experiences that disrupt the hegemony of heterosexual male experience. The cultivation of an awareness of difference, or intragroup multiplicity, can lead a black or Latin@

community of meaning to ask new questions about how the analyses of the experiences and identities of women of color and queer people of color, for example, can lead us to more inclusive, and therefore more reliable and democratic knowledge about what it means to be Latin@ or African American in the United States, as well as about what it takes to change racist structures of oppression that affect all communities of color. We believe that the epistemic effect of surfacing and scrutinizing intragroup multiplicity supports a democratizing objective of strengthening the epistemic and political growth of communities of color—here of course "communities of color" are understood as racially defined collectives who are continuously engaged in a struggle for the equal citizenship of all of their members. See also Marlon Riggs' semi-autobiographical films *Black Is/Black Ain't, Tongues Untied* and *Ethnic Notions*, and Cheryl Dunye's film *The Watermelon Woman*.

55. Obviously we are abstracting from the dynamic interaction of multiple social locations certain parts of social identity so that we may speak about the model of *communities of meaning* and about how they engage multiplicity in an epistemically and politically enabling way; thus this model is necessarily analytic. We do not mean to suggest that one experiences one's sexual identity without also simultaneously experiencing one's racial and gender identity, for example. We also do not mean to suggest that the most privileged members of an oppressed group share power with the most privileged members of dominant social groups. We are suggesting that in relation to others in the oppressed group, and in society, certain members are invested with more privilege on the basis of social class, skin color, sexual identity, gender, etc., and that this privilege is indicative of the persistence of intragroup inequalities.

References

Alcoff, Linda Martín. *Visible Identities: Race, Gender and the Self.* New York: Oxford University Press, 2005.

Alcoff, Linda Martín and Michael Hames-Gárcia, Satya P. Mohanty, and Paula M. L. Moya, eds. *Identity Politics Reconsidered.* New York: Palgrave Macmillan, 2006.

Alexander, M. Jacqui and Chandra Talpade Mohanty. *Feminist Genealogies, Colonial Legacies, Democratic Futures.* New York: Routledge, 1997.

Anderson, Elizabeth. "Integration, Affirmative Action, and Strict Scrutiny," *NYU Law Review, 77* (2002): 1195–1271.

Anzaldúa, Gloria. *Borderlands/La Frontera: The New Mestiza.* San Francisco, CA: Spinster/Aunt Lute, 1987.

Bauman, Kurt J. and Nikki L. Graf. U.S. Census Bureau. "Educational Attainment: 2000, Census 2000 Brief" http://www.census.gov/prod/2003pubs/c2kbr-24.pdf. C2KBR-24. Issued August 2003.

Bowen, William G. and Martin A. Kurzweil, and Eugene M. Tobin. *Equity and Excellence in American Higher Education.* Charlottesville: University of Virginia Press, 2005.

Bowen, William G., Derek Bok, and Glenn Loury. *The Shape of the River.* Princeton: Princeton University Press, 2000.

Chemerisnsky, Erwin, Drew Days III, Richard Fallon, Pamela S. Karlan, Kenneth L. Karst, Frank Michelman et al. "Reaffirming Diversity: A Legal Analysis of the University of Michigan Affirmative Action Cases: A Joint Statement of Constitutional Law Scholars." The President and Fellows of Harvard College. July 2003.

Collins, Patricia Hill. *Black Feminist Thought.* New York: Routledge, 1991.

Combahee River Collective. "The Combahee River Collective: A Black Feminist Statement." In *Capitalist Patriarchy and the Case for Socialist Feminism,* edited by Zillah R. Eisenstein, 362–372. New York: Monthly Review Press, 1979.

Davis, Angela. *Women, Race and Class.* New York: Vintage Books, 1981.

———. "Gender, Class, and Multiculturalism: Rethinking "Race" Politics." In *Mapping Multiculturalism,* edited by Avery Gordon and Christopher Newfield, 40–48. Minneapolis: University of Minnesota Press, 1996.

Delgado, Richard and Jean Stefancic. *Critical White Studies: Looking behind the Mirror*. Philadelphia: Temple University Press, 1997.
Delpit, Lisa. "The Silenced Dialogue: Power and Pedagogy in Educating Other People's Children." *Harvard Educational Review*. 58.3 (1988): 280–298.
Díaz, Junot. *Drown*. New York: Riverhead Books, 1997.
DuBois, W. E. B. *The Souls of Black Folk*. New York: Dover Thrift Editions, 1994.
Freire, Paulo. *Pedagogy of the Oppressed*. New York: Herder, 1972.
Giroux, Henry. "Insurgent Multiculturalism and the Promise of Pedagogy." In *Pedagogy and the Politics of Hope*, 23–57. 1992.
Giroux, Henry and Peter MacLaren. "Teacher Education and the Politics of Engagement: The Case for Democratic Schooling." *Harvard Educational Review*. 56. 3 (1986): 213–238.
Gurin, Patricia, Eric L. Dey, Sylvia Hurtardo, and Gerald Gurin. "Diversity and Higher Education: Theory and Impact on Educational Outcomes." *Harvard Educational Review* 72 (2003): 330–366.
Hames-García, Michael R. "Who Are Our Own People?" Challenges for a Theory of Social Identity." In *Reclaiming Identity: Realist Theory and the Predicament of Postmodernism*, 102–129. Berkeley: University of California Press, 2000.
———. "Which America Is Ours?: Martí's 'Truth' and the Foundations of 'American Literature.'" *Modern Fiction Studies* 49.1 (2003) 18–53.
———. *Fugitive Thought: Prison Movements, Race, and the Meaning of Justice*. Minneapolis: University of Minnesota Press, 2004.
Harding, Sandra. "Rethinking Standpoint Epistemology: "What Is Strong Objectivity?" In *Feminist Epistemologies*, edited by Linda Alcoff and Elizabeth Potter, 49–82. New York: Routledge, 1993.
hooks, bell. *Teaching to Transgress: Education as the Practice of Freedom*. New York: Routledge, 1994.
Hunter, Margaret. "Decentering the White and Male Standpoints in Race and Ethnicity Courses." In *Twenty-first-Century Feminist Classrooms: Pedagogies of Identity and Difference*, edited by Amie A. Macdonald and Susan Sánchez-Casal, 251–280. New York: Palgrave Macmillan, 2002.
Kaye/Kantrowitz, Melanie. *The Issue is Power: Essays on Women, Jews, Violence and Resistance*. San Francisco: Aunt Lute, 1992.
———. *The Color of Jews: Racial Politics and Radical Diasporism*. Bloomington: Indiana University Press, 2007.
Kincheloe, Joe L. and Shirley R. Steinberg. "Addressing the Crisis of Whiteness: Reconfiguring White Identity in a Pedagogy of Whiteness." In *White Reign: Deploying Whiteness in America*, edited by Joe Kincheloe, Shirley R. Steinberg, Nelson M. Rodriguez, Ronald E. Chennault, 3–29. New York: St. Martin's Press, 1998.
Kuklinski, James. "Review: The Scientific Study of Campus Diversity and Students' Educational Outcomes." *Public Opinion Quarterly*, 70.1 (Spring 2006): 99–120.
Lipsitz, George. "The Possessive Investment in Whiteness: Racialized Social Democracy and the 'White' Problem in American Studies." *American Quarterly*, 47.3 (September 1995): 369–387.
Loewen, James W. *Lies My Teacher Told Me: Everything Your American History Teacher Got Wrong*. New York: Touchstone/Simon and Schuster, 1995.
Lorde, Audre. *Sister Outsider: Essays & Speeches*. Freedom, CA: Crossing Press, 1984.
Lugones, María C. *Pilgrimages/Peregrinajes: Theorizing Coalition against Multiple Oppression*. New York: Rowman and Littlefield, 2003.
Macdonald, Amie A. "Racial Authenticity and White Separatism: The Future of Racial Program Housing on College Campuses." In *Reclaiming Identity: Realist Theory and the Predicament of Postmodernism*, edited by Paula M. L. Moya and Michael Hames-García, 205–225. Berkeley: University of California Press, 2000.
———. "Feminist Pedagogy and the Appeal to Epistemic Privlege. In *Twenty-first-Century Feminist Classrooms: Pedagogies of Identity and Difference*, edited by Amie A. Macdonald and Susan Sánchez-Casal, 111–133. New York: Palgrave Macmillan, 2002.

Macdonald, Amie A. and Susan Sánchez-Casal. *Twenty-first-Century Feminist Classrooms: Pedagogies of Identity and Difference.* New York: Palgrave Macmillan, 2002.

Maher, Frances and Mary Kay Thompson Tetreault. "They Got the Paradigm and Painted It White." In *White Reign: Deploying Whiteness in America*, edited by Joe Kincheloe, Shirley R. Steinberg, Nelson M. Rodriguez, Ronald E. Chennault, 148–160. New York: St. Martin's Press, 1998.

Mayberry, Maralee and Ellen Cronan Rose. "Teaching in Environments of Resistance: Toward a Critical, Feminist and Anti-racist Pedagogy." In *Meeting the Challenge: Innovative Feminist Pedagogies in Action.* New York: Routledge, 1999.

McIntosh, Peggy. "White Privilege and Male Privilege: A Personal Account of Coming to See Correspondences through Work in Women's Studies." In *Critical White Studies: Looking behind the Mirror*, edited by Richard Delgado and Jean Stefancic, 291–299. Philadelphia: Temple University Press, 1997.

Mills, Charles. *The Racial Contract.* Ithaca, NY: Cornell University Press, 1997.

Minnich, Elizabeth. *Transforming Knowledge.* Philadelphia: Temple University Press, 1992.

Mohanty, Chandra Talpade. "On Race and Voice: Challenges for Liberal Education in the 1990s." In *Beyond a Dream Deferred: Multicultural Education and the Politics of Excellence*, edited by Becky Thompson and Sangeeta Tyagi, 41–65. Minneapolis: University of Minnesota Press, 1993.

———. *Feminism without Borders: Decolonizing Theory, Practicing Solidarity.* Durham, NC: Duke University Press, 2003.

Mohanty, Satya P. *Literary Theory and the Claims of History: Postmodernism, Objectivity, Multicultural Politics.* Ithaca, NY: Cornell University Press, 1997.

———. "Radical Teaching, Radical Theory: The Ambiguous Politics of Meaning." In *Theory in the Classroom*, edited by Cary Nelson, 149–176. Urbana: University of Illinois Press, 1986.

———. "The Epistemic Status of Cultural Identity: On *Beloved* and the Post-colonial Condition." In *Reclaiming Identity: Realist Theory and the Predicament of Postmodernism*, edited by Paula M. L. Moya and Michael Hames-García, 29–66. Berkeley: University of California Press, 2000.

Moraga, Cherríe. *Loving in the War Years: Lo que nunca pasó por sus labios.* Boston: South End Press, 1983.

Moraga Cherríe and Gloria Anzaldúa. *This Bridge Called My Back: Writings by Radical Women of Color.* New York: Kitchen Table: Women of Color Press, 1983.

Moya, Paula M. L. "Postmodernism, "Realism," and the Politics of Identity: Cherríe Moraga." In *Feminist Genealogies, Colonial Legacies, Democratic Futures*, edited by M. Jacqui Alexander and Chandra Talpade Mohanty, 125–150. Routledge, 1996.

———. *Learning from Experience: Minority Identities, Multicultural Struggles.* Berkeley: University of California Press, 2002.

———. "Learning How to Learn from Others: Realist Proposals for Multicultural Education." In *Learning from Experience: Minority Identities, Multicultural Struggles.* Berkeley: University of California Press, 2002.

Moya, Paula M. L. and Michael Hames-Gárcia, eds. *Reclaiming Identity: Realist Theory and the Predicament of Postmodernism.* Berkeley: University of California Press, 2000.

Nieto, Sonia. "From Brown Heroes and Holidays to Assimilationalist Agendas: Reconsidering the Critique of Multicultural Education." In *Multicultural Education, Critical Pedagogy, and the Politics of Difference*, edited by Christine Sleeter and Peter MacLaren, 191–220. Albany: State University of New York (SUNY), 1995.

Oliver, Melvin and Thomas Shapiro. *Black Wealth/White Wealth: A New Perspective on Racial Inequality*, Second Edition. New York: Routledge, 2006.

Omi, Michael and Howard Winant. *Racial Formation in the United States: From the 1960s to the 1990s.* New York: Routledge, 1994.

Omolade, Barbara. "Quaking and Trembling: Institutional Change and Multicultural Curricular Development at the City University of New York." In *Beyond a Dream Deferred: Multicultural Education and the Politics of Excellence*, edited by Becky W. Thompson and Sangeeta Tyagi, 214–230. Minneapolis: University of Minnesota Press, 1993.

Orfield, Gary and Patricia Gándara and C. Horn. *Expanding Opportunity in Higher Education: Leveraging Promise.* Albany: SUNY Press, 2006.

"Ranking America's Leading Liberal Arts Colleges on Their Success in Integrating African-Americans." *Journal of Blacks in Higher Education*, August 2003.

Reagon, Bernice Johnson. "Coalition Politics: Turning the Century." In *Home Girls: A Black Feminist Anthology*, edited by Barbara Smith, 356–368. New York: Kitchen Table/Women of Color Press, 1983.

Rodríguez, Nelson A. "Emptying the Content of Whiteness: Toward an Understanding of the Relation between Whiteness and Pedagogy." In *White Reign: Deploying Whiteness in America*, edited by Joe L. Kincheloe et al, 31–62. New York: St. Martin's Griffin Press, 1998.

Roman, Leslie. "White is a Color! White Defensiveness, Postmodernism and Anti-racist Pedagogy." In *Race, Identity and Representation in Education*, edited by Cameron McCarthy and Warren Crichlow, 71–88. New York: Routledge, 1993.

Roman, Leslie G. and Linda Eyre. *Dangerous Territories: Struggles for Difference and Equality in Education.* New York: Routledge, 1997.

Rothenberg, Paula. *White Privilege: Essential Readings on the Other Side of Racism.* New York: Worth, 2002.

———. *Invisible Privilege: A Memoir about Race, Class, and Gender.* Lawrence, KS: University Press of Kansas, 2000.

Rothman, Stanley, Seymour Martin Lipset, and Nigel Nevitte. "Does Enrollment Diversity Improve University Education?" *International Journal of Public Opinion Research* 15 (2002): 8–26.

———. "Racial Diversity Reconsidered." *The Public Interest* 151 (Spring 2003): 25–38.

Said, Edward. "The Politics of Knowledge." *Raritan* 11.1 (Summer 1991): 17–31.

———. *Orientalism.* New York: Random House, 1979.

Sánchez-Casal, Susan. "In a Neighborhood of Another Color: Latin@ Struggles for Home." In *Burning Down the House: Recycling Domesticity*, edited by Rosemary Marangoly George, 325–350. Boulder, CO: Westview Press, 1998.

———. "Unleashing the Demons of History: White Resistance in the U.S. Latino Studies Classroom." In *Twenty-first Century-Feminist Classrooms: Pedagogies of Identity and Difference*, edited by Amie A. Macdonald and Susan Sánchez-Casal, 59–85. New York: Palgrave Macmillan, 2002.

Sánchez-Casal, Susan and Amie A. Macdonald. "Introduction: Feminist Reflections on the Pedagogical Relevance of Identity." In *Twenty-first-Century Feminist Classrooms: Pedagogies of Identity and Difference*, edited by Amie A. Macdonald and Susan Sánchez-Casal, 1–28. New York: Palgrave Macmillan, 2002.

Searle, John. "Postmodernism and the Western Rationalistic Tradition." In *Campus Wars*, edited by John Arthur and Amy Shapiro. Boulder, CO: Westview Press, 1994.

Shor, Ira and Paulo Freire. "What Is the 'Dialogical Method' of Teaching?" *Journal of Education.* 169 (1987): 11–31.

Sleeter, Christine E., and Peter McLaren. *Multicultural Education, Critical Pedagogy, and the Politics of Difference.* New York: State University of New York Press, 1995.

Steele, Claude M. "Race and the Schooling of Black Americans." *The Atlantic Monthly.* (April 1992): 68–78.

Steele, Claude M., Spencer, S. J., and Aronson, J. "Contending with Group Image: The Psychology of Stereotype and Social Identity Threat." In *Advances in Experimental Social Psychology*, edited by M. P. Zanna, 3, 2002.

———. "Contending with Group Image: The Psychology of Stereotype and Social Identity Threat." In *Advances in Experimental Social Psychology*, edited by M. P. Zanna 34: 379–440.

Srivastava, Aruna. 1997. "Anti-Racism Inside and Outside the Classroom." In *Dangerous Territories: Struggles for Difference and Equality in Education*, edited by Leslie G. Roman and Linda Eyre, 113–125. NY: Routledge, 1997.

Tatum, Beverly Daniel. "Talking About Race, Learning About Racism: The Application of Racial Identity Development Theory in the Classroom." *Harvard Educational Review*, 62.1 (1992): 1–24.

———. "Teaching White Students about Racism: The Search for White Allies and the Restoration of Hope." *Teachers College Record*, 95.4 (Summer 1994): 462–475.

———. *Why Are All the Black Kids Sitting Together in the Cafeteria? And Other Conversations about Race: A Psychologist Explains the Development of Racial Identity.* New York: Basic Books, 2003.

Teuton, Sean Kicummah. *Red Land, Red Power: Grounding Knowledge in the American Indian Novel.* Durham, NC: Duke University Press, 2008.

Trask, Haunani-Kay. *From a Native Daughter: Colonialism and Sovereignty in Hawaii.* Honolulu: University of Hawai'i Press, 1999.

US Census Bureau. http://www.census.gov/population/www/socdemo/educ-attn.html, Table A-2, "Percent of People 25 Years and Over Who Have Completed High School or College By Race, Hispanic Origin, and Sex: Selected Years 1940–2007."

Wills, John S. "Who Needs Multicultural Education? White Students, U.S. History, and the Construction of a Usable Past." *Anthropology and Education Quarterly*, 27.3 (1996): 365–389.

Wilmsen, Edward O. *Land Filled with Flies: A Political Economy of the Kalahari.* Chicago: University of Chicago Press, 1989.

Wilson, Robin. "Stacking the Deck for Minority Candidates?" *Chronicle of Higher Education* 12 (July 2002): A10.

Zinn, Howard. *A People's History of the United States: 1492-Present.* New York: Perennial Press, 2003.

2

WHAT'S IDENTITY GOT TO DO WITH IT? MOBILIZING IDENTITIES IN THE MULTICULTURAL CLASSROOM

Paula M. L. Moya

Research done over several decades in a variety of disciplines across the social sciences and humanities has shown that students and teachers alike bring their identities and experiences with them into the classroom. Identities are highly salient for students' experiences in school; they make the classroom a different place for different students. This is because students with different identities in the same classroom will face different sets of what Claude Steele calls "identity contingencies." Steele uses the term to refer to the specific set of responses that a person with a given identity has to cope with in specific settings. Indeed, *who* a student is perceived to be will affect such variables as his placement in an educational tracking system, the friends he will have to choose among, and the academic and social expectations that his teachers will have of him.[1] While these identity contingencies might seem relatively insignificant, they can have major consequences for the opportunities a person will have over the course of his or her life.

To the extent that we are genuinely interested in educating for a just and democratic society, then, we will recognize the salience of identities in the classroom. We will work to alter the negative identity contingencies that minority students commonly face, even as we find strategies for maximizing opportunities for all our students. But I will go even further than this. I argue that a truly multi-perspectival, multicultural education will work to *mobilize identities* in the classroom rather than seeking to minimize all effects of identities as part of the process of minimizing stereotypes. Only by treating identities as epistemic resources and mobilizing them, I contend, can we draw out their knowledge-generating potential and allow them to contribute positively to the production and transmission of knowledge.

IDENTITIES

What are identities? In my book, *Learning from Experience*, I define *identities* as the non-essential and evolving products that emerge from the dialectic between how subjects of consciousness identity themselves and how they are identified by others. Elsewhere in the book, I define them as "socially significant and context-specific ideological constructs that nevertheless refer in non-arbitrary (if partial) ways to verifiable aspects of the social world." I argue that identities are "indexical"—that is, they refer outward to social structures and embody social relations.[2] Insofar as identities reference our understanding of ourselves in relation to others, they provide their bearers with particular perspectives on a shared social world. They are, in the words of Satya Mohanty, "ways of making sense of our experiences."[3]

In this essay, for analytical purposes, I take the dialectical concept of identity I worked with in *Learning from Experience* and separate it into two components: ascriptive and subjective identities. I make this analytical distinction not to suggest that the two components can be, in fact, separated from one another. Indeed, identity is inescapably relational. Rather, I make the distinction because it allows me to more clearly delineate what is at stake in taking a realist—rather than an essentialist or an idealist—approach to identity. I argue that taking a realist approach to identity is critical to the project of working toward a more egalitarian and free society. Only a realist approach effectively registers the dialectical (as well as historically and culturally specific) nature of identity construction—an adequate understanding of which is essential to our ability to work toward the transformation of socially significant identities. To the extent that we are interested in transforming *this* world into a better one—insofar as we cannot get *there* except from *here*—the transformation of the identities that are central to the arrangement and functioning of society will be a necessary part of our epistemic and political project.

Ascriptive identities are what some researchers call "imposed identities," and what I sometimes call "social categories." They are inescapably historical and collective, and generally operate through the logic of visibility. Examples include racial categories such as "black" and "Asian" as well as gender categories such as "woman" and "man." Ascriptive identities come to us from outside the self, from society, and are highly implicated in the way we are treated by others. More importantly, ascriptive identities are highly correlated with the selective distribution of societal goods and resources. This is because, as a result of variable and historically specific economic and social arrangements such as slavery, employment discrimination laws, and restrictive housing covenants that unfairly advantaged some groups of people at the expense of others, different social categories have accrued different meanings and associations. These meanings and associations—many of which linger long after the economic or social arrangements that gave rise to them have been dismantled or even outlawed—are often invoked and mobilized by those in positions of relative power to justify day-to-day processes of social and economic inclusion and exclusion. These processes can range from the personally painful, as when a young black girl is refused admission to a schoolyard game by a group of white girls, to the economically debilitating, as when a Latina fails to gain a much-deserved promotion because her white male boss has trouble imagining her in a position of authority.[4]

The other aspect of the dialectical concept of identity is what we call subjective identity, or simply "subjectivity." Subjectivity refers to our individual sense of self, our interior existence, our lived experience of being a more-or-less coherent self across time. The term also implies our various acts of self-identification, and thus necessarily incorporates our understanding of ourselves in relation to others. Thus, subjective identities can refer to aspects of someone's personality, such as when we describe ourselves as being a "non-conformist," or a "joker." They can also advertise our values, such as when we identify ourselves as a "Christian," or an "ecofeminist." Finally, they can reference available social categories, such as when we self-identify as "gay" or "disabled." Although subjective identities sometimes feel as if they are completely internal, and thus under our individual control, thinkers since Hegel have agreed that subjective identities are inescapably shaped by the experience of social recognition. As Linda Martín Alcoff has argued, "the 'internal' is conditioned by, even constituted within, the 'external,' which is itself mediated by subjective negotiation. Subjectivity" she explains, "is itself located. Thus the metaphysics implied by 'internal/external' is, strictly speaking, false."[5]

REALIST VERSUS ESSENTIALIST AND IDEALIST CONCEPTIONS OF IDENTITY

I draw the distinction between ascriptive and subjective identities because how we understand the relationship between them will determine whether and when we are essentialist, idealist, or realist about identity. Essentialists about identity suppose that the relationship between the ascriptive and the subjective is one of absolute identity. They imagine, for example, that if a person can be assigned to a racial or gender category on the basis of some invariable characteristic like skin color or genitalia, then everything else of significance, including how he or she self-identifies, his or her propensity for violence, personal characteristics, and even innate mental capacity follows from being a member of that particular group. These days, there are very few scholars who claim to be essentialist about identity. Notable exceptions would be Charles Murray and Richard Hernstein, the authors of the *Bell Curve*, and the researchers who are searching the human genome for evidence that would provide a genetic basis for the sociohistorical concept of race.[6]

Idealists about identity, by contrast, claim that there is no stable or discoverable relationship between the ascriptive and subjective aspects of identity. Idealists imagine that how others regard a person should be of little consequence to the strong-minded individual who makes her own way in the world. The neoconservative minority with the "pull yourself up by your own bootstraps" mentality is one kind of person who takes an idealist approach to identity. Shelby Steele in *The Content of Our Character* and Richard Rodriguez in *Hunger of Memory* provide good examples of a neoconservative idealist approach to identity.[7] Another example of an idealist approach to identity would be that of the postmodernist who argues that we can disrupt historically sedimented and socially constituted identity categories through individual acts of parody or refusal. I am thinking here of Judith Butler's argument in her influential work *Gender Trouble*.[8] If essentialists impute too much significance to the social categories through which we receive societal recognition, idealists attribute too little. They underestimate the referential and social nature of identity. Identities, after all, refer to relatively stable and often economically entrenched social arrangements.

Such social arrangements can change, and when they do, available identities will change along with them.[9] But individuals, qua individuals, have much less power over their identities than idealists imagine.

Realists about identity, by contrast, understand ascriptive and subjective identities as always in dynamic relationship with each other. We understand that people are neither wholly determined by the social categories through which we are recognized, nor can we ever be free of them. Indeed, the intimate connection between the organization of a society and the available social categories that we must contend with in that society accounts for why no transformation of identity can take place without a corresponding transformation of society—and vice versa. This is true for everybody—black, white, male, female, gay, straight, able-bodied, disabled—but the stakes for those of us who are members of stigmatized identity groups are especially high. Because the identity contingencies we are likely to face have potentially debilitating effects on our life-chances, we ignore the dynamics of identity at our peril. To the extent that we are interested in transforming our society into one that is more socially and economically just, we need to know how identities work in order to effectively work with them.

Before I proceed, I need to make a point about the relational and contextual nature of all identities. As social constructs that draw upon available social categories, identities are indexed to a historical time, place, and situation. A consequence of this is that the same identity evokes very different associations in different places. On most mainstream news programs, a Chicana/o identity evokes associations of illegality, poverty, criminality, and delinquency. In Casa Zapata, the Mexican-American theme dorm at Stanford University, a Chicana/o identity is associated with pride, family, hard work, achievement, and solidarity. As the meanings associated with any given identity changes with the context in which that identity is invoked, the identity contingencies associated with that identity correspondingly change. There are a number of implications that follow from the contextual nature of identity, including the fact that a person can experience her identity very differently at different times, depending on the historical context and locale in which it is invoked. Claude Steele has done important work on the phenomenon of "stereotype threat," which is a particular kind of identity contingency that results from the fact that some identities are stigmatized in socially significant ways. He defines "stereotype threat" this way: "When a negative stereotype about a group that one is part of becomes personally relevant, usually as an interpretation of one's behavior or an experience one is having, stereotype threat is the resulting sense that one can then be judged or treated in terms of the stereotype or that one might do something that would inadvertently confirm it."[10] Stereotype threat is thus not only anxiety producing, but, crucially, it can measurably affect a person's performance in a realm that might alter the course of his or her future. Steele's work demonstrates empirically what most of us have known at the level of experience all along—that an identity that feels very safe in one situation can feel very threatened in another. Moreover, it helps explain why individuals who are members of certain groups might make the decisions they do—why, for example, Latina/o and African American students, who may have achieved well in elementary school, begin to dis-identify with education as adolescents and either under-perform or drop out altogether. They are responding to the myriad messages about who they are and what they are capable of that they get from the larger society. They are removing themselves emotionally, if not literally, from a very unpleasant and uncomfortable situation. Given the stereotypes about these two groups, African American and Latina/o students who care about doing well in school are almost always going to be subject to stereotype threat in the classroom—unless their teachers and fellow students work actively to alter the identity contingencies these students have to face in the classroom setting.

The relational and contextual nature of all identities reveals that the problem is not identity, *per se*, but the way in which particular identities are invoked in particular social contexts. Understanding the dialectical nature of identities helps us to avoid falling into the trap of thinking either that nothing can be done to change typical educational outcomes (women just *are* bad at Math; Latinos just *are* the type of people who drop out of school), or that individuals should be able to escape, willfully and through sheer force of character, the identity contingencies to which they are subjected. Educators who take a realist approach to identity understand the importance of changing the classroom dynamics in which people with different identities interact. By changing classroom dynamics, we transform the local social contexts in which particular identities are invoked. And because identities are dialectical, a transformation of the social context will necessarily alter the contingencies attached to particular social identities. The first step toward addressing negative educational outcomes that are identity-based, then, is understanding of the dialectical nature of identity and recognizing the fact that identities are always already invoked in the classroom—usually in pernicious ways. The next step involves figuring out a way to mobilize identities in a way that recognizes *all* identities, but especially minority identities, as important epistemic resources.

IDENTITIES AS EPISTEMIC RESOURCES

The idea that we should mobilize identities in the classroom is a somewhat unconventional idea. Identities are often thought by right-, classic liberal, and even left wing thinkers to be pernicious, or at least not conducive to rational deliberation and the public good. Some critics of identity are afraid of the difference that identities imply, afraid that an acknowledgement of cultural or perspectival difference will lead inevitably to a situation of irresolvable conflict. For others, the risk of stereotype threat and prejudice is so great as to suggest that, rather than mobilizing (and recognizing) identities, we should try to eliminate the salience of identities in the classroom completely. Such critics advocate an "identity neutral" or "color blind" approach that denies the continuing salience of certain kinds of identity for everyday interactions and experiences.

The work that those of us involved in the Future of Minority Studies project have been doing, however, suggests that seeing identities as things we would be better off without is not the most productive or accurate way to understand them. Linda Alcoff, for example, devotes a chapter of *Visible Identities* to dismantling the political critique of identities, demonstrating that such critiques are predicated on erroneous assumptions and a metaphysically inaccurate understanding of what identities are.[11] Providing careful readings of such political theorists as Todd Gitlin and Nancy Fraser, Alcoff demonstrates that their arguments against identity politics depend upon three basic assumptions about the nature and the effects of identities: (1) people with strongly felt identities are necessarily exclusivist; (2) whatever is imposed from outside as an attribution of the self is a pernicious constraint on individual freedom; and (3) identities bring with them an unvarying set of interests, values, beliefs, and practices that prevent their bearers from being able to participate in objective, rational deliberation about the common good. Such assumptions, Alcoff notes, are "hardwired into western Anglo traditions of thought"; as such, they are rarely ever made explicit and defended (31). As a way of questioning these assumptions, Alcoff examines the practices and claims of a wide range of political groups who attend to the salience of identity—from the Puerto Rican Political Action Committee (PRPAC) to the Service Employee International Union (SEIU)—to see if the picture of identity supported by these assumptions corresponds to the lived experience of identity or its politically mobilized forms. Importantly, the correspondence is not there. Alcoff argues that when we look at how identities operate in the world, we see that people with strongly felt identities are not necessarily exclusivist and that they can be capable of seeing past their own immediate interests for the common good. Moreover, we see that identity ascription is an inescapable—but not necessarily pernicious—fact of human life; it can enable, as well as constrain, individual freedom. The work Alcoff has done suggests that any dismissal of identity is, at minimum, required to begin with a metaphysically adequate understanding of it. Otherwise, dismissing identity is about as effective as dismissing gravity: you can do it, but unless you radically change the conditions that give rise to it (such as by traveling to space to achieve a condition of zero-gravity), you are not going to make much of a difference in how it works.[12]

Similarly, I have argued elsewhere that identities should be considered important epistemic resources that are better attended to than dismissed or "subverted."[13] The argument I have been making begins with the presumption that *all* knowledge is situated knowledge; there is no transcendent subject with a "God's eye" view on the world who can ascertain universal truths independent of a historically and culturally specific situation. Having recognized that all knowledge is situated, I see the importance of considering both from where a given knowledge-claim is derived, as well as whose interests it will serve, in any evaluation of its historically and culturally specific significance and truth-value. Moreover, I understand that even good, verifiable *empirical* knowledge must be evaluated in relation to a particular historical, cultural, or material context. Significantly, my view that all knowledge is situated does not lead me down the primrose path of epistemological relativism any more than my view that identities are constructed leads inexorably to the idea that they are arbitrary or infinitely malleable. I am a realist, and as such, I hold that there is a "reality" to the world that exceeds humans' mental or discursive constructions of it. While our collective understandings may provide our only access to "reality," and may imbue it with whatever meaning it can be said to have, our mental or discursive constructions of the world do not constitute the totality of what can be considered "real." The "real" both shapes and places limits on the range of our imaginings and behaviors, and therefore provides an important reference point in any sort of interpretive debate about the meaning of a text, a picture, or a social identity. The part of the "real" that exceeds humans' mental and discursive constructions of "reality" is also what occasions some "truths" to carry over across specific historical and cultural contexts.

The link between knowledge and identity stems from the fact that our identities provide us with particular perspectives on a shared social world. And while identity and knowledge are not coextensive, nevertheless, what we "know" is intimately tied up with how we conceptualize that world and who we understand ourselves to be in it. Our conceptual frameworks are thus inseparable from how we comprehend ourselves in terms of our gender, culture, race, sexuality, ability, religion, age, and profession—even when we are not consciously aware of how these aspects of ourselves affect our points of view. Our identities thus shape our interpretive perspectives and bear on how we understand both our everyday experiences and the more specialized and expert knowledge we encounter and produce through our research and teaching. They influence the research questions we deem to be interesting, the projects we judge to be important, and the metaphors we use to describe the phenomena we observe.[14] This is as true for those who have "dominant" identities as for those of us who have "minority" identities. As fundamentally social beings, we humans can no more escape the effects of our identities on our interpretive perspectives than we can escape the process of identification itself. Identities are fundamental to the process of *all* knowledge-production.

The link between knowledge and identity provides a compelling rationale for why a diverse work force, professoriate, or research team maximizes objectivity and innovation in knowledge production. People with different identities are likely (although not certain) to ask different questions, take various approaches, and hold distinctive assumptions. Insofar as diverse members of a research team conceptualize their shared social world in dissimilar ways, they may view a shared problem in discrete ways. In situations where mutual respect and intellectual cooperation are practiced, the existence of such divergent perspectives can lead to the sparking of a productive dialectic that might lead to a creative solution or advancement in knowledge. Complacency and too-easy agreement, by contrast, can lead to intellectual stultification. The presence of people who hold different perspectives but who are able to respect each other's intellect and creativity increases the possibility that a research team will come up with an innovative solution to a shared problem that looked, from one point of view, unsolvable.[15]

Solving a problem held in common is certainly not the only, and perhaps not even the best, explanation for why a diverse professoriate can lead to advancements and innovations in knowledge-production. In a disciplinary field like history or literary studies that takes as its object of study human society or culture, for example, the existence of researchers with diverse identities increases the possibility that someone might ask previously ignored research questions that open up entirely new areas of inquiry. This is essentially what has happened with such subfields as women's history and African American literature. Importantly, when the object of study is human culture or society, paying special attention to the struggles for social justice of people with subjugated identities is especially crucial to the process of investigating the functioning of a hierarchal social order such as our own. This is because subjugated identities and perspectives are often marginalized and hidden from view. Unlike the perspectives of those who have the economic means and social influence to publish and broadcast their views, the views of people who are economically and socially marginalized do not form part of the "common-sense" of the "mainstream," or dominant, culture. As I have argued elsewhere, the alternative perspectives and accounts generated through oppositional struggle provide new ways of looking at a society that complicate and challenge dominant conceptions of what is "right," "true," and "beautiful." Such alternative perspectives call to account the distorted representations of peoples, ideas, and practices whose subjugation is fundamental to the maintenance of an unjust hierarchical social order.[16] Consequently, if researchers and teachers are interested in having an adequate—that is, more comprehensive and objective, as opposed to narrowly biased in favor of the status quo—understanding of a given social issue, they will listen harder and pay more attention to those who bring marginalized views to bear on it. They will do so in order to counterbalance the overweening "truth" of the views of those people in positions of dominance whose perspectives are generally accepted as "mainstream" or "common-sense."

It is for these reasons, and one more, that I argue that teachers in multicultural classrooms would do well to recognize identities as epistemic resources and work to mobilize them in the classroom. As Michael Hames-García argues in an essay about the teaching of American Literature, an important part of educating for a democratic society involves helping students understand what is at stake in the outcome of various debates.[17] If students are to grow up to be participatory citizens in a functioning democracy, they need to see themselves as contributors to an ongoing conversation about the best way to live in the world. This will necessarily involve introducing all students—majority and minority alike—to alternative conceptions of what that "best way" might be. Whether the class is interpreting a novel or debating the merits of welfare reform, the discussion as a whole will benefit from the introduction of alternative (non-dominant) perspectives. Importantly, involving minority students in classroom discussions as privileged members—participants whose identities bring crucial (and otherwise missing) information to the discussion at hand—has the effect of changing the classroom dynamics and, by extension, the identity contingencies in that classroom. And where the teacher and students are successful at linking the perspectives expressed (in the novel, the textbook, or by the students themselves) to historically specific material interests and consequences, the stakes for students' life choices will be that much more evident. Research has shown that when education is presented as being relevant to students' lives, they will be more invested in both the discussion at hand and their education as a whole.[18] Finding ways to mobilize identities in the classroom thus serves the dual purpose of empowering students as knowledge-producers capable of evaluating and transforming their society even as it has the potential to contribute to the production of more objective, and less biased, accounts of the topics under discussion.

Educational Policy Implications

The recognition that identities are epistemic resources has implications for a wide range of policies that are external to the classroom, but that bear on what happens within it. At the most basic level, it provides a strong justification for integrated schools and classrooms. If a teacher is working in a classroom that is extremely homogeneous—along lines of race, gender, sexuality, class, religion, and ability—she will have fewer perspectival differences to exploit in her efforts to encourage her students to think critically about their own assumptions and values. Insofar as preparing students to be good citizens of a functioning democracy is an important goal of education, it must provide students with opportunities to exercise their critical capacities by reflecting on the convictions that guide their judgments about the best way to structure our common society. Students who are not encouraged to think about *why* they believe what they do will have difficulty understanding why other people believe differently. They will, moreover, be deprived of important occasions to consider changing their beliefs and transforming their identities. By contrast, a teacher whose classroom is diverse along lines of race, gender, sexuality, class, religion, and ability will have a rich variety of perspectives to draw on. She will have a greater probability of success in her efforts to encourage the sort of productive dialogue that is fundamental to the goal of educating for a multicultural democracy. Through giving her students the chance to examine their own identities, she will be training them to more adequately negotiate disagreements arising as a result of cultural, racial, economic, and class differences. Furthermore, by allowing her students to consider their own implication and agency in the structure and functioning of our society, she will be developing their critical capacities to imagine that society that could be organized differently. The epistemic and pedagogical importance of perspectival difference, then, suggests that teachers and educational policy makers should resist, in whatever ways possible, the resegregation along the lines of race and class of schools and classrooms that is currently taking place throughout this country.

A further implication of the importance of having diverse perspectives in the classroom is the need to reexamine current ability-based tracking practices. The work of educational researchers Jeannie Oakes, Mary Stuart Wells, and Irene Serna suggest that tracking, as it is currently implemented, works more to segregate along the lines of race and class than to discriminate along the lines of educational preparedness or ability. In several studies examining the decision-making processes of the people responsible for deciding how students will be tracked, these researchers demonstrate that ascriptive identities like ethnicity and gender are as instrumental in determining where a student ends up as are the student's test scores. Wells and Serna have further shown that the resistance to de-tracking is extremely strong among elite parents who perceive their children to be beneficiaries of the tracking system.[19] Such parents assume, mistakenly, that ability-based tracking is unbiased and that it ensures a more educationally challenging environment for their child. They thus fail to acknowledge the salience of identity categories for affecting educational outcomes—for their own children as well as for non-elite children. Moreover, they lack an appreciation for the potential epistemic benefits of a diverse classroom. So, while educators committed to transformative multicultural education cannot expect to easily end current tracking practices, we need to continue our efforts to develop more elaborated discourses about the economic and social salience of identity and the epistemic significance of perspectival diversity. Such discourses will be crucial to our success in affecting educational policies regarding the population diversity of our nation's classrooms.

Finally, the need for diverse perspectives and the importance of fostering dialogue in the classroom calls for a reexamination of current policies affecting the funding and oversight of our nation's public school system. As teachers know very well, it takes both time and space for us to get to know our students well, and for our students to get to know and respect each other. Moreover, it takes money to buy an adequate supply of that time and space. Without sufficient funding to hire well-qualified teachers, purchase up-to-date teaching materials, build and maintain safe and functional physical facilities, and retain the necessary administrative support staff, public schools will not be able to provide the small classrooms and interactive learning environments that are necessary for mining diverse perspectives and fostering productive dialogues.

Indeed, the steady de-funding of public schools—and the consequent rush of panicked parents toward private schools, home schooling, and school vouchers—poses a grave danger to our democratic system inasmuch as it effectively eviscerates public education's function as a shaper of civic identities. As Rob Reich discusses in his *Bridging Liberalism and Multiculturalism in American Education,* parents who pull their children out of the public school system are more likely to place them in learning environments that reinforce their beliefs rather than in environments that challenge them. This can have the effect, Reich argues, of stunting children's sense of civic responsibility and diminishing their capacity to develop what he terms a *minimalist autonomy.* Minimalist autonomy, according to Reich, "refers to a person's ability to reflect independently and critically upon basic commitments, desires, and beliefs, be they chosen or unchosen, and to enjoy a range of meaningful life options from which to choose, upon which to act, and around which to orient and pursue one's life projects." Its development, moreover, *requires* engagement with diverse perspectives and is crucial to an individual's ability to act purposefully with others in the service of creating and maintaining a democratic society.[20] So, unless we fund our public schools sufficiently to provide good, safe, educational environments that are attractive to a wide diversity of parents, we will fail to provide *all* our students with the opportunities they need to fully develop their sense of civic responsibility. Without a diversity of perspectives in the classroom, and without engaging in dialogues that challenge their sense of what is good, right, true, and beautiful, our children are highly unlikely to spend time reflecting on the best way to structure our diverse society.

Without diminishing the importance of working for large-scale school reform, I understand that teachers cannot wait for reform before they step into the classroom. Consequently, I turn my attention now to how teachers can work to mobilize identities in the classrooms they currently occupy. I begin by addressing a common mistake that teachers and students both make, that is, attributing to another student an "alternative" or "marginal" perspective that he or she does not have. I then discuss more specifically how to mobilize identities in a way that does not burden students, or stereotype them, or prevent them from growing and changing.

IDENTITY AND THE REALM OF THE VISUAL

An important part of mobilizing identities in the classroom in the way that I am proposing involves acknowledging—and then disentangling—the relationship between identity and the realm of the visual. As I indicated above, some identities appear to be visibly marked on the body. That is, they exist as social categories or ascriptive identities in part because they reference what are visual bodily characteristics (such as skin color, hair texture, limb shape, etc.) and assign to those characteristics an excess of social meaning. It is important to note that these visual bodily characteristics have no intrinsic meaning. Rather, they become imbued with meaning through the conflictive process involved in producing a social consensus about the way our society should be organized. Members of a society for whom a particular identity is especially meaningful will be socialized to select out and "see" the visual bodily characteristics commonly associated with that identity. Such socialization is necessary because such bodily characteristics are not visually obvious to everyone—especially to those people who have not been brought up to see them.

Racial identities are one example of the kinds of identities that appear to be marked on the body. Others include gender and some kinds of disabilities (such as blindness, paralysis, or limb loss). By comparison, other kinds of identities are commonly thought to be "invisible." Examples include sexuality, class, and other kinds of disabilities (such as dyslexia or chronic fatigue syndrome). Even with these "invisible" identities, though, we often behave as if we can reliably "see" identity. This is because we, as members of a society in which such identities are seen as significant, are socialized to pick up visual cues (bodily comportment, clothing, accessories) as a way of "seeing," and thus "knowing," them.

Sara Hackenberg has recently identified a process and coined a term—*visual fetishism*—that has been useful to me in thinking through our societal tendency to privilege the act of "seeing" the Other as a proxy for "knowing" the Other.[21] Even as we realize that some black people can "pass" for white, that Latina/os come in a wide range of colors and physiognomies, that some men dress and live as women and vice versa, that we cannot reliably read sexuality or class status on the body, and that many disabilities are invisible to the eye, we consistently operate in the world *as if* identities are always visible. We imagine that we can "see" difference, and that we always "know" to what racial, gender, class, or sexual orientation group someone belongs. We fetishize what is visible to us as if it contains the "truth" of the person—revealing their inner thoughts, capacities, and attitudes—even though we understand, at some level, that we may well be mistaken. We imagine not only that we can "see" race, gender, ability, and sexuality, but also that we can "know" in a reliably determinative way what those aspects of a person's identity will imply for the kind of individual that person will turn out to be.

It is important to remember that the act of "seeing" and thus "knowing" the people we come into contact with is experienced by most of us as being indispensable to our ability to act in the world. At a very basic level, visual fetishism helps orient us in the world as we act in accordance with the narratives we have internalized about who we are in relation to others. Visual fetishism can thus be a source of comfort to us as inhabitants of a rapidly changing society. But at a more problematic level, visual fetishism provides some people with an unfounded sense of superiority. This is particularly the case when such people are confronted by those racial, sexual, cultural, or bodily "others" who confound them, whose practices and values, because they are different, challenge their own. Because of the Othering it involves, visual fetishism can give some non-disabled persons a false sense of confidence about their own enduring able-bodiedness, even as it provides a measure of solace to the nativists who seek to shield themselves from the instability of values, practices, and hierarchies that racial and cultural "otherness" seems to threaten them with. In this way, visual fetishism can foster profound ignorance by preventing those who are most anxious about the existence of "others" in their midst from learning more about the "others" they know so little about, even as it can exacerbate oppression by keeping such people from interrogating their own false sense of superiority.

Even as we exercise caution with respect to judging people on the basis of how we see them, we must yet recognize that how we see them *does* matter for their experience. After all, the extent to which identities are referenced through the realm of the visual is also the extent to which they activate the pernicious aspects of visual fetishism, and thus matter to a person's day-to-day experience of oppression. In a society like ours that fears both strong women and women whose sexuality exceeds the bounds of normative heterosexuality, a lesbian who "looks" like a dyke is at greater risk of being gay-bashed than is a lesbian who is more gender conforming. Similarly, in a society like ours that has long associated skin color with status, a dark-skinned black man is at more risk of being pulled over and interrogated while driving an expensive vehicle in a predominantly white area than is a light-skinned black man. And finally, in a society like ours that, as Tobin Siebers has pointed out, has no common experience of disability, a person who has difficulty speaking is more likely to be judged by others as mentally incompetent than someone who speaks clearly—when in fact there may be no correlation between that person's ability to speak and his or her mental capacity.[22]

Mobilizing Identity in the Classroom

How can we, as teachers, mobilize identities in the classroom in a productive way? How do we avoid stereotyping students on the basis of visual fetishism even as we give due weight to the perspectives they have developed as the result of the identities they have? How do we bring our students' experiences into the classroom without either pigeonholing them as "native informants" or allowing them to be unquestioned authorities on an identity group as a whole? How, in other words, do we recognize our students as complex human beings not reducible to their ascriptive identities even as we take full advantage of the knowledge they have gained as a result of being socially situated beings?

Mobilizing identities, as I am defining the practice, involves mining our students' identity-based perspectives to see what insights into an issue they might have to offer, as well as subjecting our students' identities to evaluation and possible transformation. As educators, we want to attend to the various perspectives our students bring into the classroom, even as we give them an opportunity to change and grow. After all, if we wanted our students, upon leaving our classrooms, to be the same people they were when they entered it, we would not have accomplished very much. Moreover, because socialization as a fundamental aspect of all forms of education cannot be avoided, we need to think carefully about the values our pedagogical practices support. Education should give students the tools they need to evaluate the beliefs, conditions, and truth claims they will be exposed to throughout their lives; it should not be about merely inculcating status quo values. The purpose of a transformative multicultural education, moreover, should be to educate for democracy and social justice; it should be to help our students develop a better understanding of the structure of society and an increased sense of efficacy with respect to their own ability to influence positive social change. With these purposes in mind, I propose several principles for successfully mobilizing identities in the classroom.

Remember that every student is a complex individual with the capacity to contribute positively to the learning environment. Unless we treat our students—and, in particular, our minority students—as complex human beings with the capacity to contribute positively to the educational goals of the classroom, we risk reinforcing negative identity contingencies and creating classroom conditions that trigger stereotype threat. Since stereotype threat is activated when students fear they will be evaluated in terms of a prevailing negative stereotype about a group with which they are associated, students need to feel that their teachers, and peers, are capable of seeing them as complex individuals with the capacity to grow and change rather than as embodiments of a reductive stereotype. Although, theoretically, any student can be subject to stereotype threat, the risk for our minority students is much greater simply because they are the ones most subject to reductive and negative stereotypes in our society at large.

Work to get to know each student as a particular individual who is shaped and reshaped as a social being in and through collective identity categories and larger social structures. We can use several strategies to get to know our students as individual and complex human beings. I will suggest here a few that have worked well for me: First, ask your students to write something about themselves at the beginning of the class for you. Make the question open-ended so that you can get a sense of what aspects of their identity are most salient for each of them as individuals. Second, hold individual student conferences. This is a lot of work, but really worth it if you can make the time; there is simply no better way to get to know someone. Third, set aside a sufficient amount of discussion time, and introduce topics designed to get students talking. Think about ways to clear space for students who are too shy to talk, without forcing them to talk if they are very uncomfortable. If a student is particularly quiet during class discussions, I will ask her privately if she would like for me to call on her. Usually, she will say yes—the trouble she has in entering the discussion often has more to do with a reluctance to interrupt than with a lack of something to say. Occasionally, he will say no, and explain that he is either nervous about his language skills (this is frequently the case for ESL [English as a Second Language] students), or simply shy. In such cases, I offer alternative ways for my students to contribute to the discussion. I never want my students to be plagued by performance anxiety and I do not believe that everyone has to participate in a conversation to the same degree. The important issue for me is that everyone should have the opportunity to share his or her views in one forum or another. A number of university professors I know, myself included, have taken advantage of our universities' move toward Web-based discussion forums. I find that students who are uncomfortable talking aloud in class can be quite eloquent in online forums. Web-based discussions have not replaced in-class discussions in my courses, but they have enhanced my classroom discussions in crucial ways. Most importantly, learn to listen carefully as you allow your most die-hard assumptions to be challenged. Do not assume that an Asian student's parent pushes him too hard. Do not assume that a Latina/o student's first language is Spanish. Do not assume that your women students are not going to do well in Math. Rather, listen to what your students they say about their growing-up, their partners, their abilities and disabilities, their intellectual and social commitments. Do not expect absolute consistency and allow for contradictions. Treat each student as an individual who is shaped and reshaped by his or her changing social and economic situation.

Help your students to understand their connectedness to others by developing strategies to denaturalize your students' identities. In a society like ours that idealizes the unconstrained abstract individual, those of us who wish to mobilize identities in the classroom must help our students develop an analysis of society that allows them to understand their connectedness to others—and, in particular, to those who seem most different. This involves denaturalizing our students' customary (narrowly individualist) ways of being in the world. It means demonstrating to our students that all identities (including their own) are linked to historically, geographically, and culturally located ways of being a person in the world. Making the connection explicit will not only denaturalize the process of identity formation, but will introduce students to the complicated and far from obvious—but significant—relationship between social location, experience, and knowledge. In general, unless people's customary ways of being in the world are disturbed, their identities (and thus their interpretive perspectives) will remain un-theorized and profoundly parochial. And while even un-theorized and "inaccurate" identities can be epistemically useful to an observer for investigating the workings of ideology, they will not contribute to their bearers' ability to effect positive social change until they have been denaturalized and brought into the realm of examination and evaluation.[23]

Find strategies for denaturalizing your students' identities that are appropriate to your classroom and to your students. Denaturalizing identities in a lecture class will be a different project than in a discussion class. For example, in a lecture class I co-taught with Hazel Markus in Spring 2004, I watched as she accomplished, in an effective way, the task of demonstrating that all identities are linked to historically, geographically, and culturally located ways of being a person in the world. One day, Markus began the class by having our students fill out a short psychological survey describing themselves, their ethnic identities, as well as their attitudes about upward mobility and prejudice. In the lecture that followed, she introduced them to the large body of social science research in the United States and in Japan that describes what she has termed "self-ways."[24] In a subsequent class, Markus brought the results of the survey to share. In presenting the results, Markus demonstrated how—with some variation along gender and race lines—our students conformed to an identifiably "American" way of being a person in the world. Markus's research and pedagogical strategy effectively allowed our students to see themselves as racially and culturally located beings who have been shaped, but not wholly determined, by the values and mores of their racially and gender-stratified society. This not only disturbed our students' customary sense of themselves as self-created and wholly autonomous individuals, but it also pushed them to understand themselves as analogous to the Japanese young people who have been similarly shaped, but not wholly determined, by the values and mores of *their* particular society. Denaturalizing the process of identity formation has the advantage of helping our students understand that *everyone's* identity is complex and multiple and formed in relation to his or her situation. It helps them to avoid the pitfalls of assuming, too quickly, that they know the attitudes and assumptions of the "others" they are interacting with, even as it frees them to explore different aspects of their own identities. When students are given the tools to understand how and why they believe and value what they do, they are empowered to question their own received notions, occasionally rethink them, and, in the process, transform their identities.

Mobilizing identities in a discussion class, as opposed to a lecture, will necessarily involve the students in a more active way. Susan Sánchez-Casal has experimented with mobilizing identities in her Latina/o Studies classroom by identifying existing communities of meaning and sorting her students into small working groups based on those communities.[25] She then asks the students in each group to work together to develop arguments on issues that will be discussed in class. The beauty of Sánchez-Casal's approach is that it allows students to develop their ideas in concert with like-minded peers; it thus works against the false notion of the individual knower even as it provides students who have minority perspectives a sense of affirmation for their ideas during the crucial period of development and clarification of those ideas. I know from talking with my minority advisees that if they get no support for their ideas from the professor or even one other student in a class, they begin to withdraw from that realm of interaction by dis-identifying with it. Students need to feel that their ideas are good (i.e., valued) before they can effectively put those ideas to the test through dialogue or debate in a classroom setting. Keeping our students engaged is a prerequisite for providing them an opportunity to reorient their perspectives. Identifying preexisting communities of meaning, as Sánchez-Casal did, is thus an important strategy in the effort to mobilize identities in the classroom.

One way to identify existing communities of meaning is by noting how students sort themselves when they enter our classroom. Which students consistently sit together? Do they share a racial or ethnic background? Are they of the same gender? Do they hail from the same geographical community? Are they affiliated with a particular university club or religious group? What is the source of their identification with each other? Paying attention to where and with whom our students sit will tell us a lot about how they understand themselves relative to the other students in our classrooms. Knowing this will help us figure out how best to engage our students in the learning process. Of course, in setting up communities of meaning in the classroom, we should keep in mind the importance of avoiding polarization along one set of identity lines. While we want to give due weight to the communities of meaning into which students initially sort themselves, we also want to help students realize that they might be able to form communities of meaning that are drawn along other lines. We can do this by emphasizing the complexity of students' identities and by not letting race, or gender, or ability stand alone as the determining factor for the formation of working groups for the entire duration of the class. One possible way to address this concern is to switch up topics of discussion to allow students to see how the different aspects of their identities become salient in different situations. As we change the issue—from affirmative action to abortion, from handicapped access to online file sharing—the possible communities of meaning should alter somewhat. Changing the focus of discussion and re-forming working groups in your classroom to create new communities of meaning can reinforce the lesson that *all* people, themselves as well as others, are complex and multiple beings not reducible to their most visible ascriptive identities.

Actively cultivate an atmosphere of intellectual cooperation and mutual respect by being prepared to compensate for differences in power relations and adjudicate conflicts in values that enter the discussion. Given the hierarchical nature of our society, we are likely to be called upon to compensate or adjust for disparities in power that seep into the classroom from the larger society. Part of creating a context in which disagreements can be aired safely may thus involve interceding on behalf of a marginalized viewpoint or community. One way teachers can preempt the necessity of such intercession is to strategize ways to give marginalized perspectives and minority identities priority in the discussion. We can, for example, give students who are advocating a position that is not easily understood (or held) by the majority of students extra time to present background information necessary for understanding the issue. We can require the class as a whole to read articles, watch videos, or do research projects that excavate a minority or erased historical event or perspective. In addition, we can point to the interests historically served or denied by the social and economic structures that have privileged some identities and perspectives at the expense of others. And we can explain to our students that such apparent "imbalance" is necessary for opening up the issues under discussion and for maximizing objectivity by bringing a multitude of perspectives to bear on the issue.

Adjudicating conflicts in values can be equally as difficult but just as necessary to the project of creating an atmosphere of intellectual cooperation and mutual respect. Of course, we need to be careful to adjudicate conflicts in a way that do not close down discussion. To that end, students will need to know from us, through consistency of word and action, that we will not penalize them for taking the wrong position. Moreover, teachers should avoid having too strong a voice or position at the beginning of any debate or dialogue. In general, disagreements and strong rebukes are best voiced by fellow students, who have less real power over their peers in our classrooms than we do. This is not to say that we should stay out of the discussion entirely, or that we should tolerate any form of rudeness or disrespect. The first reason we cannot exempt ourselves from the discussion is that doing so will cause our students to mistrust us; they know we have a perspective and will feel cheated if we pretend we do not. Besides, our students expect to learn something from us (we are the teachers, after all!) and may feel that we are acting in bad faith if we expect them to lay their cards on the table while we refuse to do the same. Another crucial reason we may need to intervene in a discussion is that true dialogue can occur only in an atmosphere of mutual respect. Where real disagreements arise, we will be called upon to make sure that students show respect for each other's views. Our efforts in this vein should be directed toward fostering an atmosphere of intellectual cooperation and mutual respect while allowing for an exploration of conflict and contradiction. Our goal should not be to reach consensus (although consensus is not bad in itself!); our goal should be a respectful airing of differences and a meeting of intellectual and emotional challenges.

Remember that you are teaching the practice of critical thinking rather than a particular ideological stance. At base, remembering that we are encouraging a practice rather than delivering a product means that not every issue needs to be discussed in every classroom. Indeed, in order to effectively identify and mobilize communities of meaning in the classroom, we must be sensitive to the sorts of issues we introduce for discussion in the context of our particular set of students; it is not always safe for students to voice or champion minority perspectives. After all, if a teacher has only one gay student (or if he himself is gay) in a classroom full of anti-gay religious fundamentalists, it might not be the wisest idea to bring up the subject of gay marriage. The teacher might end up creating a situation in which his one gay student is silenced, alienated, or shamed, while his fundamentalist students are reinforced in their homophobia. Accordingly, we must bear in mind that it is neither possible nor necessary to discuss every issue in every classroom context. Just as I do not have to give my children every different kind of fish to get across the general idea that fish are in the class of things that are good to eat, so teachers do not have to discuss every hot button social issue with their students to convey the general idea that social issues are in the class of things that are good to discuss and reevaluate. Once we introduce students to the dialectic of identity and the principle of socially situated knowledge, they should be able to extend those lessons into other arenas of debate later on throughout their lives.

The key to mobilizing identities effectively in the classroom is your own identity. If we, as teachers, hold and neglect to examine and change stereotypical or prejudicial attitudes toward members of socially stigmatized groups, we are going to take those views into the classroom and mobilize them—whether we intend to or not. Because of the power dynamic inherent in every classroom situation, our identities will have a tremendous influence on classroom dynamics. As much as possible, then, we need to be aware of and understand those dynamics so that we can work with them. Whatever your identity, it is going to matter for how you interact with the students in your classroom. And because identities are relational and contextual, your identity will matter differently according to *who* and *what* you are teaching. If, for example, a teacher is an Asian man who is teaching Math to a group of white students, he is probably going to be accorded a good deal of credibility. He may be terrible at Math; he may have received a 480 on his Math SAT, and be a substitute teacher who normally teaches Art. But because of the positive stereotype our society holds about Asians and Math, the presumption he will face is that he knows what he is doing. But if she is a black woman who is teaching Math to a group of white students, she is probably going to have a hard time at first. This is not to say that she should not do it. It is to say, though, that part of her work in that Math classroom is going to involve challenging stereotypes as much as teaching differential equations.

Finally, find ways to link the issues you discuss in the classroom to your students' daily lives. The recognition that *all* identities matter in the classroom—yours as well as your students'—affirms yet again the importance of linking learning to life. Because it is not possible to check our identities at the door of the classroom, we must work to avoid the "not in my backyard," or NIMBY phenomenon that some teachers fall into when they are talking, for example, about race. Pretending that identities do not matter to in the classroom does not make them insignificant to educational outcomes. It just makes it harder to confront their very powerful effects. So, without ever accusing any of our students of being racist, or sexist, or ableist (because making such an accusations will never alleviate the problem, but will contribute to a situation of defensiveness and polarization), a teacher who is working to transform her classroom into one that meets the needs of *all* her students must find a way to acknowledge that the social dynamics we discuss and study are social dynamics that we are all a part of both inside and outside of the classroom. Even as we work to avoid the pitfalls of blaming and accusing—as well as their corollaries, guilt and defensiveness—we have to acknowledge that *we* are implicated in the production and reproduction of racist, sexist, heterosexist, and ableist ways of knowing and unknowing.

As teachers and students, we are not responsible for what our society and parents teach us, any more than we are responsible for being born into a particular situation or having an identity ascribed to us. Identities, initially, are given to us. What counts is what we do with them—whether we embrace them without question or whether we work to transform them by critically examining the dogmas of our society, thus undermining the ideologies and associations that unfairly advantage some people at the expense of others. Certainly, mobilizing identities productively in the multicultural classroom will never be an easy, or even a completely safe, thing to do. But doing so is both possible and necessary if we are to ever be successful at creating a more just and democratic society for everyone.

Notes

This essay is reprinted from: *Identity Politics Reconsidered*, eds., Linda Martín Alcoff, Michael Hames-García, Satya P. Mohanty, and Paula M. L.Moya, Palgrave Macmillan, 2006, 175 Fifth Avenue, New York, NY, 10010 and Houndmills, Basingstoke, Hampshire, England RG21 6XS, Companies and representatives throughout the world. Reproduced with permission of Palgrave Macmillan.

1. Claude Steele, "Not Just a Test," *The Nation,* 278.17 (2004): 38–40; Claude M. Steele, Steven J. Spencer, and Joshua Aronson, "Contending with Group Image: The Psychology of Stereotype and Social Identity Threat," *Advances in Experimental Social Psychology* 34 (2002): 379–440.
2. References are to my *Learning from Experience: Minority Identities, Multicultural Struggles* (Berkeley: University of California Press, 2002), pp. 86 n. 2, 13, 133. It should be clear from my definitions that I understand identities to be both *constructed* and *real*. Identities are *constructed* because they are based on interpreted experience and ways of knowing that explain the ever-changing social world. They are also *real* because they refer outward to causally significant features of the world. Moreover, because identities refer (sometimes in partial and inaccurate ways) to the changing but relatively stable contexts from which they emerge, they are neither self-evident and immutable nor radically unstable and arbitrary. Identities, in sum, are causally significant ideological constructs that become intelligible within specific historical and material contexts.
3. Satya P. Mohanty, *Literary Theory and the Claims of History: Postmodernism, Objectivity, Multicultural Politics* (Ithaca, NY: Cornell University Press, 1997), 216.
4. For concrete examples of how racial ascription works in three elementary schools, see Amanda Lewis, "Everyday Race-Making: Navigating Racial Boundaries in Schools," *American Behavioral Scientist,* 47.3 (2003): 283–305.
5. Linda Martín Alcoff, "Who's Afraid of Identity Politics?" in *Reclaiming Identity: Realist Theory and the Predicament of Postmodernism,* eds. Paula M. L. Moya and Michael R. Hames-García (Berkeley: University of California Press, 2000), 337.

6. Richard J. Herrnstein and Charles Murray, *The Bell Curve: Intelligence and Class Structure in American Life* (New York: Free Press, 1994); Vincent Sarich and Frank Miele, *Race: The Reality of Human Differences* (Boulder, CO.: Westview Press, 2004). For critiques of Hernnstein and Murray, see Bernie Devlin et al., *Intelligence, Genes, and Success: Scientists Respond to* The Bell Curve (New York: Springer, 1997); Steven Fraser, ed., *The Bell Curve Wars: Race, Intelligence, and the Future of America* (New York: Basic Books, 1995).
7. Shelby Steele, *The Content of Our Character: A New Vision of Race in America* (New York: St. Martin's Press, 1990); Richard Rodriguez, *Hunger of Memory: The Education of Richard Rodriguez* (New York: Bantam Books, 1983). For a critique of such neoconservative idealist approaches to identity, see my *Learning from Experience,* esp. Chapter 4.
8. Judith Butler, *Gender Trouble: Feminism and the Subversion of Identity* (New York: Routledge, 1990).
9. A good example of an identity that is emergent is that of "mixed-race." For more on mixed-race identity, see Ronald Sundstrom, "Being and Being Mixed Race," *Social Theory and Practice* 27.2 (2001): 285–307; Michele Elam, "Pedagogy, Politics and the Practice of 'Mixed Race,'" in *Navigating the Frontline of Academia,* eds. Deirdre Raynor and Johnnella Butler (Seattle: University of Washington Press, forthcoming).
10. Steele, Spencer, and Aronson, "Contending with Group Image," 389.
11. See Linda Martín Alcoff, *Visible Identities: Race, Gender, and the Self* (New York: Oxford University Press, forthcoming), esp. Chapter 2.
12. See also Hazel Rose Markus, Claude M. Steele, and Dorothy M. Steele, "Colorblindness as a Barrier to Inclusion: Assimilation and Nonimmigrant Minorities," *Daedalus,* 129.4 (2000): 233–259. Working with a social-psychological model of identity, Markus, Steele, and Steele argue that creating what they call "identity-safe" classrooms involves understanding the precise ways that historical identities are *already* mobilized in the classroom, generally in negative and pernicious ways. Because of this, they are opposed to "color-blind" approaches to identity.
13. See my *Learning from Experience.* See also Moya and Hames-García, *Reclaiming Identity;* Mohanty, *Literary Theory;* Hames-García, *Fugitive Thought: Prison Movements, Race, and the Meaning of Justice* (Minneapolis: Minnesota University Press, 2004); Alcoff, *Visible Identities.*
14. See Natalie Angier, *Woman: An Intimate Geography* (Boston: Houghton Mifflin, 1999), esp. Chapter 3. See also Anne Fausto-Sterling, *Myths of Gender:Biological Theories About Men and Women* (New York: Basic Books, 1992).
15. Sandra Harding, *Whose Science? Whose Knowledge?: Thinking from Women's Lives* (Ithaca, NY: Cornell University Press, 1991). See also the discussion of standpoint epistemology in *Identity Politics Reconsidered,* eds., Linda Martín Alcoff, Michael Hames-García, Satya P. Mohanty, and Paula M. L.Moya (New York: Palgrave Macmillan, 2006).
16. Moya, *Learning from Experience,* 44. See also Mohanty, *Literary Theory,* 213–214.
17. Michael Hames-García, "Which America Is Ours?: Marti's 'Truth' and the Foundations of 'American Literature,'" *Modern Fiction Studies,* 49.1 (2003): 19–53. See also Patricia Gurin, Biren (Ratnesh) A. Nagda, and Gretchen E. Lopez, "The Benefits of Diversity in Education for Democratic Citizenship," *Journal of Social Issues,* 60.1 (2004): 17–34.
18. Nadga, Kim, and Truelove, in their study of a multicultural educational initiative at the University of Washington, argue that disconnecting learning from life has the negative effect of decreasing students' sense of agency with respect to interacting with diverse peers. They show that when intergroup dialogue is combined with content-based learning, students' levels of confidence regarding their ability to make changes in the existing social structure was measurably increased. Biren (Ratnesh) A. Nagda, Chan-woo Kim, Yaffa Truelove, "Learning about Difference, Learning with Others, Learning to Transgress," *Journal of Social Issues,* 60.1 (2004): 195–214.
19. Jeannie Oakes, "Two Cities Tracking and Within-School Segregation," *Teachers College Record,* 96.4 (1996): 681–690; Amy Stuart Wells and Irene Serna, "The Politics of Culture: Understanding Local Political Resistance to Detracking in Racially Mixed Schools," *Harvard Educational Review,* 66.1 (1996): 93–118.

20. Rob Reich, *Bridging Liberalism and Multiculturalism in American Education* (Chicago: University of Chicago Press, 2002), 117.
21. Sara Hackenberg, "Reading the Seen: Mystery and Visual Fetishism in Nineteenth-Century Popular Narrative" (Ph. D. diss., Stanford University, 2004).
22. Tobin Siebers, "Passing," in *Encyclopedia of Disability* , ed., Gary L. Albrecht. 5 volumes (Thousand Oaks, CA: Sage, 2005). See also Rod Michalko, *The Mystery of the Eye and the Shadow of Blindness* (Toronto: University of Toronto Press, 1998); Tobin Siebers, "Disability as Masquerade," *Literature and Medicine*, 23.1 (2004): 1–22; Siebers, "What Can Disability Studies Learn from the Cultural Wars," *Cultural Critique*, 55 (2003): 182–216; Tobin Siebers, "Disability Studies and the Future of Identity Politics," in *Identity Politics Reconsidered*, eds., Linda Martín Alcoff, Michael Hames-García, Satya P. Mohanty, and Paula M. L.Moya, Palgrave Macmillan, 2006.
23. Because identities are indexical—because they refer outward to social structures and embody social relations—even previously untheorized identities that are brought into contact and dialogue with other interpretive perspectives have the potential to provide us with differential, and potentially valuable, access to a shared and very complicated social world.
24. Hazel Rose Markus, Patricia R. Mullaly, and Shinobu Kitayama, "Selfways: Diversity in Models of Cultural Participation," in *The Conceptual Self in Context: Culture, Experience, Self-Understanding*, eds. Ulric Neisser and David S. Jopling (Cambridge: Cambridge University Press, 1997), 13–61.
25. In chapter 1 Susan Sánchez-Casal and Amie A. Macdonald introduce the concept of *communities of meaning*, which they define as interpretive perspectives on the world that are common to people who come from similar social locations. The concept is useful for reminding us of the social nature of identity and of knowledge, and of the way in which "common sense" changes according to the social context. See also Lynn Hankinson Nelson, "Epistemological Communities," in *Feminist Epistemologies*, eds. Linda [Martín] Alcoff and Elizabeth Potter (New York: Routledge, 1993), 121–59; Susan Babbitt, "Feminism and Objective Interests: The Role of Transformation Experiences in Rational Deliberation," *Feminist Epistemologies*, 245–264.

3

Fostering Cross-Racial Mentoring: White Faculty and African American Students at Harvard College

Richard J. Reddick

Introduction

Reflecting on my own experience attending predominantly white institutions (PWIs) of higher education, I feel fortunate that I emerged from my educational experience relatively unscathed, that is, that my sense of *self* is intact. The multiple aspects of my identity survived—being an African American male of West Indian descent, growing up in Great Britain and Texas, being an Air Force "brat," labeled alternately as "gifted" and "immature" by teachers, to name a few—though I certainly recall times during which I was pushed, both explicitly and implicitly, to assimilate to the cultural mores of the dominant culture—one which was overwhelmingly white, male, American, and socially and economically privileged. I do not wish to suggest that I have rejected all these values and identity positions, or that these mores were necessarily in opposition to those with which I came to academia. However, the fact that my peers and I resisted assimilation both preserved our sense of identity, and provided a richer, more democratic schooling experience, with fulsome cross-cultural experiences for us individually, and for the campus community writ large. This imperative is precisely what Paula Moya refers to in her elaboration of a realist theory of identity (Moya, 2002). In retrospect, I am proud of our desire to "stay true" to ourselves; the truth is, it was, and continues to be, our effort to resist. Of course, this is a struggle that is shared among many people of color in the academic world.

While a graduate student at Harvard University, I was asked to help facilitate conversations about race with a group of students living in one of the residential houses. It soon became evident that the African American students in the group struggled to adjust to the housing community, as well as many other aspects of the campus environment. Their identity contingencies (Steele, cf. Moya, 2008) as African American students at a predominantly white, elite institution meant that they were often assumed to be overly sensitive when they raised objections to offensive racial comments made by white peers, comments that the white students saw as "jokes." Additional challenges that the African American students voiced to me included the assumption made by white students that black students all came from fragmented families, and had been educated in academically under-resourced urban schools. At times, these suppositions held some truth; nevertheless, being stereotyped in such a way meant African American students were forced to expend considerable energy disproving these assumptions. I instantly recognized these stressors and stereotypes from my own undergraduate experiences.

White students at Harvard, and institutions like it, are advantaged by an understanding of the subtle codes and ways of academia. By virtue of sharing a cultural and/or ethnic identity with the majority of power-brokers in the institution, intergenerational transfer (having parents who are themselves academics or graduates of highly selective colleges) or being acculturated in elite, predominantly white environments, these students are able to access what Margolis and Romero (1998) term "the hidden curriculum"—a constellation of understanding of norms and procedure, interpersonal networks, and cultural capital—leading to benefits and advantages that assist students in achieving success in a competitive environment. Though it would be erroneous to state that some African American students do not accrue these advantages at Harvard, the majority of students that I spoke to had less access to administrators and faculty who might have served as intercessors to these advantages, and were thereby not afforded the *critical access* (chapter 1) that would have assisted them in decoding the hidden curriculum, and in accessing resources and opportunities that all too often elude students of color at predominantly white campuses.

In my conversations with students, I soon found that the African American students at Harvard who were successfully overcoming these challenges had established networks of peers and staff support that provided *critical access*; these programs and cultural centers provided them a crucial link and a sense of safety (Whitla et al., 2005). But the most valued support seemed to be the relationships that some students were able to foster with faculty members at Harvard. Students spoke fondly of collaborating on research projects with interested faculty members, and dining at the homes of certain professors. Some students went so far as to label their faculty supporters as *mentors*. The mentors to whom African American students at Harvard connected did not encourage or promote the idea of their students altering or shifting their identities, but rather assisted students by giving them confidence and encouraging them to mobilize their identities (Moya, 2008) as important social and intellectual tools.

I observed that the majority of faculty mentors of African American students were also African American. Research explains why this was so: African American students look to African American faculty for support, specifically because of their belief that these faculty bring varied pedagogical techniques and perspectives to PWIs (Hurtado, 2001; Umbach, 2006). Further, research suggests that mentor-mentee relationships are more likely to occur with same-race, same-gender dyads (Turban, Dougherty, and Lee, 2002). Students informed one another about these faculty known to support young African Americans and in this manner, many African American students established an enduring network of mentors that they connected with as students, and with whom they often stayed connected, even after graduation. However, the first difficulty for African American students seeking to find African American mentors at PWIs such as Harvard is finding faculty that meet the criteria: despite increasing student diversity, the percentage of African American faculty has remained virtually stagnant—4.4% in 1975 to 5% in 1997 (Trower and Chait, 2002). The growth in African American students attending PWIs—a 56% increase since the 1980s (Harvey, 2002)—outpaces the growth in African American faculty, creating a deficit in the necessary pool of African American mentors.

Additionally, the small number of African American faculty who could serve as mentors to African American students were being unfairly burdened with their roles as mentor to students of color. This dilemma has been termed "cultural taxation" (Padilla, 1994) or the "black tax" (Cohen, 1998). This "cultural taxation" obligates African American faculty and other professors of color to take responsibility for the welfare of students of color, to mentor them, officially and unofficially, to serve on various multicultural committees, something that allows senior administrators to create the illusion of diversity at institutions that lack adequate minority faculty representation. This "hidden service agenda" places the burden of representation on the shoulders of faculty of color, far more than on white faculty (Brayboy, 2003). Junior faculty shared with me the even more intense examples of having to make hard decisions about how to allocate their time among these competing commitments. "Cultural taxation" and other forms of racial bias contribute to stress for faculty of color, who often experience the very same challenges they are asked to help students cope with (Turner, 2003). And these stressors often begin for African American faculty before they reach the rank of assistant professor. For example, as a graduate student, I often talked to my faculty mentors of all races about my commitments to younger students of color at Harvard and social justice efforts directed at inequality in education. Many times, these conversations came before, and took more time, than discussions about coursework, or research ideas.

The interactions, observations and experiences described above are how dissertation projects are born. From my own experience, I knew that I had been the recipient of excellent mentoring not only from African American faculty but also from white faculty. Indeed, scholarly narratives of faculty at historically black colleges and universities demonstrate that there is a rich legacy of cross-racial mentorship and nurturing in higher education (Allen and Jewell, 2002; Willie, Reddick, and Brown, 2005), and so I wanted to investigate if this experience was shared by African American undergraduates at Harvard. Mirroring my own experience, the research literature provides many examples of exemplary mentorship of African American scholars by white faculty (Lynch, 2002; Spence, 2005; Willie, 1986). Bringing similar examples to light among the Harvard faculty (and faculty writ large) would have two outcomes: one, providing evidence of how successful mentoring of African American students by white faculty could encourage more cross-race mentoring; and two, alleviating the cultural taxation for African American faculty by sharing the responsibility of mentoring among white faculty. I also knew from conversations with African American students and white faculty that there was often an assumption that cross-race relationships might be less fulfilling than those matching African American students and African American faculty. Yet research on this topic indicates that once students and faculty find a common understanding and form a mentoring relationship, cross-race mentoring relationships can be as successful as same-race mentoring relationships (Thomas, 2001). This finding is significant, since an increase in the number of white faculty who mentor African American students and other students of color improve conditions for faculty and students of color by widening the pool of mentors, increasing the educational success of minority students, and more fairly distributing labor among faculty of color and white faculty.

Given the myriad challenges confronting African American undergraduate students at PWIs beyond simple transitional issues—the phenomenon of underperformance, in which African American students with similar preparation and test scores perform less well than their white and Asian American counterparts (Bowen and Bok, 1998); stereotype threat, in which the simple act of identifying oneself as a member of a stigmatized group can depress performance (Steele, 1997); and racial microaggressions—I wanted to further explore which aspects of their own identity influenced white faculty to serve as mentors to African American undergraduates. Understanding how white faculty have worked successfully to provide *critical access* for African American undergraduates, helping to reify those students' identities as well as uncloak the hidden curriculum, I believe will serve to "[foster] the conditions conducive to working toward a better society" (Moya, 2002, 19). Do similar but not identical experiences of exclusion and oppression prompt white faculty to place their ethical commitments with others who are also targeted? I sought an emergent theory that could connect the different experiences of having one's identity challenged on the basis of gender or ethnicity, for instance, to the ability to empathize across racial difference with students facing challenges on the basis of race.

Background

When one discusses a phenomenon such as mentorship, it is imperative that we start from a place of common understanding. The psychologist Daniel Levinson (1978) brought the term into popular usage with the publication of his landmark study of adult development in men, *Seasons of a Man's Life*. In describing the transition from young adulthood to maturity, Levinson found a mentor served "as a host and a guide," also providing "counsel and moral support" (1978, 98). The concept of mentoring took root in management and organizational literature, most notably in Kram's seminal volume, *Mentoring at Work* (1988). Blackwell (1983) and Merriam and colleagues (1987) understood the applicability of mentoring concepts in higher education in the 1980s. However, the popularization of the concept has led to an array of definitions of what precisely constitutes mentoring. In literature reviews on developmental relationships, Jacobi (1991) and Crosby (1999) catalog respectively 15 and 18 definitions of mentoring.

Such variance has led researchers such as Sharon Merriam to comment, "The phenomenon of mentoring is not clearly conceptualized....Mentoring appears to mean one thing to developmental psychologists, another thing to businesspeople, and a third thing to those in academic settings" (cf. Jacobi, 1991, 506). However, the essential concept in mentoring is the ideal of instrumental and psychosocial support, integrating a variety of behaviors first described by Kram (1988) in table 3.1.

Table 3.1 Mentoring Functions as described by Kram (1988)

Career (Instrumental) Functions	Psychosocial Functions
Sponsorship	Role Modeling
Exposure-and-Visibility	Acceptance-and-Confirmation
Coaching	Counseling
Protection	Friendship
Challenging Assignments	—

It is important to acknowledge the significance of the individual functions of mentorship, even when such relationships do not develop into mentorship (Johnson, 2007; Thomas, 1993). Thomas has termed these interventions and functions as *developmental relationships,* "that provide[s] needed support for the enhancement of an individual's career development" (1990, 480). Thomas further notes the reciprocal nature of such relationships. While there are significant benefits to receiving any of these functions, mentorship exists only when all functions are present, and thus is a more intensive relationship (Johnson, 2007). Researchers argue that the challenges confronting African American students at PWIs demand the support that mentorship provides (Guiffrida, 2005; Reddick, 2005; Stanley and Lincoln, 2005).

Mentoring in Higher Education

In higher education, mentoring is often conceptualized in terms of relationships between graduate students and professors, or junior and senior faculty (Dixon-Reeves, 2003; Johnson and Harvey, 2002; Stanley and Lincoln, 2005). While mentoring relationships among faculty and African American undergraduates are an underdeveloped research topic, college student development researchers recognize that these relationships are a key aspect of student development (Astin, 1984, 1992; Tinto, 1987). Not every African American student will need mentoring for the purpose of navigating the PWI environment; however, given the challenges that do arise, mentoring will be valuable to a segment of this population.

Another critique of most studies of mentorship involving African Americans in higher education has been the almost exclusive focus on perspectives of the mentees at the expense of understanding the experience from the perspective of the faculty member (Guiffrida, 2005). I find that the popular conception of the professor as expert stifles our understanding that faculty, like the students they are charged to mold and develop, are often themselves struggling to understand and make meaning of their own experiences. Through sharing examples of these professors' confusion and triumph as mentors, I present a reminder that educators are not only teachers, but learners. The voices of white faculty mentors that I present in this study shift the focus to the hard work put forth by these professors in learning to be effective supporters of their students across racial lines.

Theoretical Context

In the words of biologist Sir Julian Huxley, "Ideas are too heavy to carry around in one's mind unless they are rolled on wheels of theory." To undergird the context of this study, I present here how I utilize two major concepts—Critical Race Pedagogy (CRP) and theories of cross-racial developmental relationships—and to a lesser extent, career and adult development theories.

Critical Race Pedagogy

To frame the experiences of faculty who engage in the mentorship of African American students, I look to the intersection of the legal tenets of Critical Race Theory (CRT) and education, termed Critical Race Pedagogy (CRP) (Lynn, 1999). A legal discourse emphasizing that experiences of racial oppression are not isolated, but ever-present in American life (Matsuda et al., 1993), CRT stands in opposition to modernist conceptualizations of neutrality, promoting a postmodern perspective valuing the importance of standpoint in the development of thoughts and action. When applied to education, CRT acknowledges the experiences of people of color who have encountered oppression, and further states that the consequences of racial oppression are linked to gender- and class-based bias (Delgado and Stefancic, 2001; Ladson-Billings and Tate, 1995; Solórzano and Yosso, 2000).

The next step in the evolution of CRT in the field of education came with Lynn's advancement of CRP. In his study of African American teachers, Lynn (1999) built upon the work of critical race legal scholars such as Derrick Bell and Richard Delgado, and education researchers Ladson-Billings and Tate (1995). This theoretical context locates educational settings as a site to challenge and reverse racial and ethnic oppression (Solórzano and Yosso, 2000, 2001). CRP brings together the legal framework of CRT; the concepts of care for the academic, intellectual, and cultural development of students via the work of black feminist theorists; and previous research on the practices of African American teachers (Collins, 1989; Delpit, 1988; Foster, 1990). These studies highlight the importance of cultural connectedness to students, linkages to notions of unity and self-determination, and empowering students to perform in a society that elevates individualistic values over the collective. I connect my study on mentoring to CRP through the parallels between Lynn's analysis of the experiences and pedagogical practices of African American teachers and my focus on the mentoring practices of white professors working with African American students as they respond to the consequences of students' identities in the institutional setting. Like the teachers in Lynn's work, these professors employ a variety of strategies to support their students from a psychosocial standpoint that are often perceived as being "soft" or "nonacademic"—when in fact these Maslovian approaches are often essential before one can delve into the complexities of academic work. From a realist perspective, these mentors are taking into account the function of students' ascriptive identities (Moya, 2008), and responding to them in ways that support students' academic and social success.

Theories of Cross-Racial Developmental Relationships

This section reviews the literature focusing on developmental relationships across racial and ethnic lines—typically, between white mentors and African American mentees. In a study of cross-race pairings, Thomas (1993) found that relationships that progressed beyond sponsor-protégé status (those which only featured instrumental career support) to mentor-protégé status (those which also featured instrumental career advising, but with a strong element of emotional investment in the junior person) depended greatly on both party's preferred strategies for dealing with race. Thomas categorized their responses as either *denial and suppression* (expressed discomfort about discussing race) or *direct engagement* (openly and willingly confronted issues pertaining to race in their lives and careers). When whites and African Americans in developmental relationships expressed congruent strategies (e.g., both denial and suppression or both direct engagement), mentor-protégé relationships developed. This finding suggests that mentoring relationships between African American students and white faculty can develop if both parties prefer similar strategies in their negotiation of race. In such arrangements, neither party views race as an issue that inhibits the development of the relationship. The psychosocial involvement of a mentor-protégé relationship may be more appropriate than sponsorship when the junior person encounters challenges such as perceived racism, or racial microaggressions.

Thomas' research is a promising starting point for considering developmental relationships across race among African American undergraduates and white faculty members. However, it should be noted that this study was conducted in a corporate setting, with different dynamics compared to a collegiate setting. In addition, the age range of the junior participants (ages 21 to 29) exceeds that of the typical undergraduate population. Developmentally, African American undergraduates are in a different stage—one that psychologist and progenitor of mentoring research Daniel Levinson would term *early adult transition* (ages 17–22), while the subjects in Thomas' study are mostly in the *entering the adult world* and the *age 30 transition* stages (ages 22 to 28) (Levinson, 1979). The linkages in Thomas' study, however, help to conceptualize cross-race mentoring in this study. In addition Thomas does not disclose much information about the corporations in which the subjects in either study worked—aside from the fact that the executive and management groups are predominantly white. Nevertheless, these theoretical constructs advance my analysis of how white faculty integrate their motivations and formative experiences into their mentorship of their African American undergraduate mentees by providing a basis for analyzing cross-race developmental relationships. However, in contrast to Thomas' research, my study extends cross-race mentorship theories into the collegiate context.

METHODOLOGY

Research Questions and Design Overview

There were two questions that led my research on faculty identified as mentors to African American undergraduate students at Harvard College. First, I wanted to know what aspects of their own identities influenced white faculty to serve as mentors to African American students. I also wanted to know if formative experiences with other aspects of their identity (i.e., gender and religion) informed how white faculty advised their African American mentees to respond to racially microaggressive experiences.

Table 3.2 Participant Sample, Areas of Study, and Rank

Name	Ethnicity and Gender	Area of Study	Rank	Age
Rachel Smith	white female	Applied Sciences	Advisor/ Teaching Fellow	36
Caitlin Moss	white female	Social Sciences	Assistant Professor	38
Victoria Stein	white female	Social Sciences	Professor	61
Stephen Cox	white male	Applied Sciences	Professor	60
Andy Giordano	white male	Natural Sciences	Lecturer	40
David Müller	white male	Humanities	Assistant Professor	37

I distributed a survey at Harvard College to members of African American student organizations to find white faculty mentors. The respondents to the survey numbered 64 African American students and recent graduates. Of that number, 18.8% were graduates beyond a year after their studies at Harvard. The remaining respondents were enrolled as students at Harvard College: 34.4% of the respondents were first-year students, 17.2% were sophomores, 21.9% were juniors, and 7.8% were seniors. I calculated that I had responses from 9.4% of the African American students at Harvard College.

Utilizing a purposive selection technique (Patton, 1990) from the pool of faculty identified in the survey, I selected six white faculty members or teachers who fit the criteria of having taught for at least one year, holding a teaching, advising, or administrative post at Harvard College (see table 3.2).[1]

Subsequently, I collected data that highlighted this population's perspectives on the motivations behind their mentorship, and reported their pathways in higher education and the professorate. I also sought to understand these mentors' approaches to assisting African American mentees confronting issues of race; hence, I conducted two in-depth semi-structured interviews with each participant. This approach allowed me to investigate the research questions and allowed the faculty to explore the phenomena of their own trajectory in academic life and their mentoring philosophy. The interviews were conducted in a conversational method, where I asked open-ended questions, listened actively, and pursued detail by asking the faculty follow-up questions to statements of interest (Seidman, 1998).

Data Analysis

I engaged in data analysis during and after data collection. The coding and analysis were informed by the theories of CRP (Lynn, 1999) and cross-race developmental relationships (Thomas, 1993, 2001). As this study moves beyond the realm of these frameworks, I utilized a grounded theory approach (Strauss and Corbin, 1998) to understand how faculty who serve as mentors to African American undergraduates advise and counsel their mentees when they encounter what they perceive to be racial microaggressions. Through examination of the interview transcripts, faculty responses to the questionnaire, and Web-based research of the participants, I gained an understanding of the motivations and practices involved in white faculty members' mentoring.

Site

Harvard University is a private research-intensive institution located in Cambridge, Massachussetts. The undergraduate division, Harvard College, enrolled 6,715 students in 2006–2007 in over 40 majors ("concentrations") (Harvard College Admissions Office, 2007; Harvard University Office of News and Public Affairs, 2007). Table 3.3 presents Harvard College demographic data alongside the national data for all undergraduate students enrolled in accredited degree-granting institutions the US.

University-wide, the Harvard faculty number 11,252. There has been a slight decline in the percentage of faculty of color over the past two years—21.9% to 20.8% (Harvard University Office of the Assistant to the President, 2004, 2006). Table 3.4 presents Harvard faculty demographic data alongside the national demographic data for all professors teaching at accredited degree-granting institutions the US.

It is also significant to note Harvard's classification among universities. Harvard College's undergraduate instructional program is described by the Carnegie Foundation (2007) as "arts & sciences focus, high graduate coexistence," meaning that "graduate degrees were observed in at least half of the fields corresponding

Table 3.3 Student Demographics for Harvard College, 2006 and U.S. Undergraduate Students, 2004

Category	Harvard College	United States
% female	50	48
% white	45	66.1
% Asian/Pacific Islander	15	6.4
% black/Non-Hispanic	7	13
% Hispanic	8	11.3
% Native American	1	1.1
% Unknown or other	16	N/A

Source: Harvard University Office of Institutional Research; Nat'l Ctr. of Educational Statistics (NCES).

Table 3.4 Faculty Demographics for Harvard, 2006 and U.S. Faculty at Four-Year Institutions, 2004

Category	Harvard College	United States
% female	48.3	39.3
% white	82.6	80.2
% Asian/Pacific Islander	10.1	6.5
% black/Non-Hispanic	3.8	5.2
% Hispanic	3.3	3.2
% Native American	.2	.5

Source: Harvard University Office of the Assistant to the President; NCES.

to undergraduate majors." Only 21 institutions are classified in this category, including Yale and Stanford University. Though Harvard has many unique qualities, many institutions look to Harvard for leadership on educational matters, including issues of student-faculty interaction and those pertaining to race, as evidenced by events on the campus since the early 2000s. High profile disputes between the former university president and African American faculty and administrators have also contributed to a sense that African Americans are under attack at Harvard, though there are also signs that the African American community benefited to an extent during the tenure of former president Lawrence Summers.[2] These events have forced issues of diversity to the forefront at Harvard.

Validity and Limitations

I addressed validity threats in this study utilizing several strategies recommended by experts in the field of qualitative research (Kvale, 1996; Maxwell, 1992, 1996; Seidman, 1998). To address interpretive validity concerns, I triangulated data by utilizing several data sources, including conducting multiple interviews and the faculty questionnaire. Another strategy to enhance interpretive validity was to present my analytical findings to my study group and members of my dissertation committee throughout the study. Sharing transcripts, matrices, codebooks, and memos with this interpretive community presented alternate interpretations and challenged my assumptions. I assured descriptive validity through the recording of all interviews, and entrusted their subsequent transcription to a professional.

The purposive nature of the sampling in this study precludes applying the findings to all white faculty mentors of African American students. Moreover, this study was conducted at Harvard College, a unique higher education institution. The students admitted to the college tend to come from strong academic backgrounds, so issues like remediative support rarely emerge. In addition, Harvard attracts students of all races with significant amounts of social capital, sometimes surpassing the levels of access that faculty themselves have. For such students, successful professionals outside the faculty may provide mentorship, or faculty may only serve as role models and sponsors. At an institution enrolling students with less social capital, faculty may be one of the few populations available as mentors. Findings from this study must take this exceptionality into account.

Participant Profiles

The following section provides brief biographical sketches of the faculty mentors who participated in this study. To ensure the anonymity of the professors in the sample, I have assigned pseudonyms to each participant, and disguised references to geographic places.

White Women
Rachel Smith, Former Concentration Advisor and
Teaching Fellow, Applied Sciences

A Ph. D. candidate in her mid thirties with significant teaching and advising responsibilities, Rachel describes herself as being "of European ethnic origin," and of middle-class origin from the Northeast. "We weren't like super wealthy, but we also always had everything that we needed. It was never scary," Rachel recounted. Unlike most of the professors in the study, Rachel did not receive a great amount of encouragement from her parents to attend college, but this was not a negative experience. "It didn't come from my parents. I was entirely self-driven, and they were just like, 'Okay, we'll just stand by the sidelines, and watch you go,'" Rachel recalled. "They were supportive, but they didn't *push* me." A self-described "teacher's pet" and a high-achieving high school student, Rachel attended Harvard as a premed student, but she encountered significant difficulties in her courses and switched majors. She credited a white female professor for mentoring her once she left premed, with whom she formed "a close relationship." Eventually Rachel "started taking all her classes, and I ended up really being interested in what she did." While her mentor served as her thesis advisor, the professor eventually left academia and Rachel abandoned her plans for graduate study at the time.

After some time in the workforce, Rachel eventually enrolled in a Ph. D. program in the sciences at Harvard. Rachel admitted to being somewhat in the dark about the unwritten rules in academe, reflecting that she "thought everybody else kind of knew these things, and I was the only one who didn't. I didn't understand that you have to write grants to get money to do research." Perhaps as a result of "learning by doing," Rachel was recognized by her advisor as being skilled in advising students, and was asked to work in a close capacity to the department head. Until recently, Rachel held a position as an advisor in her department, where she "interacted with just about every concentrator at least twice a year" to "help them with course choices, and general advice—making sure that nobody's about to go off the rails." Though she is no longer in the role of concentration advisor, Rachel was not surprised to have been named as a mentor "because I've gotten positive feedback from undergrads before. So I knew that I was connecting in some way." When reflecting on her contribution to students, Rachel noted that mentoring involves understanding the perspectives of African American students and breaking down barriers:

> I think we may not realize how much of a risk it feels to the student, the student may feel like it's a big risk to speak up, approach us, whatever.... It didn't always occur to me that the normal barriers of speaking to a professor are even higher for people who feel that they may not be welcomed. So, maybe we do need to be a little more proactive about reaching out to, not just students of color, but any student who seems like they are maybe a little shy, a little hesitant. I think that often gets left behind in busyness and demands and, you know, just cluelessness.... It doesn't occur to me that there may be a billion daily battles that are going right over my head.

Caitlin Moss, Assistant Professor, Social Sciences
A professor in her late-thirties, Caitlin had an initial interest and degree in a medical-related field before discovering her love for the social sciences. Caitlin is from the West Coast and describes herself as white. She describes her family as having "lots of upward social mobility desires...We weren't poor." However, Caitlin's family moved to a middle-class community, which she terms "the 'Land of Just Homogenous White Culture.' It was just relentlessly white." As a child, Caitlin was drawn to teaching and saw herself in that role at an early age. School

became an escape from a difficult home life. "It was what was sort of like the bright part of things. It was what I loved...school was a pleasant place for me."

As a college student, Caitlin studied two years at a community college and then transferred to a larger state university, where she encountered professors, white and Latino, who "just took me aside, and were telling me that I was capable of doing things that I hadn't thought I was capable of. And pointing out, 'You write so beautifully. You should go to graduate school.'" Caitlin enrolled in a masters program and discovered Emily, a white female professor, whose passion for research in Latin American societies led Caitlin to pursue a similar path. With Emily's prodding and the mentorship of a graduate student, Caitlin applied to a social science Ph. D. program with an emphasis on medical issues in Latin America.

Since arriving at Harvard four years ago, Caitlin has taught courses and worked closely with undergraduate as well as graduate students. In her opinion, Caitlin's research interests attract students who are also focused on social issues, several of whom are African American. "I work in Latin America, I work on questions of human rights, social justice, things of that sort, so I also get students who have an affinity for those kinds of topics." When I informed her that she had been identified as a mentor to an African American student, Caitlin expressed surprise. However, she thought that her willingness to listen to and spend time with students might have made a difference to some of her students:

> It's really nice to know that it made some difference for somebody. I just think it matters; I don't care who you are. Just to have somebody sit across from you sit down and listen to you and tell you [that you] have good ideas. And when you don't have good ideas, think about how you might have some....I just think it makes a huge difference, no matter who you are.

Victoria Stein, Full Professor, Social Sciences
A lifelong academic in her early sixties with experiences at several prestigious universities and various think tanks and research institutes over her career, Victoria is a professor in the social sciences from a large East Coast city who identifies as "first and foremost an American. I am Jewish....I think the identity of being a female has risen over time. I had no identity as being a female when I was younger." Victoria scored high on the citywide high school entrance exam, and chose the science magnet school where she excelled.

She then enrolled at an Ivy League institution, where she was initially dissatisfied with the social environment: "It was a terrible social environment that the university allowed, and *still* allows. I could look at someone, particularly a guy, because it was the fraternities, and I could almost instantly figure out what house they probably were in." Victoria was determined to maneuver and sought out opportunities to connect with people across the campus community. The group she eventually connected with had a commitment to civil rights and social justice. While Victoria was quick to note that she was not an activist, these friendships left an impression on her: "I would love to say I was part of it, but, I really wasn't. I was an observer. I was an appreciator."

Victoria's academic skill and promise also led her to find mentors, primarily white male professors, as a graduate student: "I had many mentors then. They're all people who won the Nobel Prize...it was an extraordinary time to be at [the graduate school]." After graduation, Victoria worked in policy institutes and was appointed to professorships at two Ivy League colleges and one prestigious Midwestern state university. A driven, intense academic, Victoria balances a position at a research institute in addition to her teaching responsibilities at Harvard, where she has taught for the past nine years. Mentorship, when it occurs, is simply a matter of her fulfilling her responsibilities as a professor: "[Mentoring is] part of my job." When asked why she mentored African American undergraduate students, Victoria connected to themes of responsibility, student engagement, and a sense of bemusement about being identified as a mentor, but an appreciation for the reciprocal nature of such relationships:

> I appreciate the fact that people think of me as a mentor, but I've never used the term. I hope that I bring out the best in people. I hope that they learn from me. I also learn from them. I guide them.... But, mainly, I allow them to be what they want to be, and hope to be, and follow their passions. I always tell them when they write a senior thesis, that they have to go to sleep with it, and wake up with it. And if they're not happy with it, they're gonna have miserable lives.

White Men
Stephen Cox, Full Professor, Applied Sciences

A career academic in his early sixties who has spent virtually all of his professional years at Harvard, Stephen describes his racial and ethnic identity as partly Jewish and Eastern European (though his name sounds English). Both of his parents came from immigrant families, who emphasized the importance of education very early in his life. Stephen said that his parents came from a "lower-class background, like one grandfather was a grocer, the other was in a lumber mill. So, I certainly got the idea from an early age that education was the way up in the world." Stephen's parents made sacrifices so that he would be able to take advantage of higher education opportunities.

Stephen showed an aptitude in mathematics and earned admission to Harvard. However, Stephen found himself struggling in his chosen concentration, eventually deciding to work in the sciences. The job was partly due to Stephen's connection to a white male professor—a "great mentor." In Stephen's recollection, this "fairly young" professor had "a great skill at getting one to do wonderful work for him." Stephen also saw this professor as a key psychosocial support during challenging times: "As my personal life started to get rocky, [he] provided a lot of personal support to me during that last year." When Stephen made the decision to enroll in a graduate program, a combination of a strong advisor, a supportive group of peers, and his wife contributed to his success.

As a white male at a highly selective school, Stephen is quick to point out "I've never experienced a disadvantage in my entire life." However, reflecting on his career, Stephen states, "I do sort of take some pride in the fact that I'm not coming from six generations of aristocracy." Recalling a statement by a friend and colleague regarding the significance of supporting the whole student, Stephen sees himself as a teacher with a responsibility to know his students, and makes an effort to help those who are in greatest need. "I tended to take the viewpoint that all students are young and haven't fully sorted out what is important to them and how to behave to achieve it," Stephen stated. "Happily this attitude never got me in trouble with students, though some might well have taken it as patronizing if they wanted to." As he is older than most students' parents at this time, his mentees appreciate such a role even more:

> The people who need [mentoring] the most are the people, very often, who aren't getting it. Partly because of this sort of sense that they don't quite belong, or they don't want to bother you.... So you've got to reassure people that it's okay to talk about these things with your professors.

Andy Giordano, Lecturer, Natural Sciences

A lecturer and undergraduate program advisor in his early forties, Andy defines himself racially as "white." Andy's ethnic background, which he describes as "Italian American," is a very important part of his identity: "It's definitely something that I nurture, and make sure I maintain." Like Stephen, he descends from two immigrant families, not far removed from the experience of living in Europe. Though he identifies as "upper middle class" by family origin, Andy is clear to note that this is a recent development in his family history. Recalling his immigrant grandparents, he states they "just had nothing." Andy's father made him aware of how his grandparents struggled: "Class is something we talk about a lot."

Andy comes from a family where higher education was valued and expected—not surprising for the son of two educators. The influence of his parents and his own personality led Andy to consider a career as an educator "because I just enjoy being in the classroom so much." Andy performed at a high level in high school and gained admission to a selective college in an urban setting on the East Coast, where he honed his teaching skills by working as a tutor for underprivileged students. Such experiences helped Andy to see the challenges for other students due to race and social class, something he himself had not been aware of: "I think I was probably pretty oblivious to it happening to other people. And it was definitely something that I never felt, you know, as a white, upper-middle class, lower-upper class, kid. I never felt it myself." However, Andy did have a strong understanding of inequity elsewhere in the world: "I think I was most concerned about people in other countries.... Central America was very big."

Andy encountered three professors, two white males and one white female, who inspired him in the research context as well as provided psychosocial support. These professors helped shape Andy's world view and strengthened his commitment to social justice. Since earning his Ph. D. seven years ago, Andy has taught courses at Harvard, as well as administering an undergraduate concentration in the sciences. Andy is responsible for advising and supporting the students in the area, a job that matches well with his passion to teach and advise students. When I asked him about his commitment to mentoring African American undergraduates, Andy said, "I think that I don't seek out African American students, but I make myself very accessible." However, by committing himself to being visible, Andy realizes that his personality is one that students, including African Americans, are drawn to.

> Harvard is an inaccessible place, with people who are not willing to talk, or give you time to talk...I don't think that I segregate my approach between African and non-African-American mentees. I would say what I do for everybody might be something

that maybe works especially well with African-American mentees. What I try to do is I try to get to know a student, in some detail, and then I try to help them find the best path through Harvard.

David Müeller, Assistant Professor, Humanities
An assistant professor in his mid thirties, David has taught at Harvard for three years and describes himself as being "bicultural," being born in Scandanavia and having grown up in Germany. Being bicultural has inspired David. "I would spend a lot of time in both countries," he stated, which helped him develop "a kind of awareness and acceptance." David's journey to higher education was self-forged, as his parents were not college educated. After a rigorous schooling experience in Europe, David spent time in the United Kingdom. He started studying the humanities at a prestigious Central European university. He worked to finance his education, but in courses David discovered several professors who both inspired him as an aspiring researcher. These professors demonstrated the importance of balance and research skills to David, and a pathway for an academic career.

David completed his doctorate and came to the United States on a post-doctoral fellowship. He admits that American racial dynamics are difficult to understand, and David was particularly appreciative of a faculty orientation session that broached the topic of racial and ethnic identity. "I think that the faculty orientation is a very good idea," David remarked. "I guess it would be different if I'd been brought up in this country. And I find it strange whenever I have to classify myself. I think I go by 'Caucasian.'" David's conceptualization of racial and ethnic identity, and inequality is significantly different when compared to the other participants in the sample, who have grown up in the American milieu and bring an understanding of societal dynamics nearly equal to their ages. However, through some social experiences in the United States, David can appreciate what the experience might be like for students in a significant minority.

> I *try* not to heed the fact that they are black. The two students I mentor are very different. I don't think that I'm doing anything differently with them. But, I think it figures in the background. It doesn't really change my attitude towards them. It doesn't really *impact* what I do *for* them. I have other students who are having a hard time as well, who happen to be white...and there's no difference in my behavior towards *any* of them. However, I do have it in the back of my mind that [being in the minority] is perhaps something that *they* are having a hard time with.

The six faculty in the study have always worked as professors and/or researchers. Mentoring at some level figured in their lives, though some of the participants received mentorship early in their career trajectory, while others encountered supportive developmental relationships as graduate students or young professionals. Through those relationships and hard work, all of the participants became successful scholars.

FINDINGS

Factors Influencing Mentorship among White Faculty

This section presents emergent themes from the analysis of factors that influence white faculty to serve as mentors to African American undergraduate students.

The themes within each of the categories are supported by emic coding[3] generated during the analysis of the interview data. In this section, I present the themes with supporting data from the interviews. This analysis also considers other aspects of identity (e.g., junior and senior status) when applicable.

Living as the "Other"

All faculty in the sample connected their mentoring of African American students to deep personal experiences at a formative period of being in the minority in the context of school, college, academia, or even at large, which I have termed "living as the 'other.'" *Othering* is a process that identifies those that are thought to be different from the mainstream, which can reinforce and reproduce positions of domination and subordination (Johnson et al., 2004). These experiences included (1) for half the sample, being a woman in an academic discipline dominated by men, (2) having a strong identity of being from working class to lower middle-class origins, and (3) having a family history of being othered (primarily in those with strong immigrant identities). It is apt to note that these aspects of identity are of great importance: for the faculty these experiences, though unique and different from the experience of being an African American undergraduate student at a PWI, placed them in a minoritized status. From their narratives, I conclude that it is, at least in part, the negative experience of having their identities sublimated by those in a dominant social position that gave the faculty mentors the opportunity to develop a meaningful empathetic connection to some aspects of what their mentees might experience in a racially stilted academic environment.

Being a Woman in an Academic Discipline Dominated by Men

The women in this study referenced experiences of isolation *as women* in their academic discipline. This was a factor in their individual disciplines as well in the professorate. As professors at a selective research university, these women also experienced isolation in their professional roles as one of a small number of women in their departments, both at Harvard and at other institutions.

For Rachel, an advisor and teaching fellow in the applied sciences in her mid thirties, her identity as a woman served as a way to connect at least partially to African American undergraduates. While careful not to draw direct comparisons gender and race, Rachel felt that her experience served to make her aware of potential issues for African American students:

> I have no idea what it's like to be a person of color. I have always been part of the majority group in all situations. So, I'm making no claim on having any idea what it's like to not be part of the majority. But that's the thing that I do understand...the gendered aspect of academia.

Caitlin, an assistant professor in the social sciences in her late thirties, provides further context into the gendered aspect of academia by invoking the term "gender tax," which can be analogized to the term *cultural taxation* (Padilla, 1994). Both terms refer to the extra burden experienced by women and faculty of color when they are implicitly or explicitly directed to take on responsibilities for their respective constituencies. Even among students of color, there is a similar burden that comes to the forefront, as they often feel pressure to represent perspectives ignored in curricular offerings, serve as "native informants," speak out against microaggressive behaviors exhibited by classmates in classrooms, cafeterias, and residence halls, and educate members of the campus community about the positive contributions of their communities, refuting stereotypes and misconceptions—perhaps best explained by an activist student of color I encountered as part of a research team: "It's like we're adjunct faculty in the college of diversity, except we don't get paid" (Reddick, 2008). In this way, Caitlin's day-to-day challenges in the department presented issues similar to those that African American students might encounter: "[There is] a gender tax that's involved. I believe it's important to give students time, but there are moments when that's difficult for me to manage."

Victoria, a full professor in the social sciences in her early sixties, worked hard to dispel gender stereotypes as a student and young academic, but she has had to confront issues directly related to her gender identity: "I think the identity of being a female has risen over time. I had no identity as being a female when I was younger. I've become, every now and then, more keenly aware of what it's like being a woman [in academia]." In Victoria's case, she strongly understood that her own responses, coupled with assumptions and expectations from her colleagues, led her to work closely as a mentor to African American students. "I think I'm probably more compassionate than many of my colleagues, but I think that that's often because I'm a woman. And it may be that students know that they can tell me things. So that's true for my African American students, as it is for my [other] students. It doesn't matter."

Having a Strong Identity of Having Working to Lower-Middle-Class Origins

Many participants in the study also discussed their own socioeconomic class identity as a strong influence in their lives. In fact, four of the participants identified as being from either working- or lower-middle-class origins. Achieving a measure of success and achievement through their own academic careers, and positions at Harvard, the participants envisioned their experiences in academia as beacons of hope for students of humble origins.

Caitlin presented stories of mentoring bonds that emerged between her and students who felt disengaged from Harvard due to their socioeconomic status. The ability to empathize with students' feelings of isolation allowed greater intimacy and an enhanced responsibility for those students' careers at Harvard College. Caitlin saw her relationship with several students established on a common experience of coming to a collegiate environment where most of their peers are solidly middle, or upper-middle class. These relationships are not established on racial or ethnic ties, but rather on class origin:

> I have a student...who is here on a full fellowship. Her mom works at an Arts and Crafts store. Her parents are divorced. I mean, she is someone that if I have a book, I give it to her, because she can't afford to buy it. I mean her situation is tough. That's how I connect with her.

For a majority of the sample, discovering that a mentee came from a humble background could accelerate and deepen their relationships. Stephen, a full professor in the applied sciences, noted that a shared class background was a point of connection: "I suppose there's a certain set of assumptions or familiarity that gives you a starting place for a conversation." However, a significant population of African American students at Harvard come from middle and upper-middle class backgrounds: a 2002 survey of 356 black students (who represent 70% of the black undergraduates at Harvard) revealed that 35.4% were upper class, 48% were middle class, and 16.6% were working and lower class (Freelon and Redd, 2002).

Echoing these findings, the faculty in the sample reported that most African American students they encountered were not from meager backgrounds. A common statement among the sample was summarized by Victoria: " I always think this is sort of funny, because my [class] background is so much lower than these students." In a similar vein, Stephen noted, "We're now getting to a point where there's enough middle class African American students here, that some of them have backgrounds that are not necessarily all that different from mine." Mentors did not indicate that being from an elite background would affect their willingness to mentor an African American student. However, as Andy discussed, working with students with fewer resources provided a sense of accomplishment: "Especially for the students who come here, with not a lot...and see them graduate—I draw a lot of satisfaction out of it."

The experience of being from a working class socioeconomic background meant that the mentors could draw parallels to their own experiences as students, and potentially help their mentees feel less isolation. Caitlin proffered an alternate strategy for her mentees from her own experience: assimilating some of the characteristics of the middle class to establish a sense of comfort. "I think, though, one thing that helped is that I learned how to don the habits of the middle class," Caitlin recalled. "Because sometimes what I see is that students don't know how to read the codes about how to act in a way that allows you to move there and feel a little bit comfortable. I think that's huge, being able to read the cultural codes."

Having a Strong Family History of Being Othered

Among the faculty in this study, over half (four of six) discussed growing up in families with a strong ethnic and cultural identity. These identities were tempered by immigrant status as well as discrimination, though not necessarily suffered by the professors themselves. When linked to social class, these professors brought an understanding of isolation and resilience in the face of daunting odds to their mentorship of African American undergraduate students.

Both Andy and Stephen discussed their strong ties to their family's immigrant past. Andy discussed his affinity for his Italian heritage. Andy also noted that his link to that heritage is different than the connection his father's generation holds:

> Some of [my father's] brothers are still there [in Italy]. And I look at how he was brought up, and just how different it was from how I was brought up. I just feel much further removed from being Italian the way he is Italian. My grandmother only spoke Italian, when I would go visit her.

Despite having a very different sense of ethnic identity, the stories of Andy's parents experiencing anti-Italian discrimination left an impression on him, though he found it absurd in a modern light. "My parents joke about how there was a time when they discriminated against Italians, which, to me, is just bizarre—no one does that anymore," Andy recalled. Though Andy's parents chose to deal with discrimination through humor, it was clear to him that his own family once was judged as inferior.

Similarly, Stephen, the son of "two parents barely born in the United States," recounted that his grandparents were lower-class European immigrants. Knowing his ancestors' struggle to succeed via hard work proved to be inspirational for Stephen. Particularly, his mother's experience getting into medical school at a time where gender was a liability served as a motivation for his own work helping students:

> My mother...got her way into [a prestigious state university medical school] from a junior college. That was ten years before Harvard Medical School would even accept an application from a woman...in 1938. And, you know, I ran across her admission letter—it's absolutely wonderful. It says her surname, and address and so on, and then it says "Dear *Sir*..."

Stephen's mother's admission to medical school changed the sense of possibility not only on an individual level, but also for her family and the next generation. Stephen linked his own mentorship of African American undergraduates to the same care that people expressed for his family, which allowed him to receive an excellent education: "You know, there was somebody who decided that an [Eastern European] woman ought to be given a chance to go to medical school if she wanted to do it. There was certainly nothing forcing anybody's hand." Through this example, Stephen witnessed the importance of intervention on the behalf of those with less advantage.

Victoria's social circle in college included her best friend, whom Victoria described as "a very political person, an activist." Through this friend, Victoria was exposed to student activism, and witnessed how people of the Jewish faith were involved in the struggle for civil rights in the 1960s:

> It was the beginning of the anti-war movement, and at the tail end of some of the important civil rights activism. [My] roommate's husband, who was several years older than us, was with the group that went to the South, and two of those were the students who were murdered in Mississippi....I was just trying to figure things out in various ways...

Victoria went further to state that she felt a great deal of empathy toward the small population of African American students and the activist community. However, she described her time in college as a "short, quiet period" as civil rights activism declined and the anti-war movement was just emerging. Victoria described how racial tensions were simmering as she neared graduation:

> There was the big student union, with this *huge* room with tables—big long bench tables. And there'd be this side here, and this side here, and then the center, and the jukebox. And one side was *known* as the Jewish side. And one side was known as the white side. And if you were white, you were non-Jewish. So, [my college] was about a

> third Jewish, two-thirds non-Jewish. And there were several African Americans in my class.... Where are they going to sit? Most of them aren't Jewish. And they're certainly not white. That was the tension in the place.

Victoria knew that the ethnic division between white and Jewish students would shift with the emergence of an African American population on campus, which she described as having "a divide, but not a nasty divide" along ethnic lines. Victoria's personal philosophy, however, was to "avoid that, pretty much." It seems that the experience of resisting the status quo gave Victoria an ability to appreciate the struggles of African American students trying to find a niche at the university.

David's cultural context was vastly different from all other faculty in this study, as he was born and raised in Europe. His family was bicultural—his mother was Scandinavian—and he was educated in Central Europe. While David never discussed his bicultural heritage as a liability, he did share a story about difficulties he encountered in school regarding language.

> I would say I was more Scandinavian than German, though my German is a lot better than my Swedish. My parents both gave up speaking Swedish with me, because I had difficulties in school for two or three years.

As a result, David is not as fluent in Swedish as he is in German. However, David viewed his bicultural roots as a highly positive experience, which greatly shaped his world view and opinions about diversity: "It *seems* to have...really bred tolerance," David recalled.

For these four white faculty mentors, being able to relate and connect to an experience of discrimination, isolation, prejudice, or simple difference served to heighten their awareness of the experience of being different in their African American mentees. This sentiment is perhaps best captured in David's words, when he reflected on the day-to-day experiences of his African American mentees, in his department, which David describes as "very white." "I have realized that it must occasionally be very hard for them," he noted. While other aspects of identity have different effects than that of race, the experience of negotiating minority status affects the perspective that these faculty bring to the mentoring relationship with African American undergraduate students.

White Male Faculty

Being Close to Someone Who Is "Othered"

An interesting finding among the subsample in this study whom might be considered the most privileged—U.S.-born white men—was their connection to close friends, or a spouse who dealt with the experience of being othered in a significant way. While both Stephen and Andy made it clear that they, as white men, had never experienced feelings of isolation, ostracization, or discrimination based on race, ethnicity, or gender, the experiences of those close to them seemed to attune both men to empathize with the challenge inherent in living or working as the other.

Stephen is very close to his wife Jane, whom he met in college and now works in academic administration. He spoke interchangeably about both his and his wife's formative experiences, stating, "I've known [Jane] for so long that I kind of merged into her, I would say." When I asked him if he himself had experienced any challenge or discomfort in his academic and professional career because of any aspect of his identity, Stephen answered in the negative, but explained how he had witnessed many challenges that his wife, Jane, confronted:

> There's so many experiences that my wife has dealt with, because she was working—it was before the expectation of women's academic equality really set in. She's had lots of experiences—which I've learned a lot from observing. But I'm not a victim of anything.

Stephen further described how Jane serves an essential role in his life, using a positive attitude and humor to confront her challenges as a woman in an academic environment. These challenges, Stephen explained, made him quite angry, but Jane's resiliency inspired him in his work as a mentor:

> In the scheme of the world anybody with [an Ivy League] diploma is, by definition, not disadvantaged, ok? But [Jane] definitely went to a college where women were not expected to work; women were not expected to really need an education. This has all changed pretty drastically. But she's much more tolerant and willing to go to bed and wake up the next morning having forgotten about them, than I am. I'm probably an angrier person than she is about this stuff. But her example has really demonstrated a good resiliency.

Jane's approach parallels what researchers seeking to understand how women scientists persist in a male-dominated field: "survivor" behaviors or resilience (Taylor, Friot, and Swetnam, 1997). From the vignette we see Jane shared humor and insight, and relied on her relationships with others to "bounce back" from negative situations. From her example, Stephen learned the importance of assisting students to move from anger to action, essential for a mentor. In Stephen's case, observing Jane has been instructive as he works with African American undergraduates encountering similar issues.

Similarly, Andy's graduate school experiences at Harvard helped to forge close friendships with diverse colleagues, but he was surprised to witness explicit gender and race bias. These episodes not only made discrimination come to life, but also, like Stephen, it angered Andy—he quickly learned that inequity also impacted the lives of his friends away from school. He was particularly affected in recounting an experience dining out with an African American friend in Boston:

> There's a couple times when we went out for breakfast, and one of the friends who came along with us was black. She went up to talk to this one guy who ran the place, and he was just ignoring her. He finally took her order, and he made her pay, and he didn't make any of us pay first. And there were several experiences like that, where I was just, again, shocked, because I thought I was coming to liberal Boston, right? And I thought these were things [that] were over with. [Bordering on emotion.] This was really, really surprising to me.

Andy described the sense as one of "unrealness. It's feeling like this stuff you read about is a million miles away. It's so shocking when it's at this restaurant, right near campus."

For the U.S.-born white men in the study, discrimination took shape not in their interactions in the professional arenas, but in the interactions that family members and friends had with faculty and members of the public. From their stories, it is clear that these situations had an intense impact on their lives as well as their work as mentors to African American undergraduates at Harvard. It appears accessibility is coupled with an empathy and understanding of what challenges African Americans encounter as students and citizens. While Stephen and Andy cannot relate directly to the experience of being othered, the close relationships with those who have served as a proxy for the psychosocial effect of such events. From a realist perspective, because Stephen and Andy have had extensive exposure to the perspectives espoused by minoritized spouses and friends, they are able to understand how their students' identities shape the lenses through which they view the world.

Heightened Awareness of the Concerns of African American Students
Participants in the study also identified students of color generally, and African American students explicitly, as a population of students for whom they had a heightened awareness. Participants theorized that African American students did not necessarily have more problems due to their race and ethnicity, as Rachel stated: "I don't think that our students of color have any higher rate of having problems than the general population." At the same time, professors did take note of students of color. From their collective perspectives, they made themselves available to all students, but especially African Americans. This attention was rooted in a respect for diversity and fairness. For instance, Caitlin's motivation for mentoring African American students is linked to her desire to see more diversity in academia:

> I don't want to sound like a do-gooder, but I don't want to work in a land of whiteness...that's not the world I aspire to....When I came here, I was like "Oh my God, there are going to be all these rich, white people." I really did think that. And that was probably provincial on my part, but I was like, "Rich, white grad students, rich, white undergraduates, and they are all going to be obnoxious, and I'm going to feel uncomfortable." Surprisingly I didn't.

Rachel similarly felt a need to reach out to African American students and students of color when she considered the potential feelings of isolation that her mentees might encounter, and the fact that not all departments appear to support all students equally:

> I do think I might make a little bit of an extra effort to make it clear that the student is welcome. And that we will support their interest in doing research if that is what they want to do...[there's an] awareness that there might be people and places and interactions that students have had where that's not the case. So I want to make it clear that we are a place where students of all sets of skill are still going to be encouraged to go as far as they want to go.

Andy felt that many African American mentees came to Harvard with not only personal goals, but also the ability to impact their families for generations with their education. For this reason, Andy discussed the extreme care he exercised when working with this student population. "I think because a lot of those students are not fourth-generation Harvard students, this is a big deal," Andy reasoned. "I can't screw up their time here. I definitely would feel that with certain African American students, I think very carefully about giving them the best possible advice and direction that I can." This sentiment was also shared by Stephen, who described a situation when he publicly admonished another professor who published a controversial article stating that black students benefited from grade inflation.

> I called [the professor], and I said, "How can you write something like that without presenting data?" And he says, "Well, you can present data to show that I'm wrong." And I said, "You're an academic! You don't posit something, and say it's true, and then say it's other people's job to prove it's false by producing data. You've got to produce data to prove it." I had an argument with him about it, and I finally said, "I'm going to write a response to the paper." But the reason I did it was that if I didn't do it, the black students would have been left with the impression that nobody who was part of official Harvard cared that [the other professor] had said that.

With the exception of Victoria, all faculty in this study made reference to the ways in which they were attentive to potential issues that African American mentees experienced. At the same time, these mentors were careful to state that their attention did not result in an assumption that all of their African American undergraduate mentees were struggling. Instead, this awareness helped the faculty detect potential issues before they escalated beyond their ability to assist the student. The professors in the sample found that the identities that African American students brought to Harvard deserved attention, rather than dismissal—as Moya (2008) discusses elsewhere in this volume.

Summary

The findings of this study highlight the central role of identity in the effectiveness of white faculty as mentors to African American undergraduate students. These aspects of identity included living as "the other," being close to someone who has experienced being othered on the basis of gender or race, and last, a sense of alliance with their students. The consequences of social identity, specifically being othered or being in close relationship to someone who was othered, seemed to be a prerequisite that allowed these professors to empathize with the experience of African American undergraduate students. Four of the six white faculty members in the study discussed their immigrant family lineage and how those experiences instilled a sense of empathy as they mentored African American undergraduates. Similarly, for the white men in this study, close personal relationships made the reality of isolation and discrimination real to them and heightened their awareness of these issues. The white faculty were further inspired to support African American undergraduates based on a sense that those students often encountered challenges to their psychosocial development at Harvard, given the sometimes challenging experiences of learning and working in a PWI. Indeed, for the faculty in this sample, their own identities as members of stigmatized groups, or allegiance to others with such identities helped shape their perspectives on how to provide support for their undergraduate mentees (Moya, 2008). While the specific experiences of oppression may differ vastly in accordance with one's identity, the effects of being oppressed or discriminated against may be commutative, thus allowing faculty to link their experiences to those of their young charges, though their experiences of exclusion and oppression are based on a different aspect of identity.

Discussion and Conclusion

Theoretical Implications

There are a number of implications of this study that move beyond the tenets of CRT, CRP, and theories of cross-racial developmental relationships. I present those implications here and discuss how this study advances a critical theory of difference that helps to analyze faculty perspectives of mentorship of African American undergraduate students.

Reconsidering CRT and CRP

I found that white faculty, though they did not necessarily personally experience racial oppression in the same manner that African American students did, were keenly aware of this reality for their mentees. For instance, Stephen and Andy discussed ethnic discrimination as an aspect of their family histories, and the connections to people of color exposed them to situations similar to those described by African Americans in the research literature. Caitlin shared her experience of being raised in a familial and community context laden with racial stereotypes, and her own efforts to challenge these beliefs. Victoria and Rachel, along with Andy and Caitlin, discussed experiences of recognizing racial inequality on a local scale (segregation on a college campus and an awareness of the isolation that many African American students experience at Harvard) and a global scale (events in Mississippi and Central America).

In their overview of CRP, Jennings and Lynn (2006) iterate the centrality of considering other constructs alongside race when reflecting on American schooling processes:

> Any form of Critical Race Pedagogy must be intimately cognizant of the necessary intersection of other oppressive constructs such as class, gender and sexual orientation. Theorizing these intersections is of high importance because individuals prioritizing one facet of their identity over another can create a false dichotomy that does not address the reality that we exist within society as subjective entities whose identities are negotiated through multiple lenses that privilege certain race, class, gender and sexual "norms." (21)

The data in the study support the importance of confronting gender and class bias in the schooling and professional experiences of faculty. Considering the challenges to women in academia noted by Andy and Stephen, and confronted by Caitlin, Rachel, and Victoria, the participants both observed and experienced how social class and gender further complicated issues of access, self-worth, and identity. Here, CRP proved to be a useful construct in analyzing the experiences of faculty by integrating other aspects of their identity as well as race, in relating to their formative and professional experiences.

CRT, and by extension, CRP, acknowledge the experiences of people of color who have encountered oppression in American society (Solórzano and Yosso, 2000). However, both theories are largely reticent on how the experiences of whites in American society might contribute to an empathetic stance or solidarity for the challenges confronting people of color. The data from the study suggests that white faculty who have experienced discrimination or exclusion due to an aspect of their identity (e.g., gender, immigrant status, or socioeconomic class) are able to empathize with African American students and in some instances, even predict when microaggressive situations might occur. However, the white participants were conscious of the differences between their discriminatory experiences and those representative of African American undergraduates. These faculty took great care to make the sensitive and critical distinction of not claiming to "understand," but as David said, experiences as the "other" gave them an idea of "what it must be like" for African American students. Such considerations are part of the process of creating *critical access,* a goal that faculty members can work toward on multiple fronts to build racial democracy in higher education (chapter 1).

The white faculty in the study, understanding this distinction, avoided positions that escalated "hierarchies of oppression" (Ladson-Billings, 2000). This distinction is of particular interest, as the other aspects of identity mentioned—socioeconomic status, immigrant identity, and even gender—varied in effect and intensity. Deigning one form of bias more intense than another creates counterproductive barriers that ultimately do little to combat the pervasive effects of prejudice. Indeed, so-called "hierarchies of oppression" or "oppression Olympics" distract from solutions and potential alliance among groups struggling on similar fronts to overcome bias. Such conceptualizations also essentialize the complexity of identity: the components of our identity often interact in myriad ways and change moment by moment, interaction by interaction.

CRP further posits that educational settings are opportune venues where educators can work to resist racial oppression (Jennings and Lynn, 2006; Lynn, 1999). The data from the study support this perspective: Stephen writing a refutation of another professor's editorial suggesting that African American students promoted grade inflation at Harvard; Caitlin supporting undergraduate students interested in the intersection of race and their respective fields of research through thesis advising; and Rachel investing significant energy to assist an African American mentee working to maintain his grades and remain at Harvard. All of these actions demonstrate the willingness of faculty in the sample to personally work against racially microaggressive situations that African American, and other students of color, might face.

Faculty in this study resonated with the theme of caring as articulated in CRP. The interview data reveal an ethos of concern for students—simply stated by Andy: "You need to actually take an interest in that whole person." Stephen's phrase borrowed from a colleague was particularly apt: "Students are more than just brains on a stick." In many ways, the root of the caring ethos emerged from the care and nurture that faculty received in their formative years from their families and mentors. A majority of the faculty discussed how their families instilled a sense of confidence and high expectations. The data clearly support the central role of caring found in CRP regarding the mentoring relationships of the faculty.

One aspect of CRP that differed from the data in the study was the concept of cultural connectedness between teacher and student. This was not evident from the perspectives shared by the faculty. On a somewhat related point, however, it did seem that academia, specifically, the Harvard undergraduate experience, exists as a cultural bond that Rachel and Stephen, both of whom attended Harvard, shared with their mentees. This connection proved helpful in putting students at ease, as the professors shared experiences in which they struggled academically. Caitlin discussed the challenges she encountered as a public school-educated student entering an elite college, and noted that she was attentive to students with similar experiences. Additionally, the Afrocentric notion of unity found in CRP was not supported by the data in the study—perhaps most obviously because none of the faculty shared identities linked to Africa with their students. Rather, they found common ground through other shared experiences and identities. Faculty discussed the importance of knowing students individually, and the significance of common interests in the growth of mentoring relationships with African American undergraduates—embracing the unique identities the students bring to the college exnvironment.

For the most part, CRT and CRP provided a useful framework to analyze the phenomenon of white faculty mentorship of African American undergraduates in the context of a highly selective higher education institution. Among white faculty, a sense of awareness regarding racial inequity tended to occur in early adulthood. The study findings also resonated with CRP's focus on the intersection of other constructs, such as class and gender, in analyzing inequity in U.S. society. The data reveal that white faculty related analogous, yet different personal experiences with discrimination and inequality, and took care not to equate these experiences to microaggressions or racism that African American undergraduates might experience. The fact that white faculty recognized the singular nature of racial discrimination, but still were able to recall their personal exposure to similar circumstances aided their ability to empathize with and demonstrate caring for African American mentees.

The study findings suggested that cultural connections between faculty and students were cursory along racial and ethnic lines: cultural commonality was simply one trait that *could* serve as a springboard to a deeper relationship. Other commonalities, such as shared research topics and attending Harvard, served as a cultural bond among faculty and their mentees. By recognizing the complexity and individuality of their student mentees, as Moya (2008) advocates, these faculty mentors found common points of interest among a broad swathe of topics. Simultaneously acknowledging the possibility that African American undergraduates might encounter microaggressive situations at the college and treating each student as an individual allowed the professors in the sample to foster *critical access* (chapter 1) and hence provide crucial opportunities and academic resources for their students' advancement—resources that are often denied to students of color at PWIs.

The most significant contribution I derive from this work is the advancement of a framework of faculty-student mentoring, which I term *critical theory of difference*. *Critical theory of difference* integrates a broader view of cultural identity, one that includes not just race and ethnicity, but also shared identity-based interests and experiences that are more common in the collegiate setting. In this study, my findings point to the need for an expansion of CRP. When examining cross-racial faculty-student mentoring relationships, we need to acknowledge the significance of analogous experiences of bias and redefine the possibility of identity connection through common educational experiences. My future research on mentoring will consider the importance of expanding theoretical conceptualizations to include *critical theory of difference*. Faculty leverage their own experiences of having experienced marginalization and threat because of their identities, or experiencing oppression—to provide a basis of understanding and empathy with African American undergraduate mentees. Though their students' conceptualization of identity differs radically from their own, it is in fact the common experience of being discriminated against that leads to alliances and ethical commitments built across race, ethnicity, gender, and multiple other aspects of identity.

Reconsidering Theories of Cross-Race Mentoring

Regarding the ways in which faculty members in this study engaged in cross-race mentoring, it was virtually impossible to ascertain their preferences in communicating about racial conflict. While the interviews gave insight in to how faculty viewed the world, it quickly became evident that while a professor articulated what appeared to be a denial and suppress preference, a contextual issue, such as a racially divisive article in the school newspaper written by a faculty colleague might inspire a race conscious response. It appears that strong mentors have multiple ways of reacting to a given situation, often taking their lead from the student regarding the preferred response. The professors in the sample discussed how racism and prejudice had affected their lives on a personal level, and could all recall instances where they experienced or witnessed racism or prejudicial behavior firsthand.

These findings bring complexity to Thomas' theories of cross-racial developmental relationships and racial conflict preferences (Thomas, 1990, 1993). Given that the students surveyed to identify mentors were members of ethnic and culturally themed organizations, it seems likely that many students may have brought a race-conscious perspective to the mentoring relationship. The resources available to African American Harvard students—administrative departments, house staffs, and peers—may be more apt and comfortable venues for students to share racial concerns. Unlike the corporate setting, where the path to advancement can be narrow and completely vertical, the collegiate setting appears to provide many paths to students aspiring to graduate and potentially explore graduate study opportunities or the workforce.

Considerations on Adult Development, Gender Issues, and Administrative Roles

Generational and age differences appear to be a significant area of focus in this study. Specifically, Stephen and Victoria were both in their early sixties and referenced certain epochal events (Vietnam and the civil rights movement) in their interviews. These historical events seem to have had a profound effect on how these faculty members viewed their roles as mentors, and their approach to working with African American undergraduate students. In addition to the historical significance of these events to their mentoring, I also noted that these professors tended to make reference to the fact that the students they mentored were close in age to their own children (they are both parents). Specifically, Stephen discussed staying in touch with the students he had worked with over the years, occasionally showing a picture from a graduate. Victoria also discussed the bond she had with former students who were now colleagues. I suspect that with age and with parenting, the skills that these professors bring to the mentoring role become sharpened. I also hypothesize that students feel comfort relating to adults who are similar in age to their parents, and mentorship in such dyads is more likely to emerge.

Levinson's (1978) theory of adult development would place these faculty between middle and late adulthood. At this stage, Levinson argues, "youthful drives" such as anger, self-assertiveness, and ambition still exist, but adults in this point in the life cycle "suffer less from the tyranny of these drives" (25). It is also a time in which mentoring can blossom, as this epoch is a time where an adult can emerge as "a more compassionate authority and teacher to young adults" (25). It is important to note that Levinson's conception of what constitutes middle and late adulthood reflects a worldview of thirty years ago, and in actuality, the descriptions for middle adulthood seem more fitting for the senior faculty in this study. The senior faculty in the study greatly reflected this sensibility. They derived a great deal of satisfaction from their work with students, and reflected on their mentoring of students by recalling letters, phone calls, and mementos much like a parent dotes on the accomplishments of an adult child.

An interesting aspect of this study was the concern about balancing professorial roles, such as teaching and mentoring, with roles that lead to tenure. While all faculty discussed the time that mentoring required, Caitlin, a junior professor, was saddened by a colleague's advice to ease her teaching load:

> She's mentoring me that I shouldn't be spending so much time with students....And that's troubling to me, because we all know that that doesn't get you tenure anywhere, it doesn't get you tenure here, and yet then where does that leave us? What kind of message is that sending?

These tensions reflect the dilemmas Levinson presents in the late settling down phase, which he demarcates as occurring between the ages of 36 to 41 (Levinson et al., 1978). There is the tension of being more independent and rejecting pressure from others; conversely, societal and professional affirmation is important. Caitlin's rejection of advice that she will need to follow to achieve tenure symbolizes the tension that late settling down faculty encounter. Academic life, with its emphasis on earning tenure, complexifies this stage because there are true disincentives to reach out as mentors. This tension is further complexified when one considers Caitlin's identity as a woman and as a junior faculty member, experiences that provide a specific social reality of the Harvard environment. Her identity based experiences lead her to know and understand aspects of social reality at the institution, in academia, and in the world, that she would not necessarily know if she were a privileged white male professor.

Another noteworthy point of discussion is how mentorship intersected with faculty holding administrative roles, from department deanships to concentration advisers. Of the six faculty in the sample, five either currently held, or had held an administrative post in conjunction with their teaching responsibilities—raising the question: Are faculty who excel at advising called into administrative roles, or does the experience of serving as an administrator enhance one's ability to mentor? From the interviews with the subjects, I feel that both explanations are true. The administrative role gave faculty a direction in which to address issues pertaining to students and their comfort in the educational setting. It was also evident that these faculty mentors brought a good deal of emotional intelligence and "soft skills" in working with students. Effective administration requires knowledge of institutional policy and resources, and the faculty in the study brought this knowledge to the mentoring role. The ability to communicate effectively with students, especially in disciplinary situations is one which both administrators and mentors must develop, and the experiences of Stephen and Andy—who developed mentoring relationships with students after difficult meetings—illustrate this point.

Advancing a Critical Theory of Difference Regarding Faculty-Student Developmental Relationships

Through the use of CRT, CRP, and theories of cross-racial developmental relationships in this study, I am able to contribute a new framework in which to conceptualize mentoring in higher education between students and faculty. I present an advancement of the cornerstone theories of CRT, CRP, and cross-racial developmental relationships—a critical theory of difference, in which faculty connect to a perception of challenges confronting African Americans at PWIs through their own lived experiences of being different by virtue of immigrant status, gender, and socioeconomic class—or those of close friends and/or family. This framework can assist future researchers in the analysis of mentoring dynamics in higher education, especially those relationships across race.

Implications for Policy and Practice

The most illustrative finding regarding policy and practice from this study is the link between formative experiences and mentoring. All participants referenced situations in their past that directly led to their interest and proficiency as mentors. If mentoring is a skill that faculties and departments wish to enhance, it appears that search committees should inquire of candidates about their experiences confronting difference. In this study, white women faculty discussed experiences as women in predominantly male environments, and white men, while stating that they had not personally confronted challenges due to race or gender, discussed how family histories of immigration, economic hardship, and relationships with significant others and friends had sensitized them to the challenges that African American undergraduates may confront. A question such as "Have you, or a person close to you, ever been in a situation in which you were different from many of your peers, and if so, how did you navigate that experience?" would allow candidates to share circumstances similar to those shared by faculty in the study. For candidates with largely homogeneous formative experiences, the inclusion of "a person close to you" allows the candidate to draw upon experiences of friends whom they may have helped. The neutrality of the question places the emphasis on the experience of being different, rather than race or ethnicity. It seems unlikely that a candidate will not be able to reflect upon some aspect of this question; however, a search committee could further pose a hypothetical scenario and ask a candidate how he or she would respond. It is also important to not assume that race or gender translates to a desire or aptitude for mentoring—candidates should be asked directly about their views on developmental relationships.

To further overcome barriers between students and professors, faculty discussed flattening hierarchical lines between themselves and students, and lessened the distance by inviting students to meet one-on-one outside of the classroom. For example, Caitlin participated in events sponsored by African American student organizations, exposing students to her research and personality:

> The Harvard black women's undergraduate association asked me to come in and talk. They were doing something on this issue of gender and violence, so I talked about my research there...and I know after that is when a couple students approached me about thesis writing, so maybe it was seeing me in another role, that I'm willing to give some time to doing something.

As this vignette shows, seeing that a faculty member is willing to engage in conversation beyond the classroom is important to students. These two examples echo the findings of researcher Ken Bain, who studied 60 outstanding college teachers in his book, *What the Best College Teachers Do*.

> Professors who established a special trust with their students often displayed a kind of openness in which they might...talk about their intellectual journey....With that trust and openness came an unabashed and frequently expressed sense of awe and curiosity about life, and that too affected the relationships that emerged. (2004, 141–142)

The findings from this study reinforce Bain's work—establishing trust with students requires a willingness to share. It seems very unlikely that the student-professor hierarchy can ever be completely eliminated; instead, faculty should endeavor to help students understand that they, too, were once students and met with varying degrees of success in different activities over the years.

Findings from this study also suggest that alongside efforts to bridge the student-professor gap, faculty can ask questions to invite African American undergraduates to share experiences of perceived racism and/or racial microaggressions. Open-ended queries can serve as springboards to deeper conversations. Faculty should listen closely for stories that suggest that a student is encountering an issue that is racially microaggressive. In such an interaction, faculty can help students not only through relating similar experiences on a personal level, but also by encouraging students to meet with colleagues that may have similar interests or expertise—a strategy employed by Stephen, and one which reinforces the concept of "constellation mentoring" (Johnson, 2007). At the same time, these questions do not presume a negative experience, and students can answer in a cursory manner if they do not yet feel comfortable sharing more intimate recollections with a faculty member. Interacting with African American students on their "turf," like Caitlin's example of attending student organization meetings as a guest speaker, can also assist faculty in learning more about the day-to-day experience of African American students, and further establishing credibility among students as someone who cares.

Another issue of particular importance in this study is the necessity of some level of engagement on the impact of race in the lives of white faculty members to effectively maintain cross-racial developmental relationships. Only one professor conducted race-related research. It seems that among this subset, students would have less of an indication that these faculty members were interested in, and able to have conversations about racial concerns. At the same time, white faculty were careful to avoid presupposing that African American students encountered racial issues. While they asserted that their approach to mentorship did not appreciably differ according to a student's race, they demonstrated an understanding of their social location as whites and that of their mentees as African Americans.

Some white faculty were better able to understand and empathize with their students by likening the experiences of African American students at Harvard to familiar experiences as immigrants and members of ethnic groups. However, they also applied problem-solving skills to the issues that African American mentees shared. This allowed the faculty to follow a protocol of sorts in responding to students' concerns. In essence, while formative experiences appear to sensitize and alert white faculty to potential issues around race for their African American mentees, it appears that those experiences are a personal reference point, evidence of the salience and epistemic component of identity (chapter 1). The professors' negotiation of identity, and the ability to understand what having one's identity under attack is like, leads to problem solving and instrumental approaches, such as Stephen's written response to the professor that wrote about grade inflation, which perhaps best address racial microaggressions.

There are more pressing concerns involving the relevance of developmental relationships in the career trajectory of faculty. As the junior participants (both in age and rank) discussed, mentoring students is a responsibility that faculty are actively dissuaded from spending time doing; in fact, senior colleagues advise their junior peers to limit the time and contact they have with undergraduate students. Yet, colleges and universities tout access and immersion with faculty as a key experience for students during their undergraduate years. Clearly, at institutions where teaching is at the core of the institutional mission, there is a stronger connection between professorial responsibilities to undergraduate students and mentoring. At institutions categorized as research focused with an emphasis on graduate education (Carnegie Foundation, 2007), it is easy to understand why faculty are the recipients of mixed messages.

Even so, deans and senior faculty can demonstrate the relevance and importance of mentoring undergraduate students. This recommendation is reinforced from the interview I conducted with Caitlin, who referenced a training session she attended at the teaching center on campus that discussed mentoring. When I asked how the university ensured that new faculty attended a workshop on mentoring, she noted, "They give you $5,000 in your research fund if you attend."

Caitlin stated that the program was well attended, even by those with little interest in teaching, because of the grant attached to their participation in the program. Although she additionally states that some in attendance "don't care," those faculty members are at least exposed to the concept of mentoring. Other "carrots" of a similar nature can both emphasize the institution's commitment to mentoring and present research that reinforces the importance of mentoring in the professor-student relationship. Mentoring awards may introduce a degree of prestige to the work of developmental relationships. Providing funds for professor-student lunches can facilitate informal meetings on campus that might lead to supportive relationships. Appointing skilled faculty members as "mentoring advisors," with a stipend, can provide a resource to less experienced faculty.

Institutions and departments can also formalize the contribution of mentors by evaluating mentoring and student support in professors' reviews. Such an evaluation might be performed by a department chair, and could consist of a brief summary submitted by the faculty member listing the students he or she advises, counsels, and mentors, and any significant efforts from mentors on behalf of students. To provide support for these claims, faculty could refer to letters of recommendation written, comments on course or adviser evaluations, and/or student evaluations. Student evaluations, while often maligned by faculty when used as a basis for promotion, can hold promise as one component of assessing faculty efficacy in developmental relationships. Instead of focusing on teaching, advising evaluations could capture student perspectives on how successfully faculty members served in the role as adviser, sponsor, and mentor. Such an evaluative tool could be fashioned to not only reflect on a professor's advising capabilities, but also to identify other faculty who served as sponsors and mentors to students.

The findings from this study additionally suggest that senior faculty members are well positioned to serve as mentors to African American undergraduate students, and thus are a key factor in providing critical access for this population. As these faculty members have considerably less professional pressure in comparison to their junior colleagues regarding tenure and promotion. Furthermore, the experience of working with diverse students over the years seems to have provided the senior faculty in the study with experiences from which to draw. Retaining senior and emeritus faculty with nominal teaching responsibilities, thus allowing them more time to interact with students, is one way to preserve the experiential knowledge of these professors. The aforementioned suggestion of having senior faculty serve as departmental advisors to other faculty members regarding developmental relationships, including mentoring, is another way in which to afford prestige and reinvigorate faculty, who have likely peaked in their organizational ascent and are particularly interested in investing time and energy in the next generation of scholars (Kram, 1988; Kram and Isabella, 1985; Levinson et al., 1978).

I write this chapter today as a faculty member at a research-intensive PWI, and I have had the experience of seeing many of the issues discussed in this essay—cultural taxation; the need for mentoring among students; and the struggle to ensure that students of all identities (especially African American and other students of color) gain *critical access* to the resources and opportunities that will help them to be successful when they leave the university setting. Knowing that there are white allies in the academy that are simultaneously sharing the responsibility of caring for and guiding African American undergraduate students is a welcome respite from the stories of racial insensitivity and indifference that are too-frequent hallmarks of the PWI experience for students of color. Through their personal experiences of being othered or minoritized and those intimate bonds that exposed Victoria, Rachel, Caitlin, Stephen, David, and Andy to the deleterious effects of bias on many levels, these professors are examples of how white faculty can forge communities of meaning with students. By linking but not privileging their own experiences over those of their student mentees, they bring diverse knowledges that help both mentors and mentees understand more fully how the apparatuses of oppression are interrelated (chapter 1). This is what education must aspire to be: part of the promise of a "better society" Moya alludes to, and a necessary step toward a racial democracy that respects all identities.

Notes

1. All names listed are pseudonyms, and their areas of research have been purposefully described in a general manner to preserve their anonymity.
2. Some of these disputes include: the departure of university professor Cornel West; the resignation of the dean of the Faculty of Arts and Sciences; and the resignation of the first African American member of the Harvard Corporation (P. Fain, "Harvard Board Member Steps Down, Citing Clashes with President." *Chronicle of Higher Education* (August 12, 2005); R. Wilson, "The Power of Professors." *Chronicle of Higher Education* (March 3, 2006), A10). All of those who departed and resigned mentioned conflicts with President Summers as a contributing factor.
3. Emic coding are categories that emerge from the data, and used to help the researcher understand the perspective, beliefs, and world view of the subject (Strauss and Corbin, 1990).

References

Allen, W. R., and J. Jewell. "A Backward Glance Forward: Past, Present, and Future Perspectives on Historically Black Colleges and Universities." *Review of Higher Education,* 25.3 (2002) 241–261.

Astin, A. W. "Student Involvement: A Developmental Theory for Higher Education." *Journal of College Student Personnel,* 25.4 (1984): 297–308.

Astin, A. W. *What Matters in College?: Four Critical Years Revisited* (1st ed.). San Francisco: Jossey-Bass, 1992.

Bain, K. *What the Best College Teachers Do.* Cambridge, MA: Harvard University Press, 2004.

Blackwell, J. E. *Networking and Mentoring: A Study of Cross-generational Experiences of Blacks in Graduate and Professional Schools.* Atlanta, GA: Southern Education Foundation, 1983.

Bowen, W. G., and D. C. Bok. *The Shape of the River: Long-term Consequences of Considering Race in College and University Admissions.* Princeton: Princeton University Press, 1998.

Brayboy, B. M. J. "The Implementation of Diversity in Predominantly White Colleges and Universities." *Journal of Black Studies,* 34.1 (2003): 72–86.

Carnegie Foundation. "Undergraduate Instructional Program Description." Retrieved March 31, 2007, *http://www.carnegiefoundation.org/classifications/index.asp?key=786.*

Cohen, J. J. "Time to Shatter the Glass Ceiling for Minority Faculty." *Journal of the American Medical Association,* 280.9 (1998): 281–282.

Collins, P. H. "The Social Construction of Black Feminist Thought." *Signs,* 14.4 (1989): 745–773.
Crosby, F. J. "The Developing Literature on Developmental Relationships." In *Mentoring Dilemmas: Developmental Relationships within Multicultural Organizations,* edited by A. J. Murrell, F. J. Crosby, and R. J. Ely, 3–20. Mahwah, NJ: Lawrence Erlbaum Associates, Inc., Publishers, 1999.
Delgado, R., and J. Stefancic. *Critical Race Theory: An Introduction.* NY: New York University Press, 2001.
Delpit, L. D. "The Silenced Dialogue: Power and Pedagogy in Educating Other People's Children." *Harvard Educational Review,* 58.3 (1988): 280–298.
Dixon-Reeves, R. "Mentoring as a Precursor to Incorporation: An Assessment of the Mentoring Experience of Recently Minted Ph. D.s." *Journal of Black Studies,* 34.1 (2003): 12–27.
Fain, P. "Harvard Board Member Steps Down, Citing Clashes with President." *Chronicle of Higher Education* (August 12, 2005).
Foster, M. "The Politics of Race: Through the Eyes of African-American Teachers." *Journal of Education,* 172.3 (1990): 123–141.
Freelon, K., and M. J. Redd. *The Black Guide to Life at Harvard.* Chicago: Ibis Communications, LLC., 2002.
Golden, D., and J. Yemma. "Tenure System Comes under Fire: Policy on Junior Faculty Leads to Early Departures and an Aging Roster." *Boston Globe,* p. A1, June 2, 1998.
Guiffrida, D. "Other Mothering as a Framework for Understanding African American Students' Definitions of Student-Centered Faculty." *Journal of Higher Education,* 76.6 (2005): 701–723.
Harvard College Admissions Office. "Harvard College Admissions Office: Prospective Students." Retrieved March 28, 2007, http://www.admissions.college.harvard.edu/prospective/academics/departments/index.html.
Harvard University Office of News and Public Affairs. Harvard at a Glance. Retrieved March 28, 2007, *http://www.news.harvard.edu/glance/.*
Harvard University Office of the Assistant to the President. *Affirmative Action Plan 2004.* Cambridge, MA: President and Fellows of Harvard College, 2004.
———. *Affirmative Action Plan 2006.* Cambridge, MA: The President and Fellows of Harvard College, 2006.
Harvey, W. *Minorities in Higher Education, 2001–2002: Nineteenth Annual Status Report.* Washington, DC: American Council on Education, 2002.
Hurtado, S. "Linking Diversity and Educational Purpose: How Diversity Affects the Classroom Environment and Student Development." In *Diversity Challenged: Evidence on the Impact of Affirmative Action,* edited by G. Orfield and M. Kurlaender, 187–203. Cambridge, MA: Harvard University Press, 2001.
Jacobi, M. "Mentoring and Undergraduate Academic Success: A Literature Review." *Review of Educational Research,* 61.4 (1991): 505–532.
Jennings, M. E., and M. Lynn. "The House that Race Built: Critical Pedagogy, African-American Education and the Re-conceptualization of a Critical Race Pedagogy." *Educational Foundations,* 19.3–4 (2005): 15–32.
Johnson, B. J., and W. Harvey. "The Socialization of Black College Faculty: Implications for Policy and Practice." *Review of Higher Education,* 25.3 (2002): 297–314.
Johnson, J. L., J. L. Bottorff, A. J. Browne, S. Grewal, B. A. Hilton, and H. Clarke. "Othering and Being Othered in the Context of Health Care Services." *Health Communication,* 16.2 (2004): 255–271.
Johnson, W. B. *On Being a Mentor: A Guide for Higher Education Faculty.* Mahwah, NJ: Lawrence Erlbaum Associates, 2007.
Kram, K. E. *Mentoring at Work: Developmental Relationships in Organizational Life.* Lanham, MD: University Press of America, 1988.
Kram, K. E., and I. A. Isabella. "Mentoring Alternatives: The Role of Peer Relationships in Career Development." *Academy of Management Journal,* 26.4 (1985): 10–32.
Kvale, S. *InterViews: An Introduction to Qualitative Research Interviewing.* Thousand Oaks, CA: Sage, 1996.

Ladson-Billings, G. "Fighting for Our Lives: Preparing Teachers to Teach African-American Students." *Journal of Teacher Education,* 51.3 (2000): 206–214.

Ladson-Billings, G., and W. F. Tate. "Toward a Critical Race Theory in Education." *Teachers College Record,* 97.1 (1995): 47–68.

Levinson, D. J., C. Darrow, E. Klein, M. Levinson, and B. McKee. *The Seasons of a Man's Life* (1st ed.). New York: Ballantine Books, 1979.

Lynch, R. V. *Mentoring across Race: Critical Case Studies of African American Students in a Predominantly White Institution of Higher Education.* Paper presented at the Annual Meeting of the Association for the Study of Higher Education, 2002.

Lynn, M. "Toward a Critical Race Pedagogy: A Research Note." *Urban Education,* 33.5 (1999): 606–626.

Matsuda, M. J., C. R. Lawrence, R. Delgado, and K. W. Crenshaw. *Words that Wound: Critical Race Theory, Assaultive Speech, and the First Amendment.* Boulder, CO: Westview Press, 1993.

Maxwell, J. "Understanding and Validity in Qualitative Research." *Harvard Educational Review,* 62 (1992):279–300.

———. *Qualitative Research Design: An Interactive Approach* (Vol. 41). Thousand Oaks, CA: Sage Publications, 1996.

Merriam, S. B., T. K. Thomas, and C. P. Zeph. "Mentoring in Higher Education: What We Know Now." *Review of Higher Education,* 11.2 (1987): 199–210.

Moya, P. M. L. "What's Identity Got to Do with It? Mobilizing Identities in the Multicultural Classroom." In *Identity Politics Reconsidered,* edited by Linda Martín Alcoff, Michael Hames-García, Satya P. Mohanty, and Paula M. L. Moya, 96–117. New York: Palgrave Macmillan, 2006.

Padilla, A. M. "Ethnic Minority Scholars, Research, and Mentoring: Current and Future Issues." *Educational Researcher,* 23.4 (1994): 24–27.

Pattillo-McCoy, M. *Black Picket Fences: Privilege and Peril among the Black Middle Class.* Chicago: The University of Chicago Press, 1999.

Patton, M. Q. *Qualitative Evaluation and Research Methods* (2nd ed.). Newbury Park, CA: Sage, 1990.

Pierce, C. "Stress Analogs of Racism and Sexism: Terrorism, Torture, and Disaster." In *Mental Health, Racism, and Sexism,* edited by C. V. Willie, P. Rieker, B. Kramer, and B. Brown, 277–293. Pittsburgh, PA: University of Pittsburgh Press, 1995.

Reddick, R. J. "Ultimately, it's about love": African-American faculty and their mentoring relationships with African-American undergraduate students (unpublished qualifying paper). Harvard Graduate School of Education, 2005.

———. "How Higher Education Can Support and Enhance Diversity Efforts for Students." In *Possibilities of Diversity: A Guide for Practitioners,* edited by H. R. Milner, 173–190. Springfield, IL: Charles C. Thomas, 2008.

Seidman, I. *Interviewing as Qualitative Research: A Guide for Researchers in Education and the Social Sciences* (2nd ed.). New York: Teachers College Press, 1998.

Smith, E. P., and W. S. Davidson, II. "Mentoring and the Development of African-American Graduate Students." *Journal of College Student Development,* 33.6 (1992): 531–539.

Solórzano, D., and T. Yosso. "Toward a Critical Race Theory of Chicana and Chicano education." In *Demarcating the Border of Chicana(o)/Latina(o) Education,* 35–65. Cresskill, NJ: Hampton Press, 2000.

———. "Maintaining Social Justice Hopes within Academic Realities: A Freirean Approach to Critical Race/LatCrit Pedagogy." *Denver University Law Review,* 78.4 (2001): 595–621.

Spence, C. N. "Successful Mentoring Strategies within Historically Black Institutions." In *Instructing and Mentoring the African American College Student: Strategies for Success in Higher Education,* edited by L. B. Gallien and M. S. Peterson, 53–68. Boston: Pearson Education/Allyn and Bacon, 2005.

Stanley, C. A., and Y. S. Lincoln. Cross-Race Faculty Mentoring. *Change* (March/April 2005): 44–50.

Steele, C. "A Threat in the Air: How Stereotypes Shape Intellectual Identity and Performance." *American Psychologist*, 52.6 (1997): 613–629.

Strauss, A. L., and Corbin, J. *Basics of Qualitative Research: Grounded Theory Procedures and Techniques*. Newbury Park, CA: SAGE, 1990).

———. *Basics of Qualitative Research: Grounded Theory Procedures and Techniques* (2nd ed.). Thousand Oaks, CA: Sage, 1998).

Taylor, M. J., F. E. Friot, and L. Swetnam. *Women Who Say "Yes" When Science Says "No": Their Lessons for Future Students and Teachers*. Paper presented at the American Association of Colleges of Teacher Education, 1997.

Thomas, D. A. "The Impact of Race on Managers' Experiences of Developmental Relationships: Mentoring and Sponsorship: An Intra-organizational Study." *Journal of Organizational Behavior*, 11 (1990): 479–492.

———. "Racial Dynamics in Cross-Race Developmental Relationships." *Administrative Science Quarterly*, 38.2 (1993): 169–194.

———. "The Truth about Mentoring Minorities: Race Matters." *Harvard Business Review*, 79(4), 98–107, 2001.

Thomas, D. A., and K. E. Kram. "Promoting Career-Enhancing Relationships in Organizations: The Role of the Human Resource Professional." *Career Growth and Human Resource Strategies: The Role of the Human Resource Professional in Employee Development*, edited by M. London and E. M. Mone, 49–66. Westport, CT: Greenwood Press, Inc., 1988.

Tinto, V. *Leaving College: Rethinking the Causes and Cures of Student Attrition*. Chicago: University of Chicago Press, 1987.

Trower, C., and Chait, R. Faculty Diversity: Too Little for too Long. *Harvard Magazine* (March-April 2002): 33–37.

Turban, D. B., T. W. Dougherty, and F. K. Lee. "Gender, Race, and Perceived Similarity Effects in Developmental Relationships: The Moderating Role of Relationship Duration." *Journal of Vocational Behavior*, 61.2 (2002): 240–262.

Turner, C. S. "Incorporation and Marginalization in the Academy: From Border toward Center for Faculty of Color?" *Journal of Black Studies*, 34.1 (2003): 112–125.

Umbach, P. D. "The Contribution of Faculty of Color to Undergraduate Education." *Research in Higher Education*, 47.3 (2006): 317–345.

Whitla, D. K., C. Howard, F. Tuitt, R. J. Reddick, and E. Flanagan. "Diversity on Campus: Exemplary Programs for Retaining and Supporting Students of Color." *Higher Education and the Color Line: College Access, Racial Equity, and Social Change*, edited by G. Orfield, P. Marin and C. L. Horn, 131–152. Cambridge, MA: Harvard Education Press, 2005.

Willie, C. V. *Five Black Scholars: An Analysis of Family Life, Education, and Career*. Lanham, MD: Abt Books, 1986.

Willie, C. V., R. J. Reddick, and R. Brown. *The Black College Mystique*. Lanham, MD: Rowman and Littlefield, 2005.

Wilson, R. "The Power of Professors." *Chronicle of Higher Education* (March 3, 2006), A10.

2

CURRICULUM AND IDENTITY

4

WHICH AMERICA IS OURS? MARTÍ'S "TRUTH" AND THE FOUNDATIONS OF "AMERICAN LITERATURE"

Michael Hames-García

MARTÍ'S "TRUTH"

According to many critics of affirmative action and curricular reform, minorities have made overwhelming gains in higher education and completely taken over English departments.[1] The reconfigured canon, as demonstrated by the new "American literature" anthologies, surely demonstrates for these critics that inequality and discrimination are things of the past and that we should get back to just reading texts as texts rather than continuing to politicize education. Such criticisms of teaching "American literature" politically from a critical, multicultural perspective rely on a false and inaccurate assessment of the current state of U.S. universities, something of which the *Chronicle of Higher Education* recently reminded me. In an article on affirmative action, the Chronicle notes that, "[t]aken together, African-American, Hispanic, and American Indian scholars represent only 8% of the full-time faculty nationwide. And while 5% of professors are African American, about half of them work at historically black institutions. The proportion of black faculty members at predominantly white universities—2.3%—is virtually the same as it was 20 years ago" (Wilson, A10). Given that "Hispanics" make up 12.5% of the total U.S. population and that blacks are 12.3%, our representation on university faculties is appallingly low (U.S. Bureau of the Census, 3).[2] Furthermore, although there were 61,000 "Hispanics" with doctoral degrees in the United States in 2000 (Newburger and Curry, 25),[3] there were well over 350,000 "Hispanics" incarcerated in the United States.[4] At the same time, there were 1,578,000 "non-Hispanic whites" with doctoral degrees and only 591,000 incarcerated (Newburger and Curry, 7).[5] In other words, there are nearly three whites *with doctoral degrees* for every one behind bars, but nearly six Latinos *behind bars* for every one with a doctoral degree. This is the world I live in, and in this world I find offensive the injunction to not teach "politically" and to leave questions of identity and power outside of the classroom, especially when the subject of my teaching is the United States.[6] The university remains a site of struggle and controversy in part because it remains reflective of the struggles and controversies surrounding identity, power, oppression, and resistance characteristic of the nation and its history. A small part of those struggles and controversies have concerned the teaching of literature, the definition of "American," and the foundations of U.S. literary history.

Those arguing for more inclusive and egalitarian definitions of "American" identity to inform teaching and scholarship in the university have turned to many sources, among them, the work of social critics in the United States such as nineteenth-century Cuban indepentista, intellectual, and exile José Martí.[7] Martí has been attractive to such scholars because of his prescient understanding of the United States' coming role of hemispheric dominance in the twentieth century and because of his astute observations about the nature of U.S. society at a crucial juncture in history. Indeed, in his 1991 book, *The Dialectics of Our America*, José David Saldívar offers a politically resistant "pan-American literary history" that begins with Martí (5). Saldívar chooses to begin with one of Martí's most famous pieces, the 1891 essay "Our America," because of its importance in signaling an awareness of the beginnings of U.S. imperialism and of a great cultural, economic, and political polarization between the two Americas of the Western hemisphere.[8] Martí urges his fellow Latin Americans not to look toward Europe and the United States for models of culture and politics, but rather to value their own indigenous traditions. He cries, "Let the world be grafted onto our republics, but we must be the trunk [*Injértese en nuestras repúblicas el mundo; pero el tronco ha de ser el de nuestras repúblicas*]" (291; 18) and urges the peoples of our America to make "common cause" with "the oppressed" (292; 19). Above all, he urges his readers to be wary of the United States, "the seven-league giant" (289; 15). "Our America" affords Saldívar an opportunity to map literary history across geopolitical borders in order to highlight critical perspectives on white, U.S. hegemony and domination. In the wake of Saldívar's invocation of Martí to open up U.S. literary and cultural studies to analyses of imperialism and to international and border studies scopes, numerous scholars have made similar efforts.[9] Others, such as Amy Kaplan and Donald Pease have also made significant contributions to this conversation. I want to begin here by taking a deeper look at another, less analyzed, essay by Martí. It is significant because of its critique of U.S. culture and the advice Martí offers to the cultural critic.

Written in 1894, "The Truth about the United States" echoes Martí's earlier work but with a more sharply honed sense of the dangers presented by the U.S. example. Martí's position as an exiled writer living in the United States gave him a vantage point from which to observe how the society functioned. Indeed, this essay represents a low-point in his estimation of the United States, a country about which he was impressively ambivalent throughout his lifetime. Rather than simply describing the country's diversity, Martí, as Rosaura Sánchez notes, links the U.S. imperial project with its "internal contradictions" (119).[10] Martí goes so far as to write that "the North American character has declined since its independence, and is less humane and virile today" (332).[11] Thus, while he acknowledges the high principles contained within the U.S. documents of revolution and the sentiments behind the abolition of slavery, he believes that the course of the nineteenth century has brought about a tragically unjust society in the United States. Martí, as usual, posits as his audience "our America," beginning by noting that "[it] is urgent that our America learn the truth about the United States" (329; 987). Martí thus seems to sound a warning to Latin America and the Caribbean, but it is not exactly on the same note as "Our America." Amplifying his call in "Our America" for Latin American nations to look to indigenous sources for models of education, culture, and government, Martí describes the "particular consequences" of the two distinct "historical groupings" of the United States and Latin America (329; 987). He wants not only to demonstrate the difference between the United States and its southern neighbors for the purpose of proving that different peoples should take the paths appropriate to their own circumstances but also to disabuse others of myths and generalizations that he feels have grown up around the United States. He makes this point sharply: "It is a mark of supine ignorance and childish, punishable light-mindedness to speak of the United States, and of the real or apparent achievements of one of its regions or a group of them, as a total and equal nation of unanimous liberty and definitive achievements: such a United States is an illusion or a fraud [superchería]" (330; 988).

In dismissing the idea of the United States as a nation of perfect liberty, Martí unites an analysis of the imperialist exploitation of native peoples on the frontier with a critique of race and class prejudice in the rest of the nation: "The hills of the Dakotas, and the barbarous, virile nation that is arising there, are worlds away from the leisured, privileged, class-bound, lustful, and unjust cities of the East" (330; 988). What is most striking to Martí is not simply that there is injustice in the United States, but that, far from a melting pot in which different cultures have fused together or coexisted harmoniously, the United States has become a country whose essence is fundamentally one of conflict and cultural clash:

> An honorable man cannot help but observe that not only have the elements of diverse origin and tendency from which the United States was created failed, in three centuries of shared life and one century of political control, to merge, but their forced coexistence is exacerbating and accentuating their primary differences and transforming the unnatural federation into a harsh state of violent conquest. [. . .] Rather than being resolved, the problems of humanity are being reproduced here. Rather than amalgamating within national politics, local politics divides and inflames it; instead of growing stronger and saving itself from the hatred and misery of the monarchies, democracy is corrupted and diminished, and hatred and misery are menacingly reborn. (330–31; 988)

Confronted by such an internally divided leviathan, Martí points to the duty of the cultural critic:

> And the one who silences this is not doing his duty, while the one who says it aloud is. For silence would be a failure of one's duty as a man to know the truth and disseminate it, and of one's duty as a good American, who sees that the continent's glory and peace can only be ensured by the honest and free development of its different natural entities. (331; 988)

The United States is thus by definition a place constituted through increasingly extreme conflict and the inability to overcome it. Furthermore, as Sánchez appropriately notes, Martí "stresses the domination and subordination forced upon particular elements within the nation-state" in his analysis of the reasons for the nation's irresolvable conflicts (119).

What Martí's essay suggests to us is that, instead of a one-sided, celebratory vision of the rise of the nation or a distortion of the reality of diverse cultures in unequal conflict, cultural critics should base the study of "American literature" on a more accurate, less celebratory vision. Martí's perspective on U.S. society enabled him to eventually come to see the nation as first and foremost a site of violence, discord, and social struggle. He therefore positions himself in this passage as a "good American" who is acknowledging the truth of difference and discord in the United States, seeing this acknowledgement as in the interest of the "glory and peace" of the continent as well as the development of the nation's own diversity.[12] He goes on in the lines that follow this passage to say that it is also his duty as "a son of *our America*" to warn his people not to take the United States ("a damaged and alien [*dañada y ajena*] civilization") as a model (331; 988, 989; emphasis added). He repeatedly acknowledges that both Latin America and the United States have faults and virtues. However, as in "Our America," he is most troubled by the unquestioning admiration for the United States of those envious of its progress and luxury, on the one hand, and those ashamed of their own mestizo origins, on the other (332; 989). Announcing a new column in his newspaper, *Patria*, Martí concludes "The Truth about the United States" by promising to "reveal the fundamental qualities that in their constancy and authority demonstrate the two truths that are useful to our America: the crude, unequal, and decadent character of the United States, and the continual existence within it of all the violence, discords, immoralities, and disorders of which the Hispanoamerican peoples are accused" (333; 990).

Notably, Martí's essay is a prism to refract the image of the United States for the benefit of viewers to the south, not a mirror intended to reflect that refracted image back to its point of origin. However, as I consider "American literature" and the teaching of multiculturalism in the remainder of this article, I want to take Martí's vision as a reflection to present back to "us" our national culture. If we, as scholars of the United States, want to be able to recognize conflict and to analyze its causes, then it would behoove us to eschew self-congratulatory approaches to U.S. culture, history, and identity in our research and pedagogy. Martí recognized that acknowledging the existence of conflict and injustice is a necessary precondition to attaining "glory and peace" and the fostering of egalitarian diversity. Both "Our America" and "The Truth about the United States" resonate with Ralph Waldo Emerson's call in "The American Scholar" and "Self-Reliance" for "Americans" to turn toward their own realities rather than those of Europe in crafting their culture. Martí refutes Emerson's image of U.S. exceptionalism, however, seeing it as standing in the way of a veracious and critical evaluation of U.S. society. Instead, he positions "our mestizo America" as the hope for a new and democratic civilization ("Our America," 292; 19). In doing so, he offers us a more objective reflection of identity in the United States and a better foundation for thinking about "American literature" than that offered by many of the traditional staples of the U.S. literary canon. I contend that one of the reasons that Martí's view of the United States has been less canonized than those of J. Hector St. John de Crèvecoeur, Emerson, or Alexis de Tocqueville is that it is less congratulatory.[13] As I will argue in the final section of this article, Martí's politically *interested* view proves to offer in the end a more objective account of the truth about the United States than Emerson's equally interested and no less partial perspective. It is therefore from Martí rather than from Emerson that I suggest we take our cue in assessing the foundations of "American literature."

Foundations of "American Literature" (1): What We Teach

Why and how is Martí's perspective useful for rethinking the foundation of the category of "American Literature" as it is embodied in major literary anthologies?[14] I want to argue that Martí's rejection of the idea of U.S. exceptionalism should prompt us, as scholars of "American literature," to do the same.[15] I want to suggest two issues that literary scholars might want to consider in this regard. First, there is a need to make explicit to ourselves and to our students the political as well as aesthetic criteria used in selecting literature for courses and anthologies. The criteria for determining what we teach as "American literature" must be based on a vision that is more objective rather than entirely one-sided, critical rather than just celebratory. Second, there is the question of how we teach. Later in this essay, I will argue that texts should be put into discussion with one another and that the social contexts in which they arise should be highlighted rather than ignored. Before dealing with that, however, I will deal in this section with the first issue—what we teach as "American literature."

Among the many questions regarding canon formation (i.e., what we teach) that have arisen for me is how literary anthology editors determine what authors and texts count as "American." Overall, editors appear to use an inconsistent combination of citizenship, geographical location, language, and subject matter (i.e., the subject of America and the United States). The first three of these become increasingly important in the contemporary period. Citizenship, in particular, has been used selectively the further back in time one looks; thus, the national citizenship of John Smith or Samuel de Champlain is irrelevant. Several anthologies have broken with the trend of requiring citizenship for authors after the eighteenth century, and the most notable new inclusions are American Indian authors and orators, as well as some other "outside observers" such as Martí and Sui Sin Far.[16] For these authors, subject matter and geographical location suffice. Geographical location usually involves a projection of the present-day borders of the contiguous 48 states anachronistically into the past. While geographical location and subject matter allow Martí, Sui Sin Far, and Native Americans into the anthologies, Christopher Columbus and others like him are included by virtue of subject matter.[17] In addition, while the inclusion of early authors is less constrained by a language requirement, more recent authors not writing in English, such as Tomás Rivera and Isaac Beshevis Singer, are almost never included in "American literature" anthologies, despite citizenship, geography, and subject matter.[18] Finally, subject matter has proven to be the most consistent (although far from universal) criterion. Indeed, much of the recent "canon busting" in U.S. literature has been in the name of making the canon more inclusive and representative of the diversity of experiences and perspectives on what "America" and "an American" are. The highly charged ideological component inherent in defining the category explains why science fiction has rarely been viewed as "American literature"—despite the great number of wonderful science fiction authors on the U.S. scene (e.g., Isaac Asimov, Samuel R. Delaney, Ursula K. Le Guin, H. P. Lovecraft, Marge Piercy, and Kurt Vonnegut, Jr.).[19] Unless it can be shown to "actually be about" the United States, speculative fiction is rarely considered to be either "serious" enough or sufficiently topical to merit inclusion. Whereas much of the romantic and gothic fiction of Nathaniel Hawthorne and Edgar Allan Poe has accreted a long interpretive legacy establishing its concern with U.S. history and identity, most twentieth-century science fiction has yet to achieve this kind of resonance among critics.

In light of such considerations, one can see that, even in the case of "undisputed classics," the canon of "American literature" has not been created on the sole basis of aesthetic value. It exists partly due to ideological determinations about the relative importance of citizenship, geography, language, and subject matter. Conversely, many works of great aesthetic value are not taught as "American literature" because they do not fit the ideological and political requirements of academics and publishers.[20] Anthologies prior to the 1980s focused on portraying a civic sense of what it is to be "American." They did this almost exclusively through works by white men, and they sought to put these texts into a continuous line of development that demonstrated the creation of "the American character" as an exceptional entity. Much has been made of how this involved, in the early twentieth century, the removal both of many women writers and of a number of "New England gentlemen" poets from the canon; in their place, such previously neglected writers as Melville and Thoreau were ensconced. Black, Latino, American Indian, and immigrant writers were left out altogether. The high-point of this period saw the elevation of *Moby Dick* to national epic in a sometimes explicitly ideological move of the cold war.[21]

Beginning in the 1970s and gaining strength through the 1980s was a new phase of canon revision. Whereas earlier curriculum changes had eliminated the Greek and Latin classics in favor of English and American texts, or displaced "genteel" literature with literature reflecting the brash "spirit of the frontier" and rugged U.S. exceptionalism, this new phase was motivated by feminist and minority identity politics. It grew out of a critical wave of scholarship that analyzed the historical and social connections between power and identity in society and described the links between social oppression and the devaluation of cultural contributions by women and people of color. Changes happened most quickly with regard to contemporary writers, but over the course of 40 years, from the early 1960s to the present, tremendous recovery efforts unearthed a wealth of neglected or suppressed literature, much of which was critically acclaimed even by white male critics in its day.[22]

In the wake of calls for curricular reform, anthology editors have seriously rethought many aspects of their job.[23] In addition to introducing women and minority authors, the new editions of most anthologies include unprecedented selections from oral culture (e.g., Native American oratory, slave songs, and work and folk songs), working-class and labor writers (e.g., Rebecca Harding Davis, Hamlin Garland, and Upton Sinclair), white immigrant writers (e.g., Emma Lazarus, Anzia Yezierska, and Abraham Cahan), non-citizen writers in the United States (e.g., Sui Sin Far and José Martí), writers from "the West" (e.g., Louise Amelia Smith Clappe, Bayard Taylor, and Harriet Prescott Spofford), and even writers outside the area of the 48 contiguous states (e.g., Aleut tales,

Bartolomé de las Casas, and James Grainger).[24] Today, the latest edition of the *Norton Anthology of American Literature*—appearing in five volumes and totaling well over 5,000 pages—has dramatically changed from its first edition. Beyond its division into five volumes, the current state of the *Norton* is typical of other anthologies, which have scrambled to incorporate women and minority authors, but still lag behind the sweeping changes introduced in the 1980s and controversially extended by the first edition of *The Heath Anthology of American Literature* in 1990.[25] The changes are particularly noticeable with regard to gender. However, much of the change seems more dramatic than it is, simply because earlier editions were almost entirely white and male.[26] The 2002 edition of the *Norton* still devotes 4,432 pages to white and European authors (82.53%), but 572 pages to black authors (10.65%), 229 pages to Native American authors (4.26%), 73 pages to Latino authors (1.36%), and 64 pages to Asian American authors (1.19%). Tallies that count selections, rather than pages, also make the degree of change seem greater than it is. White authors make up 68.80% of the total, but their selections occupy 82.53% of the pages, while Latinos make up 2.99% of the authors, but receive 1.36% of the pages.[27] Thus, for example, volume A's "Native American Trickster Tales" includes six selections by Native Americans, but these combine to only 36 pages, compared to William Bradford's one selection (from *Of Plymouth Plantation*) at 40 pages. Furthermore, 180 of the 572 pages devoted to black authors are made up of selections from just two writers: Frederick Douglass and Nella Larsen. Excluding Spanish explorers, no Latino or Latin American authors appear in the first three volumes—which cover the periods of both the Mexican American and Spanish American wars! The only Latino author to appear prior to 1945 is William Carlos Williams, who accounts for 18 of the 73 pages devoted to Latino authors (another 25 are devoted to Gloria Anzaldúa).[28] (By contrast, *Herencia: The Anthology of Hispanic Literature of the United States*, edited by Nicolás Kanellos, offers well over 300 pages of material dating before 1945.)[29]

Clearly, different teaching constituencies require different approaches. Yet, I want to briefly consider how unexplored assumptions about who our students are influence both what we teach and the reasons we provide for teaching it. In the critical material on the teaching of multiethnic U.S. literature, a great deal of space is taken up by considerations of the best way to teach literature by nonwhites to white students. Of course, the majority of college students nationally (and at my own institution) are white. Still, many essays begin with the disclaimer that most students are white and then go on to consider them exclusively as the subjects of multicultural education, as if there were no need to consider students of color or as if the only purpose for teaching multiethnic literature were to sensitize our white students.[30] In *Cultivating Humanity*, for example, Martha Nussbaum criticizes both conservative approaches to teaching literature and what she calls "the spirit of identity politics": "an approach to literature that questions the very possibility of sympathy that takes one outside one's group, and of common human needs and interests as a basis for that sympathy" (109).[31] Of course, some dominant trends in literary criticism do tend toward a relativist approach to cultural understanding; however, Nussbaum misrepresents the case when she implies that "most" literary criticism takes the position that African American literature is only for African Americans or that "the argument in favor of *Invisible Man* [is] that it affirms the experience of African-American students" (110).[32] Indeed, Nussbaum's argument, which stresses understanding others across differences, runs the risk of placing whites at the center of the multicultural curriculum.

For Nussbaum, people of color are always an object to be understood by "us," although this happens through implication rather than directly in her text:

> [I]t is hard to deny that *members of oppressed groups* frequently do know things about their lives that other people do not know. [. . . I]n general, if we want to understand the situation of a group, we do well to begin with the best that has been written by members of that group. We must, however, insist that when we do so it may be possible for us to expand *our own* understanding—*the strongest reason for including such works in the curriculum*. We could learn nothing from such works if it were impossible to cross group boundaries in imagination. (111; emphasis added)

The referent in this passage for "groups" remains "oppressed groups," so that the referent for "we" becomes implicitly those outside such groups. Multiculturalism, for many critics like Nussbaum, is about understanding the other, without much consideration for what it is like to be the other within the context of multicultural education. While no one would assert that the strongest reason for including Jane Eyre in the English syllabus is so that African American students (or any students, for that matter) can come to feel sympathy toward the experiences of nineteenth-century English women, a parallel argument about *Invisible Man* is commonplace. Much can be learned about the situation of nineteenth-century English women from Jane Eyre, but a more plausible reason for learning this would be to understand how gender, nation, and class have come together in shaping the world we inhabit today. This is something like Ellison's argument about *Invisible Man*: "our nation has a history of racial obtuseness and [. . .] this work helps all citizens to perceive racial issues with greater clarity" (Nussbaum 110). While Nussbaum sees Ellison's point as incompatible with "affirm[ing] the experience of African-American students" (110), *Invisible Man*, among other things, affirms African American students' experience of being perpetually misperceived by whites, of being made simultaneously indispensable and invisible to the project of multicultural U.S. democracy. This affirmation is a necessary part of more clearly perceiving racial issues in the United States. It enables one to understand, for example, how a theorist like Nussbaum can unintentionally turn a discussion about the inclusion of African Americans in the curriculum into a consideration of the best way to educate students (implicitly white) to become "world-citizens" (110). Nussbaum's mistake announces pressingly the alarum to rethink not only what is taught in the classroom, but how one teaches it, to whom, and why.

Foundations of "American Literature" (2): How We Teach

I would like to avoid Nussbaum's error through an incorporation or awareness of structures of oppression within the teaching of multiethnic literature; I therefore suggest that courses should focus on portraying diverse experiences in U.S. literature in relation to oppression and resistance. Martí's "truth" causes me to wonder what would happen if one were to take oppression and resistance as well as cultural diversity as the basis for reconstructing the canon. An honest and more truthful account of the United States, such as that provided by Martí—one that acknowledges the discord and violence at the heart of the U.S. national self—can enable a much more radical reconstruction of the U.S. literary canon than any suggested solely from the argument for diversification. Martí's vision of U.S. society as first and foremost a site of violence, discord, and social struggle demands a transformed U.S. literary landscape and methodology. This transformation accords with the opinions of many others, such as W. E. B. Du Bois, or William Carlos Williams, who wrote in a 1925 essay, "The Fountain of Eternal Youth," that "History begins for us with murder and enslavement, not with discovery" (39). Including perspectives like Martí's without making explicit the incommensurability between them and those that have organized the canon into which they are now being introduced results in a de facto cultural relativism. Cultural relativism discourages students from evaluating different positions as better and worse, as more and less truthful. As a result, students are not encouraged to take perspectives other than their own "seriously," that is, to incorporate them, or to engage with them at an emotional level.

Thus, one recurring problem with teaching multiethnic literature is that non-canonical texts are simply brought into the "American literature" classroom to set the traditional canon into relief.[33] This pluralist approach rarely makes a central issue out of the connections between cultures, the reasons why some cultural perspectives have been excluded from the canon, or new texts' possible incommensurability with the traditional ideological underpinnings of "America" and "American literature." In addition, it promotes tolerance and risk-free diversity without acknowledging the difficult examination of exploitation and injustice (as well as questions of blame) that have to be addressed openly before students can make decisions about the sources of racism, oppression, and exploitation or the current consequences of colonization, genocide, and imperialism. Many scholars have launched powerful critiques of pluralist approaches to multicultural education and argued for alternative *critical* versions of multiculturalism.[34] Educational theorists Michael Apple and Linda Christian-Smith, for example, have deplored what they call domination "through compromise and [. . .] 'mentioning'" (10). This approach incorporates some material into the curriculum by or about nondominant groups in society (incorporations won usually after lengthy political struggle), but in such a way that the power of its political critique is diffracted or diffused.

Similarly, Paula M. L. Moya criticizes an additive approach to multiculturalism that decontextualizes texts and ideas (146). According to her proposal for a realist approach to multicultural education, an essential aspect of successful multicultural education is an understanding, not merely of different cultures' experiences, but also of how cultures are interrelated and connected by historical practices and systems of domination (156–58). Multicultural education must therefore proceed from the understanding that diverse cultural formations exist within a differentiated structure of hierarchical relations. It must also, Moya argues, acknowledge how diverse cultural "experiments" in the realms of knowledge and ethics are indispensable for the greater flourishing of humanity. This also means that students and teachers have to be able to make critical evaluations about culturally different conceptions of human flourishing (160–66).[35] Furthermore, Moya makes the point that the study of different cultures will rarely be either easy or comfortable. Instead, she writes that conflict is "inevitable and necessary, as a potentially creative, and not always destructive force" (171). Indeed, this final and most important of Moya's points is crucial for the present essay and resonates with my appropriation of Martí's "truth" about the United States as a starting point for understanding "American literature." Students are already aware of the conflict and oppression in the world; they live in the world, after all. I contend that we as teachers do them no favors by encouraging them to think of literature as a utopian place apart from the violence and discord of the real world. Not only do we shield them from the "controversies" that make the study of literature interesting to many of us as scholars, but we deprive them of the educational and epistemological value of discomfort that arises from confronting new ideas and being forced to defend or let go of old ones.[36] As Moya writes, "conflict [. . .] is absolutely necessary for epistemic and moral growth. [. . . W]e need to learn to work through it rather than attempting to cover it over or trying to avoid it" (171).

In *Beyond the Culture Wars: How Teaching the Conflicts Can Revitalize American Education*, Gerald Graff argues that teachers need to not make "a focused curriculum out of [the university's] lively state of contention" (11). Rather than a liberal pluralist approach to multiculturalism that would be "content to let cultural and intellectual diversity proliferate without addressing the conflicts and contradictions that result," Graff believes that pedagogy should bring different cultural and intellectual perspectives into conversation, "teaching the conflicts" (10). One thing that I find particularly useful about Graff's work is that he resists the impulse of many defenders of multiculturalism to conclude by mitigating the very real threat to one's sense of self sometimes posed by encountering difference. Rather than seeing diversity as a step toward a greater sense of unity, Graff prefers fostering a common debate to imposing a common culture: "We need to distinguish between a shared body of national beliefs, which democracies can do nicely without, and a common national debate about our many differences, which we now need more than ever" (45). He ends his book by looking at several examples of institutional efforts to create communities of conflict rather than commonality.[37]

My own first experience with a class based on the controversies in U.S. literature was as an undergraduate in a course titled "Race and Ethnicity in American Literature," taught by Frann Michel at Willamette University. Michel began the course by teaching excerpts from Dinesh D'Souza's *Illiberal Education* (just published) and an article on the U.S. literary canon by Paul Lauter (general editor of the *Heath*).[38] I do not know whether she was familiar with Graff's arguments (*Beyond the Culture Wars* was not yet published, but some of its main points had already appeared in journals). I do remember class discussions about these readings being highly charged and uncomfortable. However, in part because they were uncomfortable, I approached them with seriousness and took the often difficult and painful step of rethinking some of my own ideas. Knowing the disciplinary controversy also enabled me to see the texts we went on to read (by such writers as Nella Larsen, William Faulkner, and Ralph Ellison) as themselves participating in *debates* about the meaning and role of race, gender, and class in U.S. society and letters rather than just espousing particular points of view. Again, I was invested in these texts at an emotional as well as intellectual level in part because the earlier readings had given me a sense of their relevance to contemporary, heated debates over education and social justice of which I saw myself as a part.

One limitation of Graff's project is that he at times focuses too closely on cultural debate over "differences" rather than material conflicts over, for example, unequal distribution of wealth, land, resources, and political power. For this reason, I think that his model risks lending itself to a reasoned "democratic discussion" in which participants do not feel themselves or their sense of self and wellbeing called into question (45). This comes through in several of Graff's hypothetical examples, in which professors holding differing positions simply argue for and/or revise their theories (47–52, 53). What is missing from these is what we all (students and faculty alike) go through whenever we learn something worthwhile: emotion, pleasure, pain, anger.[39] In addition to playing down the emotional character of conflict, especially necessary for the project of learning, Graff seems to undervalue the importance of an active teacher, demonstrating a suspicion of "radical pedagogy" and dismissively associating consciousness raising with "authoritarianism" (25). As a result, although he would certainly not endorse such an outcome, one could imagine a teacher simply presenting conflict and controversy in a way that lacks personal investment, thus encouraging students to participate disinterestedly. The examples he endorses in his final chapter veer away from this direction, but, as someone with a focused goal to teach against oppression, I remain wary of the liberal pluralist overtones that linger in his proposal.[40] A more critical approach to multicultural pedagogy must address material discrepancies in society and not shy away from students' and instructors' own investments (physical and emotional) in those discrepancies, that is, how many of us benefit from social inequality.

Moya's considerations about how knowledge is actually arrived at (not only through "positions" to be debated, but also through less coherent, more emotional "thinking through" at the individual and collective levels) should therefore be an important addition to Graff's model of controversy and conflict in multicultural education. Another important element here—also explicit in Moya's proposal—is the centering of students' and teachers' own identities in the educational process. As Moya makes clear, attention to identities in the classroom is not the same thing as cultural relativism or the uncritical celebration of identity (160–61, 169–70). The right kind of attention to identity can resolve the difficulties with teaching texts that Stephen Railton warns about in his essay on white audiences' reception of Harriet Beecher Stowe's *Uncle Tom's Cabin*. Railton discusses how students, in particular white students, can respond to this novel in ways similar to how some white readers responded to the novel in its day. He charts how white readers can use an identification with Eliza's suffering or Tom's self-sacrifice to feel greater sympathy toward blacks, but also (even perhaps to a greater degree) can use this identification in a "self-interested, excursionary way" to gain their own redemption: "Tom is not allowed to become a person who might want to live next door but remains a personification of what is missing in the Anglo-Saxon inner life" (108). Railton's map of possible white responses to *Uncle Tom's Cabin* records the danger I see in Nussbaum's discussion of *Invisible Man* as well. Again, one sees the white subject central to a project that was originally to be about the liberation of nonwhite subjects from racist oppression. Stowe's text presents African American culture as more virtuous than European American culture, suggesting that whites have something to learn from blacks. Such a gesture is laudable, but can obscure material questions about equality and social justice, allowing "whites to see themselves as the disadvantaged race" (Railton, 109). Questioning this assumption through a constant reference from texts to material contexts is a first step, but students and teachers must also see themselves as materially invested in the outcome of debates over culture, identity, and power.[41]

WHICH AMERICA IS OURS?

Everything I have argued for up to this point resists the impulse to attempt to bring about a better society merely by presenting students with a positive image of diversity in our course syllabi.[42] This impulse is either an ultimately ineffective shortcut or an easy out. Given the fact of an extremely rancorous diversity in society, the university, and the classroom that is punctuated by both personal and structural injustice, I want to consider what kinds of approaches to the teaching of literature (specifically "American literature") might work against oppression and injustice. Two primary implications follow from my argument at this point: first, students must be involved and feel something personally at stake in the conflicts motivated in and about literary texts in order for their education to truly challenge oppression and exploitation; second, texts must be put into relation with one another (historically, rhetorically, and ideologically), not simply included side by side in a smorgasbord of diverse experiences.

Consider an example of a text recently included into the canon: Du Bois's *Souls of Black Folk*. In the *Instructor's Guide* to the *Heath*, contributing editor Frederick Woodard makes numerous excellent suggestions for teaching Du Bois's text in relation to images of blackness, gender, African American folk music, Hegelian thought, and terminology ("Negro," "black," and "African American"). What's missing from these suggestions is the pain and anger directed toward white people in Du Bois's text. My white students certainly pick up on that anger, which is why many come into class after reading the assignment disliking Du Bois and thinking of him as "racist" or "antiwhite." To downplay the anger in Du Bois's text or to avoid the ways in which he is, indeed, "antiwhite" or to ignore his claims for epistemological privilege for blacks in the United States would be a disservice to both white and black students. All students need to hear and to engage with that anger and with Du Bois's claim that blacks understand something about white America that many whites do not. Chances are they have heard similar claims before from the Reverend Al Sharpton or from rap lyrics. Du Bois presents an opportunity to make sense of that claim and its limits, to give it context (for example through considerations of the lynching epidemic of the time), and (for white students) to understand why an intelligent, antiracist person might make it. The contributors to the *Instructor's Guide* are fine scholars and the intent of the *Heath Anthology* in including Du Bois is admirable; I think what can go wrong in the process of inclusion is that many critics and educators attempt to shield classrooms from anger and conflict and to predigest controversy and emotion, making sense of it before it is felt. Works intended to be radical additions to the curriculum can thus end up sanitized.

The *Harper* anthology has perhaps gone the furthest in rethinking the nature of the anthology itself so as to avoid the pattern of isolated "great works" presented as points along a continuum of tradition. Through numerous short sections (many called "Cultural Portfolios"), this anthology presents competing voices together, as part of a debate. These sections present brief excerpts from U.S. Supreme Court decisions, popular journalism, diaries, speeches, and poetry. While the *Heath* certainly has the greatest diversity of authors, the *Harper* shows the least reticence to present "American literature" as a series of material conflicts; it includes, for example, racist texts, loyalist texts opposing the War of Independence, anti-American texts, and nonliterary and visual sources such as illustrations and political cartoons.[43] These texts, worth studying in their own right, also provide invaluable context for understanding the writings of abolitionists, revolutionaries, patriots, and other, "literary" authors. What I am really interested in, however, is not just context but debate—not debate for its own sake, but with the understanding that different positions can and must be evaluated for what they can tell us about what is "American," what is "literary," and what is "right." In this, the *Harper Anthology* surpasses the *Heath* by breaking with the tradition of organizing texts solely around "great" authors. Thus, selections by Du Bois or Franklin appear in multiple locations, as parts of different conversations. Aesthetic, social, and political issues play as much of an organizing role in the *Harper* as do authors and chronological concerns.

None of the anthologies I examined for this study, however, includes the highly moving text of John Brown's statement to the court at his trial for treason or his rhetorically brilliant "Declaration of Liberty by the Representatives of the Slave Population of the United States of America," yet they all discuss the importance of Brown for other writers. How useful might it be to present not only poems about Brown by other authors, but also his own words? How might his defense of violence to free slaves as a patriotic duty rather than treason offset Lincoln's defense of war a decade later in his "Second Inaugural Address"?[44] After assigning Brown's text, I asked my students why they thought Brown was not included in "American literature" anthologies. They hypothesized that his inclusion would make the government look bad for executing someone we would today all agree was in the right and therefore risk undermining students' faith in their country, rather than making them "feel good" about it. (I think that a desire to make students feel good about their country is the same reason, ironically, that proslavery texts are today usually omitted.) Also, my students suspected that someone who was executed for treason might not seem to anthology editors sufficiently "American" to be included. Both of these hypotheses seem to me to point to a bigger problem with cultural pluralism as an ideology guiding the conceptualization of "American literature" anthologies. Through the avoidance of controversy and conflict, a pluralist approach hopes to give an impression of consensus and community in the anthology and, ideally by the end of the term, in the classroom. The idea that the answers to questions such as "what is American literature?" can come tied up in a bow at the end of one or two semesters is ludicrous and does not do justice to the complexity of the human mind or the complexity of literature.

Martí's *Letters from New York* can provide an excellent teaching opportunity regarding the "messy" intractability of conflict, although the method in which they are usually presented can stand in the way. Martí offers a superb example of the conflicted nature of belief on the personal level. It is not the case that Martí was an unequivocal opponent of the United States. Many leftist critics writing about Martí do their best to ignore or to explain his views on race (often problematic from a contemporary viewpoint) and his contradictory stance toward the United States. Perhaps the most common way of acknowledging Martí's shortcomings is to compartmentalize them into the "early" period of his exile in the United States.[45] When critics, even those who present Martí's complexities, encourage us to think of him as having or developing *a position* (albeit one that is partly right and partly wrong)—rather than contradictorily trying to sort out several incompatible beliefs to which he partially adheres—they do us and students a disservice. A linear developmental narrative (Martí gradually comes to consciousness through his experiences in the United States) may be comforting, but it is not entirely accurate. This narrative gives us a great anticolonial hero, but hides from us the conflicted and self-contradictory human intellectual. James W. Loewen, criticizing the tendency for history textbooks to create heroes out of historical figures, attributes heroification to, among other things, "the wish to avoid ambiguities, a desire to shield [students] from harm or conflict, the perceived need to control [students] and avoid classroom disharmony, [and] pressure to provide answers" (35). Resisting this tendency with Martí could enable us to ask difficult questions about his portrayal of "ignorant or wild Indians" (not in 1880, but in 1894, the year before his death) or, in "Our America," the persistent glorification of indigenous cultures while minimizing the far vaster inheritance from Africa in his native Cuba.[46] If we teach Martí as an intellectual who sorts through his ideas throughout his life, students could see, for example, how he became a victim of the press-induced panic over anarchism following the 1886 Haymarket bombing, wholeheartedly supporting the death sentence given to those who were tried for the incident, and how he flip-flopped (along with the majority of American popular sentiment) the following year.[47] Even on the level of the individual author, then, foregrounding conflict, controversy, and debate can tremendously enrich the study of "American literature."

What I most like about Martí's characterization of the United States as a model for teaching "American literature" is that it keeps in view all the things that traditional approaches to literature and "American identity" suppress: violence, conflict, discord. Far from making students feel bad about the nation, this approach enables them to have a more accurate assessment of what it has accomplished and how. Do teachers benefit students by pretending that these things do not exist, that the United States is a harmonious "salad bowl" of difference or that injustice and violence do not have identifiable agents, victims, and resisters? Take Martí's numerous discussions of lynching, for example. I have found that students have very little sense of the importance of lynching in the nineteenth and twentieth centuries. Often, students believe that a lynching was "just" a hanging, that only black men were lynched, that lynching only occurred in the South, and that a lynching was an isolated incident perpetrated by a few white men. Although this matches the common Hollywood portrayal of a lynching by masked Ku Klux Klan marauders in the woods in the dead of night, the truth about lynching is somewhat different. Men, women, and children (of many races, ethnicities, and religious backgrounds, although overwhelmingly black) were lynched from Oregon to Minnesota to California to New York to Florida. In addition, lynchings were often huge spectacles, sometimes announced in advance by the press. Horrible torture, mutilation, and burning of the victims were common, and crowds of white men, women, and children would collect "souvenirs" from victims' bodies and pose (wearing no masks) for photographs with the corpses. Such photographs were sometimes made into souvenir cards. Furthermore, throughout the final decade of the nineteenth century and into the twentieth century, lynching occurred with alarming frequency: according to Ida B. Wells-Barnett, at least 241 blacks were lynched in 1892 alone ("Red Record," 157).

Martí's writings on lynching attempt to convey some of the irrationality and horror of U.S. chauvinism and racism, as well as their quotidian character. In an 1891 essay, for example, Martí describes the lynching of 11 Italians in New Orleans. He is at pains to stress the fact that the men who participated in the lynching are ordinary and unremarkable members of U.S. society:

> From this day forward, no one who knows what pity is will set foot in New Orleans without horror. Here and there, like the last gusts of a storm, a group of murderers comes around a corner and disappears, rifles on their shoulders. Over there another group goes by, made up of lawyers and businessmen, robust blue-eyed men with revolvers at their hips and leaves on their lapels, leaves from the tree where they have hung a dead man. ("The Lynching of the Italians," 297)

Another central concern for him appears to be the official and public approval of the lynching. After noting that four of the Italians had, a few hours earlier, been found innocent of the crime for which they had been accused, he describes the events leading up to the lynching:

> [A] committee of leading citizens named by the mayor to assist in punishing the murder, a committee led by the chief of one of the city's political factions, convokes the citizens in printed and public appeals to a riot to be held the next day . . . then attacks the parish jail with only the most minimal interference, meant only to preserve appearances, from the police, the militia, the mayor, or the governor . . . rushes bellowing through the corridors in pursuit of the fleeing Italians, and with the butt-ends of its revolvers smashes in the heads of the Italian political leader, the banker, and the consul—consul of Bolivia. . . . Three more of those who, like the banker, had been absolved, along with seven others, are killed, against the wall, in the corners, on the ground, at point-blank range. Returning from this task, the citizens cheer the lawyer who presided over the massacre and carry him through the streets on their shoulders. (297)

This description evokes a sense of surreal horror in a contemporary reader, as it probably also did for many of Martí's Latin American readers. Clearly, however, the lynching was not seen as an unusual break with decorum by the population of New Orleans.[48]

In "A Town Sets a Black Man on Fire," Martí goes even further to portray the extent to which racism pervades every aspect of U.S. society. This brief essay describes three scenes. The first is a cakewalk in New York City, a spectacle put on by black couples dressed up in "elaborate getups with patent-leather shoes, the women in dancing pumps and the men in frilled shirts, so they can be mocked, ridiculed, whistled and shouted at, and have coins thrown at their heads by frenetic, curly haired players from the gaming dens, by the gamblers on the stock market who are called brokers, and by students from the two great colleges" (311). The last scene is that of a lynching in Arkansas that culminates with a white woman approaching a black man she has accused of assaulting her, after he has been tied to a tree and doused with kerosene:

> "Get back, everyone, get back, so the ladies can see me." And when Mrs. Jewell, in a triangular scarf and hat, came out from among the crowd, on the arms of two relatives, the crowd burst into a round of cheers: "Hurrah for Mrs. Jewell!" The ladies waved their handkerchiefs, the men waved their hats. Mrs. Jewell reached the tree, lit a match, twice touched the lit match to the jacket of the black man, who did not speak, and the black man went up in flames, in the presence of five thousand souls. (313)

Included between these two accounts is one of blacks "fleeing" the United States for Liberia. A powerful portrayal of the "nation of immigrants," Martí's account of the United States dramatically challenges the ideology of idealism that was being forged by writers like Emma Lazarus during his own time. The example of blacks emigrating to Africa hones the point established by the other two examples: the racism of the United States is pervasive and possibly incorrigible to the point that the only solution for blacks is to flee for their lives.

In my experience teaching American literature, background knowledge of lynching and racial conflict in the late nineteenth-century United States is invaluable for understanding such otherwise stale academic debates as that between Du Bois and Booker T. Washington. Yet, while most American literature anthologies include both Du Bois and Washington, none include the kinds of graphic accounts of lynching contained in Ida B. Wells-Barnett's "Southern Horrors" or "A Red Record" or in Martí's writings on lynching. As a part of my classes, I have found that teaching Wells-Barnett before Du Bois and Washington enables students to place the two male writers into a social and historical context that reveals the serious stakes implicated in their rhetoric. (It also introduces students to a preeminent nineteenth-century, black, female intellectual, activist, and journalist, disabusing them of the belief that the great black intellectuals of the time were all men.) One possible reason why Martí's and Wells-Barnett's accounts of lynching are not included in anthologies may be that they are not "pro-American," in the sense that they do not argue that the true nature of U.S. society lies in egalitarian ideals that have been betrayed by Southern racists. They thus contrast with many abolitionist writers who see themselves as defenders of the true U.S. character against corrupt slaveholders. Instead, Wells-Barnett and Martí present an unflattering view of the United States, suggesting that its true nature is violent and racist, and that there is no simple way to correct this by appealing to its own noble legacies. Such selections make for difficult reading, especially for white students, providing no "easy way out" and no way to shunt the responsibility for racism onto "un-American" social outsiders.

What is therefore necessary in the "American literature" classroom is to teach about social divisions, knowing that they already exist in the classroom (so there is no real need to worry about "replicating" them there through teaching about them). The emphasis in some multicultural programs on always "ending on an upbeat" can obscure the downside of U.S. life both in the past and in the present. Many critics have stressed that an incomplete canon that does not highlight conflict is not only politically suspect but also quite simply inaccurate.[49] There is, furthermore, no reason to think that focusing on conflict and "political" questions in texts should obscure questions of "literariness." For one thing, the question of what is literary itself is one of the conflicts at the heart of "American literature." For another, as Graff discusses in his consideration of teaching Conrad's *Heart of Darkness*, putting texts within social contexts of debates over what humanity and universality mean, for example, can actually increase students' appreciation of the aesthetic depth (or lack thereof) of classic texts, as well as nonclassics (25–33).[50]

The nineteenth-century United States presented in Martí's *Letters from New York* or Wells-Barnett's "A Red Record" is completely unlike that presented in "American literature" anthologies. Yet Martí's perspective on the United States can help us to restructure "American literature" not by suspending aesthetic evaluation of texts as better and worse, but by requiring that we reconsider the criteria by which we make such evaluations. Satya Mohanty argues in "Can Our Values Be Objective?" with regard to aesthetic knowledge, that new possibilities for aesthetic experience become possible "[w]hen new social relations become imaginable" (825). As with ethical knowledge, we are therefore wise to open ourselves up to the widest number of possibilities for what might count as beautiful. Furthermore, he makes a case for believing that, if "aesthetic responses are not simple but complex, and even the accurate detection of beauty is itself dependent on feelings and ideas that are in themselves not aesthetic [. . .] then] the traditional aesthetic isolation . . . of beauty blinds us to the objective nature of beauty" (827). According to this model, "aesthetic experiences are unavoidably linked to ethical and metaphysical values and perspectives, and they can enlarge our conception of what it means to be more fully human—that is, they can radically deepen and alter our existing conceptions of human flourishing" (830). Given the inextricability of aesthetic and ethical perception, then, one can see why a more *truthful* account of U.S. history, society, and identity must be a factor in making not only political but also aesthetic decisions about "American literature."

Rethinking "American literature" should, therefore, entail the kinds of judgments of which cultural relativism does not permit. According to *political* criteria, the essays of William Apess, Du Bois, Martí, or Wells-Barnett may be better than those of Emerson, insofar as Emerson's ethnocentric, partial perspective with regard to American identity impairs his ability to accurately perceive the texture of American letters.[51] However, one might also argue that "An Indian's Looking-Glass for the White Man," *The Souls of Black Folk*, "Our America," and "A Red Record" could be demonstrably superior in aesthetic terms to Emerson's essays once one no longer perceives these essays apart from the material human misery and joy they refer to, ignore, or obscure. Once we have freed ourselves from the fallacy that a text's aesthetic value lies *in the text itself*, then social context and what texts tell us about "possibilities for human flourishing" become essential knowledge for evaluating them rather than external considerations to be banished to the headings and footnotes.[52]

I would like to end this essay by briefly considering a poem whose aesthetic value could be entirely missed without a deep understanding of the conflictive and violent history of U.S. society. Michael Harper's poem "American History" begins with an invocation of one of many famous acts of white-on-black terrorism in the United States, the bombing of a church in Birmingham, Alabama during the civil rights movement. Harper immediately connects this incident to a much longer history of racial oppression. What is most striking to me about the poem, however, is how he then makes the submerged nature of African American history itself the subject of the poem, ironically suggesting that the reason the middle passage is less known than the War of Independence is because murdered blacks have hidden *themselves*:

> Those four black girls blown up in that Alabama church
> remind me of five hundred
> middle passage blacks
> in a net, under water
> in Charleston harbor
> so *redcoats* wouldn't find them. (lines 1–7)

The title of the poem ensures that readers do not miss the irony. It demands that they acknowledge not only that blacks have been victimized by whites for centuries in the United States, but also that the history of that victimization has been suppressed. The poem ends with another, playfully ironic twist: "Can't find what you can't see / can you?" (lines 8–9). It thus repeats a central claim of my essay, that an understanding of the reality of U.S. history is critical for accurately perceiving the present. It also demonstrates my claim with regard to the inclusion of literature by racial and ethnic minorities: not only can you not find what you can't see (the question of inclusion in the canon), but you can't learn from what you can't understand. For this reason, it is imperative that we engage in the pursuit of more truthful and enabling accounts of the foundations of "American literature." Otherwise, we will fail to perceive the truth—and beauty—even of what is directly in front of us.

NOTES

Hames-García, Michael. "Which America Is Ours? Martís 'Truth' and the Foundations of 'American Literature.'" *Modern Fiction Studies* 49.1 (2003): 19–53. Copyright The Purdue Research Foundation. Reprinted with the permission of the Johns Hopkins University Press.

1. Two useful overviews of these charges (the "culture wars") are Gerald Graff, *Beyond the Culture Wars: How Teaching the Conflicts Can Revitalize American Education* (New York: Norton, 1992), 16–25 and Gregory S. Jay, *American Literature and the Culture Wars.* (Ithaca, NY: Cornell University Press, 1997), 1–12.
2. These percentages are overlapping and do not include those persons indicating more than one race. On distinguishing between representation of minority writers in the curriculum and representation of minority persons in the university, see Henry Louis Gates, Jr., *Loose Canons: Notes on the Culture Wars* (New York: Oxford University Press, 1992), 178–183, and John Guillory, *Cultural Capital: The Problem of Literary Canon Formation* (Chicago: University of Chicago Press, 1993), 4–5. See also David Palumbo-Liu, "Introduction," in *The Ethnic Canon: Histories, Institutions, and Interventions*, ed. David Palumbo-Liu (Minneapolis: University of Minnesota Press, 1995), 5–9, on the co-optation, accommodation, and "containment" of multiculturalism.
3. I am counting here only those who are native-born or naturalized citizens; 42,000 of these are men; Eric C. Newberger, and Andrea E. Curry, *Educational Attainment in the United States (Update)* (Washington, DC: Dept. of Commerce, Economics and Statistics

Administration, U.S. Census Bureau, 2000, <http://www.census.gov/population/socdemo/education/p20-536/tab10.pdf>), 27.
4. Since the U.S. Census Bureau has not made available incarceration demographics for 2000, this number is a rough extrapolation from different sources: according to the Bureau of Justice Statistics' *Profile of Jail Inmates* 1996, 18.5% of the approximately 567,000 inmates of local jails in 1996 were "Hispanic" (see Caroline Wolf Harlow, *Profile of Jail Inmates 1996* [U.S. Department of Justice: Bureau of Justice Statistics. Rev. 1998. http://www.ojp.usdoj.gov/bjs/pub/pdf/pji96.pdf], 3), and according to its *Sourcebook of Criminal Justice Statistics 2000*, 47,023 "Hispanics" were in federal prisons in 2000 and 17% of the over 1,059,000 state prisoners in 1997 were "Hispanic" (U.S. Dept. of Justice, 524, 519).
5. In 1996, 37.3% of local jail inmates were "non-Hispanic white" (Harlow, 3) and, in 1997, 33.3% of state prisoners and 29.9% of federal prisoners were "non-Hispanic white" (U.S. Dept. of Justice, 519). No numbers are available for "non-Hispanic whites" with regard to federal prisoners in 2000.
6. Numerous critics have discussed how teaching "American literature" has always been exceptionally political rather than ever "merely" aesthetic. See, among others, William E. Cain, "Opening the American Mind: Reflections on the 'Canon' Controversy," in *Canon vs. Culture: Reflections on the Current Debate*, ed. Jan Gorak (New York: Garland, 2001), 3–14; Graff, 145–163; Judith Fetterley, "Introduction: On the Politics of Literature," *The Resisting Reader: A Feminist Approach to American Fiction* (Bloomington: Indiana University Press, 1978), xi–xxvi; Frances Smith Foster, "But, Is It Good Enough To Teach?" *in Rethinking American Literature*, eds. Brannon and Greene (Urbana: National Council of Teachers of English, 1997), 194–195; Jay 1997, 146–57; Paul Lauter, *Canons and Contexts* (New York: Oxford University Press, 1991), 22–47; Satya Mohanty, *Theory and the Claims of History: Postmodernism, Objectivity, Multicultural Politics* (Ithaca: Cornell University Press, 1997), 5–9; Palumbo-Liu 1995, 3–9; Amy Kaplan, *The Social Construction of American Realism* (Chicago: University of Chicago Press, 1988); William V. Spanos, *The Errant Art of Moby-Dick: The Canon, the Cold War, and the Struggle for American Studies* (Durham: Duke University Press, 1995); Kenneth W. Warren, *Black and White Strangers: Race and American Literary Realism* (Chicago: University of Chicago Press, 1993), 71–108. See also Harriet Pollack's brief but superb testimonial in "Eudora Welty and Politics: Did the Writer Crusade?" in *Eudora Welty and Politics: Did the Writer Crusade?* eds. Harriet Pollack and Suzanne Marrs (Baton Rouge: Louisiana State University Press, 2001), 1–5.
7. Born in 1853 in Cuba, José Julián Martí y Pérez came of age during Cuba's first war of independence, the Ten Years War (1868–1878). Martí spent much of the 1880s and early 1890s in New York, chronicling continental affairs for numerous newspaper readers throughout Latin America and assisting in the development of a revolutionary movement among Cuban exiles.
8. All parenthetical citations to "Our America" refer first to the English translation, followed by the Spanish original. This will enable readers who would like to check the citation in the original text to do so easily. In addition, when I found the translation ambiguous, I have provided a quotation from the Spanish original in brackets.
9. See, for example, Jeffrey Belnap and Raúl Fernández, eds. *José Martí's "Our America": From National to Hemispheric Cultural Studies* (Durham, NC: Duke University Press, 1998). Another important collection of essays examining the United States in relation to imperialism is Amy Kaplan and Donald E. Pease, *Cultures of United States Imperialism* (Durham: Duke University Press, 1993); see especially Kaplan, "'Left Alone with America': The Absence of Empire in the Study of American Culture," Kaplan and Pease, 3–21.
10. Martí's account of the United States could, therefore, answer some concerns that view multiculturalism as opposed to an internationalist framework (see, e.g., Spears).
11. All parenthetical citations to "The Truth about the United States" refer first to the English translation, followed by the Spanish original. "The Truth about the United States" does not appear in the Cuban edition of Martí's complete works. Spanish citations to this text

are therefore from the Venezuelan edition. In addition, when I found the translation ambiguous, I have provided a quotation from the Spanish original in brackets.

12. Interestingly, critics disagree over Martí's view of national identity in general and whether or not he advocated a unified, coherent (nation-type) identity or a looser, more diffuse notion of identity for "our America." Pease, e.g., argues that Martí is an antinationalist, resisting the power of the state to create subjects by the fusion of "interest groups" into a national core. Pease casts Martí as a kind of proto-poststructuralist, opposed to state power and national identity in all their forms (Kaplan and Pease, 44). By contrast, Susana Rotker in "The (Political) Exile Gaze in Martí's Writing on the United States," Belnap and Fernández, 58–76, sees Martí as an exile preoccupied with the construction of a "home" in the remembered nation. She writes that "Martí . . . was able to assemble a sort of unity out of fragmentation in a space of condensation" through his *Letters from New York* (66).

13. Crèvecoeur's "What Is an American" and Emerson's "The American Scholar" and "Self-Reliance" are reprinted in all major anthologies of "American literature." I discuss Emerson at greater length later in this essay. Although Tocqueville is generally not included in "American literature" anthologies, Pease discusses the influence of his views on the United States, particularly in contrast to those of Martí.

14. Anthologies I have consulted for this study include the following: *The Norton Anthology of American Literature*, edited by Nina Baym et al.; *The Heath Anthology of American Literature*, edited by Paul Lauter et al.; *Anthology of American Literature*, edited by George McMichael et al.; *Harper American Literature*, edited by Donald McQuade et al.; and *Anthologie de la Littérature Américaine*, edited by Daniel Royot et al.

15. On the politics of this particular vision of "American literature," see the works cited in note 6. Cain, in particular, argues that this vision presents a view of "American literature" that is flawed, inadequate, and wrong, and demonstrates how the inclusion of noncanonical texts and contexts can force us to revise our opinions about traditional texts (3–5, 7–8).

16. American Indians were specifically denied citizenship in the U.S. Constitution and did not receive the ability to claim citizenship until the passage of the Indian Citizenship Act by congress in 1924. José Martí is included only in the *Heath*, while Sui Sin Far, resident of the United States from 1898 to 1912, appears in both the *Heath* and the *Norton*. Richard S. Pressman notes that there have been experiments since World War II at "giving an international perspective through the use of foreigners (mainly British) to comment on American life" (58), although these voices were eventually eliminated along with those of women and minority writers.

17. Authors from Alaska, the Aleutian Islands, Hawaii, Guam, Puerto Rico, and the past possessions of Cuba and the Philippines are nearly always ignored; as, traditionally, have those from entire regions of the mainland United States, such as the Pacific Northwest. Both the *Heath* and the *Harper* have begun to break these traditions: the *Heath*, by including Inuit poetry, the *Harper*, by including an Eskimo story. Beyond U.S. borders, Prentice-Hall's *Anthology of American Literature* offers only Columbus, while the *Norton* includes numerous European explorers who did not reach the present-day United States. Going further, the current *Heath* and *Harper* editions add writings and oral narratives by indigenous peoples in Mexico and Central America. The *Harper* even includes Icelandic and Norse sagas about the Viking arrival in the Americas.

18. Rivera is now included in the *Heath*, as is much bilingual poetry by Latinos. Nobel prize-winning Yiddish-language writer Singer is found only in the *Harper*. Marc Shell's and Werner Sollors's new *Multilingual Anthology of American Literature: A Reader of Original Texts with English Translations* (New York: New York University Press, 2000) is a monumental challenge to the concept of the United States as an English-language nation with an English-language literature. Unfortunately, the anthology is not easily teachable to undergraduates and has, in my opinion, an undue focus on European languages. See also Sollors and Shell's *Multilingual America: Transnationalism, Ethnicity, and the Languages of American Literature* (New York: New York University Press, 1998).

19. Notably, the *Norton* has come to include both Vonnegut and Le Guin, although the selections, while fine, are not representative of the science fiction work both authors are best known for.

20. See note 6, above. Beverly Peterson has described her experiences introducing students to works like Poe's "Why the Little Frenchman Wears His Hand in a Sling" in order to disrupt their expectations of how canons are formed, including the role of individual beliefs, entertainment, and "social values" ("Inviting Students to Challenge the American Literature Syllabus," in *Teaching English in the Two-Year College*, 28 [2001]: 379–82).
21. See Spanos's excellent study of the mid-twentieth-century appropriation of *Moby-Dick* as an allegory for U.S. exceptionalism in response to the cold war with the Soviet Union.
22. Lauter provides an excellent account of some of the changes that took place in the canon in the early part of the twentieth century (*Canons and Contexts*, 22–47). For a useful chronicle of changes from the late nineteenth century to the end of the twentieth, see Richard S. Pressman "Is There a Future for the *Heath Anthology* in the Neo-Liberal State?" *symplokē*, 8 (2000): 57–67.
23. Karen L. Kilcup, in "Anthologizing Matters: The Poetry and Prose of Recovery Work," *symplokē*, 8 (2000): 36–56, provides an excellent overview of the various pressures influencing editors of anthologies, from aesthetic criteria to political concerns to the demands of the publishing industry. See also the thoughtful forum reassessing anthologies and "American literature," "What Do We Need To Teach?" especially Martha Banta, *American Literature*, 65 (1993): 330–34 and Warren, "The Problem of Anthologies," *American Literature*, 65 (1993): 338–342.
24. Writing in 1993, Jay Fliegelman provides one of the best and most concise overviews of the kinds of literature that are still underrepresented in "American literature" anthologies, from pro-slavery and temperance literature to nineteenth- and twentieth-century sermons to illustrations and manuscript facsimiles ("Anthologizing the Situation of American Literature" *American Literature*, 65 [1993]: 334–338). Some of his concerns have been (albeit unevenly) addressed by recent editions of the *Heath* and *Harper*.
25. Carla Mulford notes that *Heath* issued an anthology in four volumes in the early 1970s, but that the experiment proved unmarketable (Carla Mulford, "Seated amid the Rainbow: On Teaching American Writings to 1800," *American Literature* 65 (1993): 348).
26. Lauter is useful for considering the dire state of anthologies into the 1980s:

> In the three best-selling anthologies of 1982, women were 3 of 41 nineteenth-century writers in the Norton text and 6 of 40 in both the Macmillan and Random House volumes. . . . [I]n its 1985 second edition, [the Norton] included the work of thirty-five women and sixteen black authors. . . . [W]omen are here 22 percent of the authors included and occupy 15 percent of the page space; blacks are 10 percent of the authors and occupy 4.5 percent of the pages. [. . . The Norton and Macmillan anthologies] concentrate on presenting the work of eleven white men: they take up nearly 41 percent of the total pages in the Norton text and, in the Macmillan volumes, a whopping 49.2 percent. . . . Franklin, James, and Emerson each have about as many pages devoted to them as all the black writers combined; both Cooper and Twain are allotted more space than all the women writers (apart from Chopin) taken together. (Lauter, *Canons* 100)

Lauter also notes that the 1990 *Heath* first edition included "work by 109 women writers of all races, twenty-five individual Native American authors . . . fifty-three African-Americans, thirteen Hispanics . . . and nine Asian-Americans" (101).
27. Black authors account for 12.24% of the total, Native Americans for 12.82%, and Asian Americans for 2.14%. For the purposes of these statistics, I have included Indian-European mixed-race authors of Latin America prior to 1848 as "Native American" and counted Spanish explorers and missionaries as "white."
28. Williams's mother, Raquel Hélène Hoheb, was Puerto Rican, of the same generation as the Cuban Martí. Lisa Sánchez González discusses the implications of considering Williams as a poet shaped by the Puerto Rican diaspora (*Boricua Literature: A Literary History of the Puerto Rican Diaspora* [New York: New York University Press, 2001],

42–56). It is probable that the propensity not to think of Williams as Puerto Rican is related to his long-established presence in the canon.
29. *Herencia* will surely be the standard bearer in the field of U.S. Latino literature. Its 638 pages of multilingual writings from wide-ranging ethnic groups and five-century-plus scope towers over *The Prentice-Hall Anthology of Latino Literature's* mostly contemporary 544 pages of primarily English-language selections from the three dominant U.S. Latino ethnicities. Herencia's association with the Recovering the U.S. Hispanic Literary Heritage Project and respected editorial board also lends it an academic legitimacy that cannot be claimed by Norton's forthcoming anthology.
30. See, e.g., Foster, 195–196.
31. Although Nussbaum claims that "[m]uch teaching of literature in the current academy is inspired by the spirit of identity politics" that "celebrates difference in an uncritical way and denies the very possibility of common interests and understandings, even of dialogue and debate, that take one outside one's own group," she fails to offer an example (110). Her gesture is not at all uncommon. Many critics, both left and right, use the bogey man/straw man of "Balkanization" or "tribalism" to discredit their opponents. Usually, this claim is either asserted as fact without argument (as with Nussbaum) or established through a slippery slope fallacy (e.g., Paul Bryant "Diversity of What, for What?" *CEA Forum*, 24.2 [1994]: 10). When examples are given, they are often colloquial or anecdotal (as in, "I have a colleague who once said . . .") rather than scholarly books or essays (e.g., Spears "Multicultural versus International?" *CEA Critic*, 59 [1996]: 6). One of the many problems with the gesture toward Balkanization/tribalism is that it conjures a specter of the United States degenerating to the state of ethnic violence found "elsewhere in the world." This specter presupposes an innocent United States that is free of violent ethnic conflict. The threat to peace is then made out to be minorities rather than the dominant groups who have exercised an unremitting legacy of ethnic violence against people of color for more than 200 years. If Martí's nineteenth-century observations can teach us anything, it is that the United States prior to "multiculturalism" has always been among the most Balkanized, tribalist, and violent societies in the world.
32. For my own critiques of cultural relativism in contemporary theory, see "Who Are Our Own People? Challenges for a Theory of Social Identity," in *Reclaiming Identity: Realist Theory and the Predicament of Postmodernism*, eds. Paula M. L. Moya and Michael Hames-García (Berkeley: University of California Press, 2000), 102–129 and "How to Tell a Mestizo from an Enchirito®: Colonialism and National Culture in Gloria Anzaldúa's Borderlands/La Frontera," *diacritics*, 30.4 (2000): 102–122.
33. E.g., Graff's otherwise excellent model can lend itself to slowing down change through a "teach one classic text, teach one nonclassic text" approach. His "teach the controversies" model, in fact, is presented as a possible compromise between the left and right insofar as it guarantees the inclusion of both conservative and radical voices in the debate (25–33, 51–52). Lois Rudnick, with the best of intentions, also embraces a "one classic, one nonclassic" approach. One might also usefully contrast such models of contesting the classics externally with Eve Sedgwick's justification for preserving the Western canon so that one might critique the internal contradictions that hold it together (*Epistemology of the Closet* [Berkeley: University of California Press, 1990], 48–59). On the use of multiculturalism to bolster the traditional canon, see Palumbo-Liu, 16–17.
34. On critical versions of multiculturalism, see, among others, James A. Banks "Approaches to Multicultural Curriculum Reform" in *Multicultural Education: Issues and Perspectives*, eds. James A. Banks and Cherry A. McGee Banks, 3rd ed. (Boston: Allyn, 1997), 237–243; Chicago Cultural Studies Group, "Critical Multiculturalism," *Critical Inquiry*, 18 (1992): 530–555; David Theo Goldberg, "Introduction: Multicultural Conditions," in *Multiculturalism: A Critical Reader*, ed. David Theo Goldberg (Oxford: Blackwell, 1994), 1–41; Peter L. McLaren, "White Terror and Oppositional Agency: Towards a Critical Multiculturalism," in *Multicultural Education, Critical Pedagogy, and the Politics of Difference*, eds. Christine E. Sleeter and Peter L. McLaren (Albany: State University of New York Press, 1995), 33–70; Paula M. L. Moya, *Learning from Experience: Minority*

Identities, Multicultural Struggles (Berkeley: University of California Press, 2002); Palumbo-Liu 2, 5, 9–14; Marjorie Pryse, "Teaching American Literature as Cultural Encounter: Models for Organizing the Introductory Course," in Brannon and Greene 186–190; Ella Shohat, and Robert Stam, *Unthinking Eurocentrism: Multiculturalism and the Media* (London: Routledge, 1994); Sleeter, "Introduction: Multicultural Education and Empowerment" and *Multicultural Education as Social Activism*, 91–134 (Albany: State University of New York Press, 1996).

35. Also important for Moya's realist position is the work of Mohanty (see esp. *Literary Theory* 210–211, on the epistemic importance of emotion, and 240–247, on multiculturalism as "epistemic cooperation"). Sean Teuton, drawing from Mohanty's conception of cultures as laboratories in which different experiments are worked out, also stresses the importance of political sovereignty and equality as a precondition for cross-cultural exchanges of knowledge ("Internationalism and the American Indian Scholar: Native Studies and the Challenge of Pan-Indigenism." *Identity Politics Reconsidered*, eds. Linda Martín Alcoff et al. New York: Palgrave Macmillan, 2005).

36. Some commentators are very critical of the use of anger in the classroom and even use their feelings of discomfort and/or the discomfort of students as a justification for refusing to teach about groups other than their own to members of that group (see Spears 4–5, 2–3). Clearly instructors should not set out to traumatize students and should take care to avoid trauma themselves. Avoiding all discomfort and unpleasantness, however, harms the educational process more than it helps.

37. He argues that multiculturalists "are not rejecting the idea of a common culture so much as asking for a greater voice in defining it" (45), but later notes that "The current attack on 'divisiveness,' 'Balkanization,' and so forth is really an attack on the unpleasant fact of social conflict itself while fobbing off the responsibility for it on somebody else" (46). Many conservative and liberal critics seem to be most fearful of losing some (imagined) sense of community, glossing over the fact that such community has never existed and has typically been posited through domination. Schlesinger's and Gitlin's titles (respectively, *The Disuniting of America* and *The Twilight of Common Dreams*) illustrate this, implying as they do the reality of a "good old day" when unity and commonality reigned, despite their fictional nature (see also Bryant, 10; Spears, 3). For refutations of this myth, see Cain, 3–11; Graff, 47–51; Jay, 39–43; Pollack, 1–5.

38. Lauter's essay was an earlier version of the second chapter of *Canons and Contexts*.

39. This may partly have to do with the fact that his vision of the teaching of literature often remains bound to texts and authors, rather than issues. As James S. Laughlin notes, the emotional aspect of learning is most present in Graff's autobiographical anecdotes (Brannon and Greene, 231–248, 233–234). Laughlin criticizes Graff for not recognizing sufficiently the need to introduce controversies in a way that affects students viscerally and gives them a reason to be invested in the conflicts taught. His point is partly a Freirian one about the need for students to generate and explore questions from their own initiative, rather than having information (even controversy) brought to them by a teacher. I agree with Laughlin's prescription for pedagogy, but think that Graff's method may be in fact closer to his own than he admits.

40. See, e.g., note 33.

41. Several critics have raised questions about the possibility for aligning multiculturalism with materialist or class critique. Pressman, e.g., sees multiculturalism, identity, and "identity politics" more pessimistically than I do. His concern with class difference and material inequalities develops into a late-Marxist critique of identity that lacks nuance. While I believe there is much of value in his final conclusions that anthologies and canons may not ultimately be salvageable and that diversification of the canon is at high risk of co-optation by neoliberalism, I think that he moves too quickly to an anti-identitarian gesture. This includes making the absurd claim that ethnicity is not a great issue in our society (because on the campus on which he teaches "hardly an eye bats when an Hispanic and an Anglo pair off") but that class should be our primary focus (64). Guillory also argues for an increased attention to class, claiming that "the critique of the canon does

indeed belong to a liberal pluralist discourse within which . . .] the category of class has been systematically repressed" (14). I of course agree that more attention should be paid to class in the study of U.S. literature, especially in its interrelation with other factors of identity, but Pressman and Guillory present too reductive and unitary a notion of "multiculturalism" and "identity politics." For a better and subtler consideration of multiculturalism's complex relationship to capitalism and to class, see Palumbo-Liu, 9–14.

42. See Graff's account of how the university fosters its self-image as a "conflict-free ivory tower" (6). Sleeter also rejects models of multiculturalism that simply seek to create "an unoppressive, equal society which is also culturally diverse" in the classroom without addressing structural oppression and inequality in society ("Introduction," 10).
43. See Fliegelman's criticisms of the *Heath* in note 24. For these reasons, I am in agreement with Banta that the *Harper* presents the most advantageous "American literature" anthology currently available (333).
44. Brown's writings can be difficult to find, but are worth the effort. See John Brown, "A Declaration of Liberty by the Representatives of the Slave Population of the United States of America," in *John Brown and His Men*, ed. Richard J. Hinton (New York: Arno, 1968), 637–643; and Richard D. Webb, ed. *The Life and Letters of Captain John Brown* (Westport, CT: Negro University Press, 1972), 216–219.
45. This is true of a number of otherwise fine essays (see Pita, 130–133; Rotker, 69; Sánchez, 118).
46. On Martí and race, see Ferrer; Gillman, 96–104; and Helg.
47. See Martí, "Class War in Chicago: A Terrible Drama."
48. See also Wells-Barnett's "Mob Rule in New Orleans: Robert Charles and His Fight to the Death," 253–322.
49. See, e.g., Cain, 3–5 and Moya, 174.
50. See also Cain, 7–8. Mulford notes how student interest in early American literature has revived with the inclusion of Native American literature and questions of colonialism and cultural conflict alongside the traditional Puritan canon (342–348).
51. On Emerson, see Jenine Abboushi Dallal's excellent essay exploring the continuity between 1840s expansionist discourses and Emerson's individualist and aesthetic ideologies: "American Imperialism UnManifest: Emerson's 'Inquest' and Cultural Regeneration," *American Literature*, 73 (2001): 47–83. On the further relation of American letters to expansionist discourses, see also Frederick Wegener, "'Rabid Imperialist': Edith Wharton and the Obligations of Empire in Modern American Fiction," *American Literature*, 72 (2000): 783–812.
52. Although I am not in complete agreement with his conclusions, in "Why Does No One Care about the Aesthetic Value of Huckleberry Finn?" *New Literary History*, 30 (1999): 769–784, Jonathan Arac offers a thoughtful consideration of these issues, taking *Huckleberry Finn* as an occasion to reconsider the meaning of the aesthetic. His essay offers somewhat of a counter-point to Mohanty's essay on aesthetic value. Cain offers the insightful observation that conservative critics of multiculturalism rarely define "literary value" (9).

References

Allen, Esther, ed. and trans. *Selected Writings*. New York: Penguin, 2002.
Apple, Michael W., and Linda K. Christian-Smith. "The Politics of the Textbook," in *The Politics of the Textbook*, ed. Michael W. Apple and Linda K. Christian-Smith. New York: Routledge, 1991, 1–21.
Arac, Jonathan. "Why Does No One Care about the Aesthetic Value of Huckleberry Finn?" *New Literary History*, 30 (1999): 769–84.
Banks, James A. "Approaches to Multicultural Curriculum Reform," in *Multicultural Education: Issues and Perspectives*, ed. James A Banks and Cherry A. McGee Banks (3rd ed.). Boston: Allyn, 1997, 229–250.

Banta, Martha. "Why Use Anthologies? or One Small Candle Alight in a Naughty World." *American Literature*, 65 (1993): 330–334.

Baym, Nina Wayne Franklin, Ronald Gottesman, Philip F. Gura, Jerome Klinkowitz, Arnold Krupat et al., eds. *The Norton Anthology of American Literature* (6th ed.). New York: W. W. Norton, 2002.

Belnap, Jeffrey, and Raúl Fernández, eds. *José Martí's "Our America": From National to Hemispheric Cultural Studies*. Durham, NC: Duke University Press, 1998.

Brannon, Lil, and Brenda M. Greene, eds. *Rethinking American Literature*. Urbana, IL: National Council of Teachers of English, 1997.

Brown, John. "A Declaration of Liberty by the Representatives of the Slave Population of the United States of America," in *John Brown and His Men*, ed. Richard J. Hinton. New York: Arno, 1968, 637–643.

Bryant, Paul T. "Diversity of What, for What?" *CEA Forum*, 24.2 (1994): 7–11.

Cain, William E. "Opening the American Mind: Reflections on the 'Canon' Controversy," in *Canon vs. Culture: Reflections on the Current Debate*, ed. Jan Gorak. New York: Garland, 2001, 3–14.

Chicago Cultural Studies Group. "Critical Multiculturalism." *Critical Inquiry*, 18 (1992). 530–555.

Dallal, Jenine Abboushi. "American Imperialism UnManifest: Emerson's 'Inquest' and Cultural Regeneration." *American Literature*, 73 (2001): 47–83.

del Río, Eduardo R., ed. *The Prentice-Hall Anthology of Latino Literature*. New York: Prentice, 2001.

Ferrer, Ada. "The Silence of Patriots: Race and Nationalism in Martí's Cuba." Belnap and Fernández, 228–249.

Fetterley, Judith. "Introduction: On the Politics of Literature." *The Resisting Reader: A Feminist Approach to American Fiction*. Bloomington: Indiana University Press, 1978. xi–xxvi.

Fliegelman, Jay. "Anthologizing the Situation of American Literature." *American Literature*, 65 (1993): 334–338.

"Forum: What Do We Need To Teach?" *American Literature*, 65 (1993): 325–361.

Foster, Frances Smith. "But, Is It Good Enough To Teach?" Brannon and Greene, 193–202.

Gates, Henry Louis, Jr. *Loose Canons: Notes on the Culture Wars*. New York: Oxford University Press, 1992.

Gillman, Susan. "*Ramona* in 'Our America.'" Belnap and Fernández, 91–111.

Gitlin, Todd. *The Twilight of Common Dreams*. New York: Metropolitan, 1995.

Goldberg, David Theo. "Introduction: Multicultural Conditions," in *Multiculturalism: A Critical Reader*, ed. David Theo Goldberg. Oxford: Blackwell, 1994, 1–41.

González, Lisa Sánchez. *Boricua Literature: A Literary History of the Puerto Rican Diaspora*. New York: New York University Press, 2001.

Graff, Gerald. *Beyond the Culture Wars: How Teaching the Conflicts Can Revitalize American Education*. New York: W. W. Norton, 1992.

Guillory, John. *Cultural Capital: The Problem of Literary Canon Formation*. Chicago: University of Chicago Press, 1993.

Hames-García, Michael. "How to Tell a Mestizo from an Enchirito®: Colonialism and National Culture in Gloria Anzaldúa's *Borderlands/La Frontera*." *diacritics* 30.4 (2000): 102–122.

———. "Who Are Our Own People? Challenges for a Theory of Social Identity," in *Reclaiming Identity: Realist Theory and the Predicament of Postmodernism*, ed. Paula M. L. Moya and Michael Hames-García. Berkeley: University of California Press, 2000, 102–129.

Harlow, Caroline Wolf. *Profile of Jail Inmates 1996*. U.S. Department of Justice: Bureau of Justice Statistics. Rev. 1998. <http://www.ojp.usdoj.gov/bjs/pub/pdf/pji96.pdf>.

Harper, Michael S. "American History." Baym et al., 3008.

Harris, Trudier, ed. *Selected Works of Ida B. Wells-Barnett*. Oxford: Oxford University Press, 1991.

Helg, Aline. *Our Rightful Share: The Afro-Cuban Struggle for Equality, 1886–1912*. Chapel Hill: University of North Carolina Press, 1995.

Jay, Gregory S. *American Literature and the Culture Wars.* Ithaca, NY: Cornell University Press, 1997.

Kanellos, Nicolás et al., eds. *Herencia: The Anthology of Hispanic Literature of the United States.* Oxford: Oxford University Press, 2002.

Kaplan, Amy. "'Left Alone with America': The Absence of Empire in the Study of American Culture." Kaplan and Pease, 3–21.

———. *The Social Construction of American Realism.* Chicago: University of Chicago Press, 1988.

Kaplan, Amy, and Donald E. Pease, eds. *Cultures of United States Imperialism.* Durham, NC: Duke University Press, 1993.

Kilcup, Karen L. "Anthologizing Matters: The Poetry and Prose of Recovery Work." *symplokē* 8 (2000): 36–56.

Laughlin, James S. "Beyond *Beyond the Culture Wars*: Students Teaching Themselves the Conflicts." Brannon and Greene, 231–248.

Lauter, Paul. *Canons and Contexts.* New York: Oxford University Press, 1991.

Lauter, Paul, Jackson Bryer, Anne Goodwyn Jones, King-Kok Cheung, Wendy Martin, Charles Molesworth et al., eds. *The Heath Anthology of American Literature* (4th ed.). Boston: Houghton, 2002.

Loewen, James W. *Lies My Teacher Told Me: Everything Your American History Textbook Got Wrong.* New York: Simon and Schuster, 1995.

Martí, José. "Class War in Chicago: A Terrible Drama." Allen, 195–219.

———. "La verdad sobre los Estados Unidos." *Obras completas.* Vol. 2 Caracas, 1964. 987–990.

———. *Letters from New York.* Allen, 89–254.

———. "The Lynching of the Italians." Allen, 296–303.

———. "Nuestra América." *Obras completas.* Vol. 6. La Habana: Editorial Nacional de Cuba, 1963.

———. "Our America." *Selected Writings.* Allen, 288–296.

———. "A Town Sets a Black Man on Fire." Allen, 310–313.

———. "The Truth about the United States." Allen 329–333.

McLaren, Peter L. "White Terror and Oppositional Agency: Towards a Critical Multiculturalism," in *Multicultural Education, Critical Pedagogy, and the Politics of Difference,* ed. Christine E. Sleeter and Peter L. McLaren. Albany: State University of New York Press, 1995, 33–70.

McMichael, George, Frederick Crews, J.C. Levenson, Leo Marx, and David E. Smith, eds. *Anthology of American Literature* (7th ed.). Upper Saddle River, NJ: Prentice, 2000.

McQuade, Donald, Robert Atwan, Martha Banta, Justin Kaplan, David Minter, and Robert Stepto, eds. *Harper American Literature* (3rd ed.). New York: Longman, 1999.

Mohanty, Satya P. "Can Our Values Be Objective? On Ethics, Aesthetics, and Progressive Politics." *New Literary History,* 32 (2001): 803–33.

———. "Colonial Legacies, Multicultural Futures: Relativism, Objectivity, and the Challenge of Otherness." *PMLA,* 110 (1995): 108–118.

———. *Literary Theory and the Claims of History: Postmodernism, Objectivity, Multicultural Politics.* Ithaca, NY: Cornell University Press, 1997.

Moya, Paula M. L. *Learning from Experience: Minority Identities, Multicultural Struggles.* Berkeley: University of California Press, 2002.

Mulford, Carla. "Seated amid the Rainbow: On Teaching American Writings to 1800." *American Literature,* 65 (1993): 342–348.

Newberger, Eric C., and Andrea E. Curry. *Educational Attainment in the United States (Update).* Washington, DC: Dept. of Commerce, Economics and Statistics Administration, U.S. Census Bureau, 2000, <http://www.census.gov/population/socdemo/education/p20-536/tab10.pdf>.

Nussbaum, Martha. *Cultivating Humanity: A Classical Defense of Reform in Liberal Education.* Cambridge, MA: Harvard University Press, 1997.

Palumbo-Liu, David. "Introduction," in *The Ethnic Canon: Histories, Institutions, and Interventions*, ed. David Palumbo-Liu. Minneapolis: University of Minnesota Press, 1995, 1–27.

Pease, Donald. "José Martí, Alexis de Tocqueville, and the Politics of Displacement." Belnap and Fernández, 27–57.

Peterson, Beverly. "Inviting Students to Challenge the American Literature Syllabus." *Teaching English in the Two-Year College* 28 (2001): 379–382.

Pita, Beatrice. "Engendering Critique: Race, Class, and Gender in Ruiz de Burton and Martí." Belnap and Fernández, 129–144.

Pollack, Harriet. "Eudora Welty and Politics: Did the Writer Crusade?" *In Eudora Welty and Politics: Did the Writer Crusade?* Ed. Harriet Pollack and Suzanne Marrs. Baton Rouge: Louisiana State University Press, 2001, 1–18.

Pressman, Richard S. "Is There a Future for the *Heath Anthology* in the Neo-Liberal State?" *symplokē*, 8 (2000): 57–67.

Pryse, Marjorie. "Teaching American Literature as Cultural Encounter: Models for Organizing the Introductory Course." Brannon and Greene, 175–192.

Railton, Stephen. "Black Slaves and White Readers." *Approaches to Teaching Stowe's* Uncle Tom's Cabin. Ed. Elizabeth Ammons and Susan Belasco. New York: MLA of America, 2000. 104–110.

Rotker, Susana. "The (Political) Exile Gaze in Martí's Writing on the United States." Belnap and Fernández 58–76.

Royot, Daniel Jean Béranger, Yves Carlet, and Kermit Vanderbilt, eds. *Anthologie de la Littérature Américaine* (2nd ed.). Paris: Presses Universitaires de France, 1995.

Rudnick, Lois. ". . . And (Most of) the Dead White Men Are Awful: Reflections on Canon Reformation." *Radical Teacher* 59 (2000): 36–40.

Saldívar, José David. *The Dialectics of Our America: Genealogy, Cultural Critique, and Literary History*. Durham, NC: Duke University Press, 1991.

Sánchez, Rosaura. "Dismantling the Colossus: Martí and Ruiz de Burton on the Formulation of Anglo América." Belnap and Fernández, 115–128.

Schlesinger, Arthur M., Jr. *The Disuniting of America* (Rev. ed.). New York: W. W. Norton, 1998.

Sedgwick, Eve. *Epistemology of the Closet*. Berkeley: University of California Press, 1990.

Shell, Marc, and Werner Sollors, eds. *Multilingual America: Transnationalism, Ethnicity, and the Languages of American Literature*. New York: New York University Press, 1998.

———. *Multilingual Anthology of American Literature: A Reader of Original Texts with English Translations*. New York: New York University Press, 2000.

Shohat, Ella, and Robert Stam. *Unthinking Eurocentrism: Multiculturalism and the Media*. London: Routledge, 1994.

Sleeter, Christine E. "Introduction: Multicultural Education and Empowerment," in *Empowerment through Multicultural Education*, ed. Christine E. Sleeter. Albany: State University of New York Press, 1991, 1–23.

———. *Multicultural Education as Social Activism*. Albany: State University of New York Press, 1996.

Spanos, William V. *The Errant Art of Moby-Dick: The Canon, the Cold War, and the Struggle for American Studies*. Durham, NC: Duke University Press, 1995.

Spears, Beverly. "Multicultural versus International?" *CEA Critic*, 59 (1996): 1–7.

Teuton, Sean. "Internationalism and the American Indian Scholar: Native Studies and the Challenge of Pan-Indigenism," in *Redefining Identity Politics*, ed. Linda Martín Alcoff, Michael Hames-García, Satya P. Mohanty, and Paula M. L. Moya. New York: Palgrave Macmillan, 2006.

U.S. Bureau of the Census. *Overview of Race and Hispanic Origin: 2000*. Washington, DC: U.S. Dept. of Commerce. 2000. <http://www.census.gov/prod/2001pubs/c2kbr01-1.pdf>.

U.S. Department of Justice, Bureau of Justice Statistics. *Sourcebook of Criminal Justice Statistics 2000*. Washington, DC: USGPO, 2000.

Warren, Kenneth W. *Black and White Strangers: Race and American Literary Realism.* Chicago: University of Chicago Press, 1993.

——. "The Problem of Anthologies, or Making the Dead Wince." *American Literature*, 65 (1993): 338–342.

Webb, Richard D., ed. *The Life and Letters of Captain John Brown.* Westport, CT: Negro University Press, 1972.

Wegener, Frederick. "'Rabid Imperialist': Edith Wharton and the Obligations of Empire in Modern American Fiction." *American Literature*, 72 (2000): 783–812.

Wells-Barnett, Ida B. "Mob Rule in New Orleans: Robert Charles and His Fight to the Death." Harris, 253–322.

——. "A Red Record: Tabulated Statistics and Alleged Causes of Lynchings in the United States, 1892–1893–1894." Harris, 138–252.

——. "Southern Horrors." Harris, 14–45.

Williams, William Carlos. "The Fountain of Eternal Youth," *In the American Grain: Essays.* New York: New Directions, 1956, 39–44.

Wilson, Robin. "Stacking the Deck for Minority Candidates?" *Chronicle of Higher Education* (July 12, 2002): A10.

Woodward, Frederick. "W.E.B. Du Bois," in *Instructor's Guide for the Heath Anthology of American Literature* (4th ed.), ed. John Alberti. Boston: Houghton, 2002, 482.

5

THE MIS-EDUCATION OF MIXED RACE

Michele Elam

> The emergence of university courses and student organizations by, for, and about biracial and multiracial people strongly indicates that this identity deserves social and academic legitimacy and institutional sustenance.
> Teresa Kay Williams et al., "Being Different Together in the University Classroom: Multiracial Identity as Transgressive Education"[1]

> ...what I am always cautious about is persons of mixed race focusing so narrowly on their own unique experiences that they are detached from larger struggles, and I think it's important to try to avoid that sense of exclusivity, and feeling that you're special in some way.
> Barack Obama, from the documentary *Chasing Daybreak: A Film about Mixed Race in America*[2]

The national education industry has emerged as one of the most powerful vehicles through which mixed race is currently manufactured and marketed. Anthologies, collections, pedagogical manuals, and educational materials in print, media, and web form have popularized, propagandized, and institutionalized particular ideas and ideals of mixed race. The "Mis-education of Mixed Race" explores the ways in which K-12 through college curricula have begun canonizing the emergent field of "mixed race studies." This canonization often occurs with the explicit rationale of inclusiveness and equity of representation, and sometimes lays claim to a revolution that shares much with radical challenges to education that the black and brown power movements initiated in 1960s and 1970s. These earlier reforms came with critiques of the way education normalized traditional racial and social hierarchies. Mixed race educational reforms often evoke this civil rights tradition, and sometimes the curricula do challenge certain social assumptions, particularly promoting the acceptability of interracial unions and transracial adoption.

But like the earlier bids for curricular change, this more recent move to integrate the study of mixed race into schools shares as much with traditional educational systems that normalize certain acceptable forms of cultural literacy over others, and in that sense, mixed-race education functions as both a vehicle of change and not-change, of challenge and yet of accommodation. In this case, the cultural literacy being taught involves and presumes the acceptance of mixed race as a unique and distinct type of racial experience deserving of its own recognition and thus requiring its own special cultural instructions and social prescriptives. Just as educational systems usually pose as neutral "information delivery" mechanisms, so do these; the fact that analysis of their claims is often unwelcome—even, at times, represented as politically backward—makes study of them all the more pressing. As mixed-race education is increasingly mainstreamed, and continues to gain influence in shaping perception and social practice, it essential to examine its particular educational mandates and requisites, to explore advocates' assumptions about race and identity that inform their policy recommendations. I argue that the cultural instructions in the materials I examine here are distinctive to the degree they are more than just "youth-oriented," at least to the degree that expression is used to refer merely as an innocuous bookstore classification or age-specific demographic. Rather, these textbooks and curricula posit mixed-race education as for and about the "next" generation. This is education as script and illustration for a projected future, one which both models and prescribes ideal social relations.

Their project does not differ from most children's books or educational programming, of course, most of which also tends towards this didactic impulse. But mixed-race education is often unself-reflective of the ways it sometimes replicates traditional prejudices as part of the very process it uses to overturn others. For in the progressive effort to normalize the "atypical" family of multiply-raced individuals, targeted consumers are almost invariably cast as an imagined community of light-skinned children of suburban middleclass, heterosexual parents, as well as educators enlightened enough to recognize their peculiar needs. As I will argue, almost all of these recent educational mandates are tacitly couched in these conceptual frames, frames which both enable and yet complicate the goal of what has been called "oppositional" and "transgressive" pedagogy. Articles such as "Challenging Race and Racism: A Framework for Educators" by Ronald David Glass and Kendra R. Wallace, "Being Different Together in the University Classroom: Multiracial Identity as Transgressive Education" by Teresa Kay Williams, Cynthis L. Nakashima, George Kitahara Kich, and G. Reginald Daniel, and "Multicultural Education" by Francis Wardle are all excellent contributions to the study of mixed race. But they also tend to take as a given that the educational project of institutionalizing the concept and practice of mixed race is, in and of itself, "progressive"—that is, to assume it ipso facto challenges the status quo.[3] My argument here involves a provisional critique of some of the notions of mixed race that are currently emerging in curricula. This critique is meant as a way of clearing space for alternative pedagogies that potentially encourage more politically complex and self-reflective understandings of mixed-race identification, hopefully allowing students' experiential claims to provide the basis for epistemic social insight.

THE CURRICULAR INSTITUTIONALIZATION OF MIXED RACE

In part the undergraduate classroom has become a locus of advocacy because demographic study of those who identify as mixed race suggests that the vast majority are people under 25.⁴ It is no accident, therefore, that there is a nationwide mushrooming of undergraduate courses on mixed race in the humanities and social sciences at institutions such as Yale, New York University, Vassar, Smith, Stanford, University of California at Berkeley, and many others. Many have hosted student-sponsored national conferences, educational workshops, and leadership summits on the subject. The national organization, EurasianNation, offers a link on its website to "The Top 19 Mixed Race Studies Courses" in the United States and Canada. On that same site is an article on "The Explosion in Mixed Race Studies" by Erica Schlaikjer (April 2003) that refers to "the new generation of academics…pushing the boundaries of ethnic studies."⁵ Many of the mixed-race organizations, websites, affinity and advocacy groups, magazines and journals that have emerged in the last few years have begun aggressive campaigns for educational reform, particularly regarding reading lists for and representation of the "mixed-race experience" at the college level.⁶

The explicit goal of much of this work is to educate a constituency; but in fact mixed-race constituencies are as often generated by these efforts. The classroom is a hub for some of the most active work in creating populations who identify as mixed race. In fact, the new crop of undergraduate courses on mixed race around the country are, to a great extent, preceding and anticipating the emerging body of critical literature on mixed race. These courses, often requested by the students themselves, have become the developing ground for nascent political identities and social organizing, both the result of and the inspiration for student clubs, youth leadership summits, and national student conferences devoted to the issue of mixed race. That is, while there certainly has been scholarly work, much of what is happening in the classroom is less obviously also driving our critical understandings of mixed race.⁷ That educational environments are a crucible for defining and refining what it means to be "mixed race" should remind us that pedagogy is not the mere effluvia of research, that these classroom events and practices are more than the realization and application of theoretical models and principles, and certainly more than an inevitable effect of changing demographics.⁸

COVERING: THE EYE'S INSTRUCTION

Covers reveal and conceal; they provide the prefatory function of visually glossing the pages within, of implicitly sanctioning certain ways of reading over others. Their edifying orientation for readers and viewers becomes especially important when the genre or topic is novel. But literary covers can unintentionally narrow as well as open perspective. How to represent mixed race when some expressions of ethnic identity are socially encouraged and approved of, and others are not? When the social performances of ethnic identity occur in the context of a society which rigorously delimits and monitors expressions of ethnicity. Kenji Yoshino argues in *Covering: The Hidden Assault on Our Civil Rights* that "covering," the coerced pressure to hide crucial aspects of one's identity, provides an adaptive strategy for the ethnic minorities that deploy it, but that the conditions for it are necessarily repressive.⁹

And in fact, covers have quietly played a critical role in the creation of certain restrictive, normative models of mixed race: *The Sum of Our Parts*,

Mixed-Race Literature, and *Mixed: An Anthology of Short Fiction on the Multiracial Experience*[10] people their covers with stylized facial abstractions; *The Multiracial Child Handbook, What Are You: Voices of Mixed-Race Young People, Racially Mixed People, The Multiracial Experience: Racial Borders as the New Frontier, New Faces in a Changing America: Multiracial Identity in the 21st Century*[11] feature pictures of "real" people, suggesting, similarly that the mission of the volume is to represent a snapshot of a population previously invisible. In both cases, stylized or realistic, the images function as typological, as marking a diverse but distinct people. *The Multiracial Child Resource Book, New Faces in a Changing America, Racially Mixed People*, and *The Multiracial Experience*, in particular, are filled to the margins with middle-class studio portraits of interracial couples and school-pictures of their light-skinned, well-groomed children not only seek to domesticate cross-racial sex, to visually purge its historical stigma and taboo; they also trigger the realist commitment to photo-documentary and the putative unimpeachability of the seen. We are encouraged to take for granted the idea that the photo-shopped families on the page are but a synecdoche: they imply there must be millions of others similarly miscegenated in middle America.

The covers equate visibility with not only with social recognition but also with political representation; indeed, civil rights campaigns associated with mixed-race advocacy often have as their goal that mixed-race people must be seen on the census, on stage, in media, in office.[12] But this act of rendering mixed race intelligible to the eye can sometimes interrupt political engagement in the name of it. The anthology covers, for instance, in the admirable service of making visible one marginalized population, effectively—and not accidentally—marginalize another: the images work together to codify anew the already-iconic status of the heteronormative unit at the expense of other family formations. The photos are not merely reportage of neutral demographic phenomenon, but the graphic naturalization of a particular political representation of a people. The fact that only heterosexual couples appear (and appear over and over again) on these "family album" covers extends the presumption of heterosexuality to the other images of solitary mixed race children—if they are the biological or adopted offspring of same-sex or intersex couples, we never see it. By implication, they are not deemed "representative" of the mixed race constituency, and thus silently omitted from the field of representation. To borrow Toni Morrison's insight in "Unspeakable Things Unspoken": "invisible things are not necessarily 'not there,'" and that "certain absences are so stressed, so planned, they call attention to themselves...like neighborhoods that are defined by the population held away from them.".[13] If we take Morrison's cue, then the covers, in this way, can teach us what to see what is left unseen. But there are historical and social pressures not to see the unseen.

These pressures are perhaps made most evident if we remember that the strategy of agitating for racial rights by conforming to a normative sexual model has a disturbing precedent. As Roderick Ferguson argues "canonical sociology—Gunnar Myrdal, Ernest Burgess, Robert Park, Daniel Patrick Moynihan, and William Julius Wilson—has measured African Americans' unsuitability for a liberal capitalist order in terms of their adherence to the norms of a heterosexual and patriarchal nuclear family model. In short, to the extent that African Americans' culture and behavior deviated from those norms, they would not achieve economic and racial equality."[14] Through the compromises of such imagery, the representations of the mixed-race nation risk making palatable race-mixing under cover of conservative "family values." Once beyond the legal and social pale, interracial marriage is rehabilitated as a model for the American Way.

A version of this impulse appears as early as the 1993 *Time Magazine* special issue, "The New Face of America: How Immigrants are Shaping the World's First Multicultural Society," which explicitly stated that the mulatta represents America as the melting pot of the world. The glossy-lipped twenty-something sepia ingénue on the cover is the cybergenetic fantasy of the male editors who computer-mated men and women from supposedly all races and regions, and hailed as the Daisy Miller for the new century. Created as the "beguiling...symbol of the future, multiethnic face of America," her feminine high-yellow appeal seems at first at odds with fears about immigrant "rising tide of color" that is the subject of the series of articles inside that edition.[15] But, in fact, the morph-mulatta is the sexual accommodation of white fears of immigration, suggesting that mixed race women, like all women of color, are rendered sexually available to white men, and through this historically hypergamic tradition of miscegenation achieve what more easily colonization, for the race may be "blanched," as William Byrd suggested in the 1700s and thus incorporated into the body politic without threatening the status quo.[16] Stripped of all ethnic markers (hair, clothes, cultural context) and, we are to assume, any politics, this new mulatta is no threat to nation; she functions in perfect concert with the anthologies' similarly appeasing heteronormative photomontages of mixed race. The risk that race mixing will lead to other taboo transgressions and civil rights petitions—same-sex or immigrant rights protests, for instance—lurking in the pages following the cover both occasions the New Face of America and explains its function as antidote.

MARKETING FOR THE NEXT GENERATION: EDUCATIONAL PRODUCTS FOR THE SPECIAL CHILD

This re-packaging of mixed race as nationalist expression is echoed and enhanced in its commercialization. Kimberly McClain DaCosta's excellent analysis of the market targeting to and profiling of multiracials in the 1990s notes that the advertisements featuring images of mixed race people and interracial couples "requires no knowledge about multiracials and their putatively unique habits and needs. Rather, their impact, and the advertisers' motivation for using such images, lies in their symbolism—the ability to evoke for a viewer positive qualities, feelings, or desires. Of course, multiracialism's capacity to evoke such desirable qualities, and even what is considered 'desirable,' are historically and context specific."[17] These desirables include the branding of multiracials as a distinct population associated with a hip, young, new people, "drawing on existing culturally resonant narratives of the meaning of racial mixedness for the purposes of selling stuff. In doing so, he shape social perceptions that multiracials exist as such. Through the marketplace, multiracials are being constituted as subjects" (DaCosta, 2007, 172). Those narratives of mixed race are culturally resonant, I argue further, because they dovetail with American tenets of individualism, iconoclasm, and forward-looking modernity. It should be no surprise, then, that this creation of a mixed race subjecthood is increasingly targeted at and for both the very young and the parents of youth, those stewards of America's next generation.

Consuming mixed race is represented as expressing oneself, a form of national obligation to self-realization—no better fulfilled than through the active support of one's children's efforts to "find wholeness."[18] The related and growing niche market associated with educational mixed race products, products pitched as necessities for a "healthy" multiracial identity: parent resource handbooks for "mixed race" childrearing, specialized toys and books for biracial children, signature clothing for mixed-race-identified youth, "unique" hair care products for "blended" tresses, and new lines of healthy "multiracial" skin lotions and creams. These products advertise good self-esteem through proper care of one's specialized epidermis, package a step-by-step program to happier family life in "mixed" hair instruction manuals, and promise a better world through a multiracial literacy achieved through subscription to a children's library of "mixed race"-friendly picture books. The Web-based company, Like Minded People: Clothing for the Conscious, Inspired by Life, which offers items that create a sense that the socially enlightened can and should wear garb that reflects their evolved perception, t-shirts with scripts across the front like "[Not] Other," "Everyone Loves a Mixed Girl," and "What Are You?" These messages signal in-group code references that take a stand on or " talk back" to putatively mixed race issues and interests—in this case, the statements take up positions, announcing the refusal of a Census category of "Other," which was an option considered briefly by the Office of Management and Budget during the 2000 Census deliberations; offering social affirmation to a girl who might feel denied it; and turning on its head the most commonly asked question of people perceived to be racially ambiguous, asking the would-be offender instead, "what are you?" These t-shirts are represented as forms of social bonding through social intervention—those who wear them form the "like-minded," a community of belief more than blood.

Of course in the effort to represent the similarly-minded, some sheep might be lost. What if one identifies as mixed race but would have found perfectly acceptable the "other" Census option? What if the claim that everyone loves a mixed race girl, in its effort to counteract the implied view that someone somewhere does not love her, slides into the historic tendency towards colorism and the privileging of light-skinned females as especial beauties? What if the challenge "what are you?" does *not* lead to the kind of "conscious" self-reflection on the part of the viewer who might be tempted to ask that question that the company suggests it hopes for, but instead leads only to a straightforward non-ironic interpretation of the question: that is "what am I? oh well..." and then goes on to list his or her own genealogy. In other words, what if someone reading this t-shirt is not prompted to think "hmmm, why do I not ask that question of myself? Why would I ask what are you versus who are you? Why would I ask that question of this racially ambiguous person and not someone else? What are my own presuppositions about who is what or who belongs where? What is it about my own uneasiness about being unable to racially classify this person?" What if, instead, he or she misses the intellectual challenge and the question only reinforces the practice of asking others "what are you?" My point is not that the messages ought to be honed to defend against such misinterpretation, but rather to note the ways in which these are all posited as sites of education for the "like minded" that tend to advocate positions more than to raise questions, to exclude those unlikedminded, and, even more dangerously it seems to me, to create a consumerist climate in which, as DaCosta cogently puts it, "'recognition' (be it in the form of representation in advertisements or in the census) is substituted for a politics of civic and economic equality" (2007, 169).

Literary History in the Making: Canonizing Mixed Race

One of the most significant moves in mixed race education has been the revisioning of literary history, and, in turn, the surveys and period courses usually based on it. This revisioning is in many ways billed as a great awakening to a lost past, a truth to put to the lie to monoracial identity. But the new ideal of mixed race is also about forgetting as much as it is any recovery, for the recent project of finding a "mixed race literary tradition" involves what Mireille Rosello calls "amnesiac creolity." Identity, Rosello suggests, is always

> the product of forgetting or repressing the inherent hybridity or creolity that constitutes it. The politics of hybridity then calls for the remembering of the truth that was always already there, for the recovering from the amnesiac sleep of identity. It is almost as if the error or illusion of our misplaced faith in identity can now be replaced with a new creed of hybridity: I once was lost but now am found, was blind but now I see. But is not this teleological narrative of redemption similar to modernity's promise of history as the gradual emergence of humanity into the dazzling light of the truth? The narrative of the beyond—beyond identity, beyond race, beyond racism—is in many ways a revision of the Enlightenment narrative of the universal subject which gradually shed all particularity and contingency to emerge into the light of its true being, with the signal difference that this has now been recast as essentially hybrid rather than essentially singular.[19]

Rosello's insight holds, similarly, for the proposals for some of the new literary histories canonizing mixed-race literature. Their advocates celebrate the "truth" of lost-but-now-found hybrid literatures putatively eclipsed by monoracialist biases, but end up, just as Rosello suggests, merely making the hybrid subject a new kind of Enlightenment universal ideal. This is particularly ironic because mixed-race advocates often claim to challenge the political processes of literary canonization, yet remain wed to the same processes by which certain writers get placed in certain categories—they may change the racial calibrations but the system itself, in which writers of one color are categorized by race not genre or style, arguably changes little.

The United States canon has always been as dependent upon racial categorization as it has aesthetic values, arguably beginning with W. D. Howells, the "Father" of American Realism, who in the late nineteenth century, couples race and genre when he praised Paul Laurence Dunbar as most true to his race when writing poems in dialect.[20] In their challenge, however, multiracial advocates have begun a problematic reinterpretation of African American literary history by redefining authors previously identified (or self-identified) and anthologized as "black" according to the racial discourses of the day, ascribing to them a new multiracial identity. Through this form of presentism, imposing the values and standards of the present moment to the past, these writers and their texts are being "saved," redeemed and relieved of their blackness, celebrated and canonized through a process in which bi- and multiraciality becomes an index of heroic self-definition.

Central to these reinterpretations is what used to be called the "mulatto/a," now reconceived not as a "neither/nor" or even a "yet both" figure but as a new-age pioneer who seeks to "transcend" racial categorization through the synergism beyond "the sum of our parts," to name the title of one recent anthology. Charles Chesnutt, Jean Toomer, W. E. B. Du Bois, Nella Larsen, and many others are all re-presented in this latest canon as misunderstood trailblazers. Jonathon Brennan in the Introduction to his edited *'Mixed Race' Literature*, approvingly cites Jean Toomer's comment that people like him were "in the process of forming a new race" (3) and praises him for breaking free of the "corral" (4) of monoracialism imposed on him by his black contemporaries and current black critics—namely Henry Louis Gates, Jr., in this case. Werner Sollors argues in *Neither White nor Black yet Both* that " 'race mixing' has its [own] tradition, an interracial tradition that needs to be explored."[21] In particular, he takes to task scholar Richard Bone for his interpretation in 1975 of Charles Chesnutt's story 1898 story, "The Wife of His Youth," in which Bone treats the light-skinned protagonist, who belongs to the exclusive Blue Vein Society, as black rather than mulatto (13). Sollors lays especial responsibility for what he sees as a critical miscalculation in scholarship that emerges during black arts era, stating that it was upon the reissue of the book in the "1960s...[where it] received...readings as a 'black text' " (12). Indeed, some critics suggest that the putative misrepresentation of those of mixed race is the legacy of tyrannizing forces of literary history circa 1960s and 1970s, a literary history that was crudely forged, according to literary critics like George Hutchinson and Paul Spickard, in the fires of anti-white sentiment and racial resentment. Both argue that narratives of the mixed race experience do not fit comfortably within any known literary traditions, though many of these narratives clearly extend rhetorical and racial acts of self-definition that appear across texts as varied as antebellum slave narratives or late-twentieth-century black bourgeois memoir. Although mixed-race literature shares in the ethnic bildungsromane, my point here is simply that an insistence on mixed-race distinctiveness can eclipse the ways many of these texts directly participate in genres and literary practices historically associated with ethnic literary traditions. This is not to make them less "special," but to suggest that their specialness does not require separation.

It is important to note that Sollors' anthologizing a literary history of interracialism, which understand texts as cohabiting within Anglo and African American traditions, is quite different from the new and separate literary history proposed by Hutchinson and Brennan. If his is descriptive, theirs is acquisitive, seeking to claim as its own not only past narratives that might be redefined as mixed, but also the recent plethora of mixed-race memoirs, quasi-fictionalized autobiographies, essays and fiction. The new literature I refer to includes, to name only a very few published since 1994:

Shirlee Taylor Haizlip, *The Sweeter the Juice: A Family Memoir in Black and White* (1994); Lisa Jones, *Bulletproof Diva: Tales of Race, Sex, and*

Hair (1994); Barack Obama, *Dreams from My Father: A Story of Race and Inheritance* (1995); Judy Scales-Trent, *Notes on a Black White Woman: Race, Color, Community* (1995); Gregory Howard Williams, *Life on the Color Line: The True Story of a White Boy Who Discovered He Was Black* (1995); Marcia Hunt, *Repossessing Ernestine: A Granddaughter Uncovers the Secret History of Her American Family* (1996); James McBride, *The Color of Water: A Black Man's Tribute to His White Mother* (1996); Scott Minerbrook, *Divided to the Vein: A Journey into Race and Family* (1996); Toi Derricotte, *Black Notebooks: An Interior Journey* (1997); Edward Ball, *Slaves in the Family* (1999) and *The Sweet Hell Within: A Family History* (2001); Danzy Senna, *Caucasia* (1999) and *Symptomatic* (2004); Lalita Tademy, *Cane River* (2001) and *Red River* (2007); Rebecca Walker, *Black, White, Jewish: Autobiography of a Shifting Self* (2001); Neil Henry, *Pearl's Secret: A Black Man's Search for His White Family* (2002); Ronne Hartfield, *Another Way Home: The Tangled Roots of Race in One Chicago Family* (2004); Emily Raboteau, *The Professor's Daughter* (2005); Essie Mae Washington-Williams, *Dear Senator: A Memoir by the Daughter of Strom Thurmond* (2005); June Cross, *Secret Daughter: A Mixed-Race Daughter and the Mother Who Gave Her Away* (2006); Angela Nissel, *Mixed: My Life in Black and White* (2006); Bliss Broyard, *One Drop: My Father's Hidden Life—A Story of Race and Family Secrets* (2007); Dave Mathews, *Ace of Spades* (2007); Judith Stone, *When She Was White: The True Story of a Family Divided by Race* (2007). We can also include collections of autobiographical narratives that contain pieces by those who variously identify as "mixed": Carol Camper, ed., *Miscegenation Blues: Voices of Mixed Race Women* (1994); Lise Funderberg, ed., *Black, White, Other: Biracial Americans Talk About Race and Ethnicity* (1995); Claudine Chiawei O'Brian, ed., *Half and Half: Writers on Growing Up Biracial and Bicultural* (1998); Pearl Fuyo Gaskins, ed., *Who Are You? Voices of Mixed-Race Young People* (1999). Recent biographies also emphasize and explore mixed-race identity: Kathryn Talalay's *Composition in Black and White: The Life of Philippa Shuyler, The Tragic Saga of Harlem's Biracial Prodigy* (1997), and George Hutchinson's *In Search of Nella Larsen: A Biography of the Color Line* (2006).

Perhaps the coup de gras came with *Mixed: An Anthology of Short Fiction on the Multiracial Experience*, published by W. W. Norton, the ultimate stamp of institutional validation.[22] Major academic as well as popular presses have begun producing and marketing collections and literary surveys of mixed race writing; these volumes and anthologies quite literally institutionalize and commercialize an alternative literary history.

Many literary critics rehearse a common mantra of the mixed race advocates: we merely wish to value equally our parents, both white and black antecedents. When transposed to the literary realm, however, this bid for filial interraciality risks constructing a literary history that is at best ahistorical and at worst a distortion of black literary accomplishment in a playing field that was and is not equal. I am less interested in simply locating extant representations of racial hybridity or of finding some lost mixed-race literary tradition than in the larger project of analyzing the politics of this new literary history in-the-making. The heated academic debates provide a crucial glimpse into what is emerging as the profound political and institutional implications of this new mixed-race movement.

Part of my critique is of the circular tautology at work in most arguments for a distinct literature by mixed-race peoples. Brennan, for instance, both calls for—but then also presupposes—a "mixed race tradition" (17) in which to place literature. He insists we must consider the mixed race writer's work within every single ethnic literature tradition corresponding to the writer's "mix"—Langston Hughes, he says, is "African-French-Cherokee-European American" (29) and thus one should see his work as a "hybrid text" which is the sum of all those traditions. Susan Graham of Project RACE argues that, were he alive today, Langston Hughes, the oft-called poet laureate of the race, would have self-identified as "mixed"; Brennan simply bypasses Graham's hypothetical scenario and asserts as self-evident fact that Hughes *is* a "mixed writer."[23] Hence we find only twenty pages into the book the kind of conflation of blood ancestry and literary category—and reification of both—that he earlier says he wants so much to avoid.

Most importantly, Spickard's interpretation of the idea that "mixed race" is constituted by narrative does not address the way in which mixed race is socially or politically situated both with and without individual consent, or the way in which one cannot simply doff and don racial and cultural identities. He argues, for instance, mixed race peoples "draw their life-force from fashioning and refashioning the story of the ethnic self" (92), but he then turns what could have been a move towards literary analysis into a taxonomy of (multi)racial traits and tropes. Citing Reginald Daniel, Spickard claims that, "For multiracial people, you live your racial narrative by creating it. The created element is particularly strong in the case of multiracial people. There is an element of fictionalizing to it, but it's not falseness. It is choosing the proportions and the proper fit of the various ethnic elements one possesses."[24] The fact that fictionalizing is not falsity nicely addresses the common charge that people of mixed race are "inauthentic" or racial posers. That said, Spickard's sartorial metaphor for self-fashioning suggests identity can be selectively styled from pieces of clothing chosen for the correct "proportion" and "proper fit," implying mixed race identified people can and should creatively suit themselves up in whatever racial garb they choose.

If for the select few perceived as racially ambiguous by a dominant culture, race does involve a heightened ability to make situational choices about one's racial identity, then by choice we must not merely mean a willingness to buy into, literally, the commercialization of race affects; that is, "choosing" race, one hopes, ought not to translate into, for example, simply purchasing hiphop wear. Such an exercise of choice reduces "crossover dreams,"[25] as Noel Ignatiev puts it, into a point-of-purchase sales gimmick that markets race as apolitical and endlessly portable. As Harry J. Elam, Jr., suggests in "Change Clothes and Go: A Post-Script to Post-Blackness," this free-market view of race reinforces the idea that one can just slip in and out of identities without political commitment or ethical consequence.[26]

Thus I find both the thematic and multitraditional approach useful but limited in teaching "mixed race," because it risks resuscitating the very racial categories it says are on their last gasp, risks dehistoricizing the literature, risks inattention to the literary and cultural specificity of this peculiar form of literary production, and risks participating in the global commodification of race. So for instance, it is important to analyze this literature in terms of its genre of choice (autobiography and memoir) not as transparent reflection or sociological description of the "mixed race" experience, but as a function of ethnic narrative and practices of literary self-representation. Thus the critical race scholarship by David Palumbo-liu, Ramon Saldívar, Amritjit Singh, Robert Young, and others might better serve analyses of "mixed race" narrative than a reinvention of literary history. The latter approach would allow us to see how such literature can be situated in, does in fact participate in respective ethnic literary traditions, which far from bastions of racial tribalism, usually already account for hybridity. Rather than see mixed-race individuals as *sui generis* and "outside history," they must be considered as racialized subjects *in* history. Thus even if novelist Jean Toomer loved the writer Georgia Douglas Johnson's literary salons as a sanctuary "out of time" where his "new friends...were not concerned with being either coloured or white," such a claim says less about a unique identity movement that emerges out of history than one deeply embedded in its historical time and place.[27] Toomer identified over time across a racial spectrum, but his preference for Johnson's salon must be understood, Elizabeth McHenry argues, in terms of the concerns of the Washington, DC upper-class, light-skinned elite, the center of black aristocracy from the end of Reconstruction to World War I (261), an elite that distinguished itself sharply from black masses until the Red Summer of 1919 and the race riots throughout the country sharpening the race line. In other words, the desire to be outside history has its own history.

Cultural Instruction and Pedagogies of Mixed Race

Much recent fiction and drama tries to account for multiple racial affiliations within any one group and to work against the notion of homogenous and totalizing racial categories implicit in, for instance, both black cultural nationalism and what Harold Cruse calls white "racial particularism."[28] These texts beg the question: does the "mixed-race" category productively complicate all racial boundaries or does it risk instituting and reifying yet another kind of racial categorization—in effect, does the designation of "mixed race" dissemble or merely replicate reductive models of "race?" Prompted by the recent work on a "post-positivist realist" politics of identity,[29] I would like to ask: What *would* a politically progressive and theoretically sophisticated mixed-race politics and aesthetics look like? To that end, even as I am cautious about the popular and scholarly appeal of mixed race, I am curious about what multiple racial identification *does* allow for intellectually, experientially and artistically. In short, it is a given that "mixed race" literature will be taught as these debates rage on about its literary status and its social implications, so the answer is not to resist teaching it but to teach it critically, to allow the course content to participate, and take a defining lead, in these debates.

To that end, I want to argue that inclusion of so-called mixed-race literature alone is not a satisfactory pedagogical response. Appropriate here is Michael Hames-García's critique of canon-inclusion in which he notes that, ironically, incidental representation of marginalized literature in an anthology can function to highlight and heighten the status of the traditional literatures which dominate it in terms of page numbers and length of selection. Furthermore, inclusion which occurs in the absence of considerations of "oppression and resistance"[30] (versus the safer and more commonly preferred theme of innovative "discovery" and sanitized "contact") leads merely to "risk-free diversity" (Hames-García, "Which America Is Ours?" 30). This "additive" (30) approach to multicultural education also fails to be much use if it advances only an understanding of different cultures' putatively discrete experiences; rather it can encourage appreciation and analysis of "how cultures are interrelated and connected by historical practices and systems of domination."[31] Considering cultures and socially-located identity in these more charged terms in the literature will surely raze the ideal notion of polite and "uninvested democratic discussion" (31) in which the protocols of civility can unwittingly coerce silence. Teachers need to anticipate—and set the context for—more vigorous and sometimes agonistic but ultimately much more provoking and productive conversations.[32]

One of the important and yet rarely discussed issues related to the teaching of mixed race in the context of cultural interaction, including oppression, is whether or not mixed race individuals and communities can be considered oppressed, marginalized—often even whether they can be considered minorities. On the one hand, mixed race peoples can to some degree lay claim to being among the most *representative* of oppressed populations. They are representative not because those of "crossed blood" bear a uniquely "heavy cross"(vii) as Berry Brewton puts it, elevating their victimage to something akin to divine sacrifice. For Brewton, these are the über-outcast, "pathetic folks of mixed race ancestry...raceless people, neither fish not fowl, neither white, nor black, nor red, nor brown (vii). Yet to argue that they are victimized primarily by racial categorization itself because it cannot reflect their experience lays blame rather conveniently at the foot of theoretical abstraction—blames "race" rather than racism. As Kimberlé Crenshaw cogently argues,

> The embrace of identity politics...has been in tension with dominant conceptions of social justice. Race, gender, and other identity categories are most often treated in mainstream liberal discourse as vestiges of bias or domination—that is, as intrinsically negative frameworks in which social power works to exclude or marginalize those who are different. According to this understanding, our liberatory objective should be to empty such categories of any social significance. Yet implicit in certain strands of feminist and racial liberation movements, for example, is the view that the social power in delineating difference need not be the power or domination; it can instead be the source of social empowerment and reconstruction. The problem with identity politics is not that it fails to transcend difference, as some critics charge, but rather the opposite—that it frequently conflates or ignores intragroup difference."[33]

Crenshaw's theory of the need for identity politics to account for intragroup difference is especially useful in recasting the oppression of mixed-race people: rather than see mixed race people as oppressed by racial categorization itself (the liberal argument, according to Crenshaw), I argue that mixed-race people bear oppression because, historically, their bodies have borne physical testimony to sexual violation as an exercise of racial privilege, what Hortense Spillers has called "the will to sin"[34] by a dominant culture. They represent the reference point, the very nexus of cultural collisions, conflicts and conjoinings; those of mixed race heritage can be the issue of loving relations, surely at times, but also, overwhelming from a historical perspective, they are the result of hypergamic relations, in which one party, usually the woman, occupies a significantly lower social and racial status.

So in deciding whether or not mixed-race people qualify and identify as oppressed minorities, it is not a matter of assessing historical injury based on some impossible measurement of blood quantum, which only leads to unproductive debates about whether some fictive percentage of "black" or "Latino" heritage outweighs the white blood, with its historic dominance. As social scientists have long argued, racial advantage or disadvantage in terms of income, health care, social leverage, likeliness to be incarcerated and a host of other indicators, is not, and ought not to be, determined anew with each individual case but by the "class" of people to whom individuals most likely belong.[35] That is, race, *and this holds for mixed-race identification as well*, is socially and politically salient even if the particular experiences of individuals vary. Historically, so-called mulattos, mestizos, hapa individuals have been singled out as part of the threat of the "growing tide of color" in the United States even if any particular person identifying as mixed race does not consider herself oppressed (or does not, as might very well also be the case, recognize the ways in which he or she is oppressed). In sum, "mixed race" people can most certainly be considered among those minorities historically oppressed.

Yet, some argue, with much justification, that a temporal shift has occurred in the last decade which has lent great cultural prestige to the "mulatto."[36] Those who are light-skinned, who do not identify with any particular racial community, or who are aligned with civil rights agenda, are held up high as exemplary symbols of the melting pot, representatives of racial harmony, ambassadors of cultural and racial rapprochement. Many mixed-race individuals claim to be oppressed by—through their sense of exclusion from—monoracial communities and relate more with white people, indeed have the racial profile of white people to the extent they do not fully recognize themselves as racialized and thus are oblivious to color hierarchies from which they benefit socially. This, to me, nevertheless does not change my view of mixed-race individuals as having a valid association with marginalized people. Rather, I see many of them as participating in what Paula Moya calls "neoconservative minority identity politics" (2002, 132). They bear all the typological features: "ambivalent relations with the minority communities with which they are identified by others" (132) although allowing themselves to be exploited as "exemplary" representatives of that community, even serving occasionally as "native informants" (133) for those "outside" the group. Those who engage in neoconservative identity politics also tend to "overlook the structural and inegalitarian nature of society," ignoring the "structural inequities that contribute to the correlation between the likelihood of incarceration and nonwhite racial status, and between poverty and the female gender." Instead, inequalities are attributed to the "cultural character of the subordinated individual or group," and thus, consistent with this focus on culture in this sense, "neoconservative minorities have a liberal understanding of individual agency"(133). Still, Moya sees them as "fellow travelers" even as she sees their ways of negotiating racial discrimination and inequality as problematic, and I would argue that to the degree mixed race advocacy is all about identity—and racial identity in particular—they can be counted as fellow travelers.

Nonetheless, that does not mean that someone who identifies as mixed race has, necessarily, especial critical insight or epistemological acumen regarding either their experience or regarding race in general.[37] Thus it is important in teaching mixed race literature to discuss the conditions and circumstances behind most historical race mixing, which in Brazil, Cuba, Jamaica, South Africa, Suriname, and the American South, to name a few, leads us directly to a history involving crossracial exploitation.[38] Too often students' historical memory extends only to the 1967 Loving Decision, the Supreme Court decision which rendered interracial marriages legal: with that as the originary marker, students tend to associate race mixing only with marital free will, with the brave, free choice of partners, with what looks like the transgressive personal politics of interracial love. But to shrink the meaning of mixed race to the present historical moment creates the impression that racemixing has always been or is now always necessarily "progressive." Even in post–civil rights era liaisons, power relations are never absent, although the repeated invocation of the Loving Decision as page one of a heroic American narrative about state discrimination overcome often suggests just this. The retroactive scripting of the Loving Decision as the foundational moment in a narrative of love transcending race and nation can distract us from analysis of how interracial unions and the children born of them do, in fact, necessarily participate in the racial, economic and social economies of a nation.[39] Students might be better served if we presented the phenomenon of race mixing within a larger international context and if we weaned ourselves from triumphalist or romantic historical plotlines in order to see how mixed race may carry forward past inequities even though it has come to represent a post-Civil Rights escape from the past. It may also help students to appreciate that identity politics is not "an end in itself," as Moya puts it (131), that they can grant the viability of their experiences as a way to gain useful knowledge about the world and yet more self-critically and self-reflective assess those experiences.

To this end, I suggest adapting several "realist proposals" (158) for multicultural education that seem well-suited to teaching mixed race from both historically-informed and social-justice perspectives. In "Learning How to Learn from Others: Realist Proposals for Multicultural Education," Moya encourages first the study of culture not simply as related to food, dress, anthems (which too often translate into the bland multicultural days for K-12 programs in which families serve up representatively "ethnic" dishes to share), but of culture as "practices involving habits of interaction, communicative codes, norms of behavior, and artistic expressions" (2002, 158). Coming at culture through this critical lens would allow much more analysis of the kinds of synergistic dynamics cross-culturally for mixed race experiences in more complex ways than simply the usual framework of the mixed race, which is posed in terms of either/or race loyalties rather than generative interactions. One could then more usefully ask: how are these interactions put in relation in the experience of a person who identifies as "mixed"? What choices does one make, and are made for one, when some practices are engaged, performed, compete with or give way to each other at any given event or moment? To begin considering these and other questions about the relation of culture to identity is to begin also to appreciate, track, and even participate in "cultural change" (150), which is to set *culture in motion*, to view it diachronically instead of as frozen into some synchronic tableaux in which "culture" is a static portrait of traits.

Further, understanding culture and cultural groups as already always heterogeneous and evolving would preempt the red herring of race identity as subscription to a monolithic "race," or worse, as an antiquated investment in fixed categories, or as a belief in a musty, misguided cultural nationalism based on some quaint and vaguely Marxist notion of "unity." Many advocates complain that the mixed race experience is defined by this desire to reject homogenous and totalizing race categories, which grossly simplifies the diverse lived experiences of those within minority communities. This means that mixed race advocates, so concerned with being boxed in, accidentally box themselves into a notion of identity that erroneously and unnecessarily gives to undue weight to descent, origins, and heterodox genealogies, as if one could never grow up and into cultures, or that cultures themselves are malleable, permeable, responsive. At worst, mixed race positions like these tend to demonize the minority communities (as with the case in which mixed race advocates publicly chastised the NAACP for resisting the "Mark All That Apply" census box), and blames their dissatisfactions on race and "identity politics" generally, rather than interrogating the socio-economic factors at play in their frustration with the status quo. When mixed race advocates see other minority and monoracial groups as their primary problem, interethnic coalition-building and broad-based civil rights mobilization seems out of the question, which may be precisely why Newt Gingrich and others on the political Right were early in favor of the Census revision. It also means that occasionally mixed race advocates, rather than constructing a subject that claims "forms of cultural belonging" that encourage "human flourishing" (159), instead construct a defensive identity for themselves. This embattled sense of identity limits critical self-reflection and thus does little to discourage commercial and political co-optation of their "mixed" image, at the expense of their darker-skinned brethren, in often unwitting support of a color-struck status quo. In this way, not more carefully theorizing their experience can lead to complicity with the very status quo that mixed race advocates claim to challenge, and to perpetuate and extend unknowingly institutionalized systems of inequity—both in the social world and, as I have suggested, in the field of literary studies.

I suggested at the outset of this chapter that the classroom is pivotal in driving our understanding of "mixed race." The classroom is social microcosm, as Andrea Lunsford, Paolo Friere and many others have argued, and in many ways it never was a room in an insular "ivory tower" but, rather, already existed and exists as an active site for social engagement. As those participating in "mixed race" identity projects continue to bring those interests to academic institutions, both faculty and students will no doubt find "educational prescriptives" (2002, 171) through the sometimes necessarily contestatory cross-cultural communication that can take place in university classrooms. Hopefully those identifying as mixed race will continue to find legitimacy in their experience of both personal and social realities, and yet work to develop more theoretically-nuanced and politically astute understandings of the identities they claim.

Notes

Many thanks to Harry Elam, Jr., Susan Sánchez-Casal, and especially to Amie A. Macdonald for their helpful editorial suggestions, unflagging goodwill, and always generous collegial support. I am also grateful to the larger Future of Minority Studies (FMS) community. This chapter is drawn from a longer chapter in *Mixtries: Mixed Race in the New Millenium*, Michele Elam (Stanford University Press, forthcoming). Copyright © by the Board of Trustees of the Leland Stanford Jr. University.

1. Teresa Kay Williams, Cynthia L. Nakashima, George Kitahara Kich, G. Reginald Daniel, "Being Different Together in the University Classroom: Multiracial Identity as Transgressive Education," in *The Multiracial Experience: Racial Borders as the New Frontier*, ed., Maria P. P. Root, 359–379.
2. A group sponsored by MAVIN, the mixed-race advocacy organization, met with Senator Barack Obama, in Washington, DC at his offices (April 25, 2005). The interview is included in the documentary, *Chasing Daybreak: A Film about Mixed Race in America*, directed and edited by Justin Leroy, produced by Matt Kelley (MAVIN Foundation, 2006). More on the documentary can be found at www.chasingdaybreak.com. Last accessed January 14, 2009.
3. All collected in Maria P. P. Root, *The Multiracial Experience: Racial Borders as the New Frontier* (Thousand Oaks, CA, 1995), 341–394.
4. Stanford's Center for Comparative Studies in Race and Ethnicity's Executive Summary of "The Population of Two or More Races in California," *Race and Ethnicity in California: Demographics Report Series*—No. 4 (Nov 2001) by Alejandra Lopez makes note that the percentage of people identifying as "mixed-race" or marking multiple racial designations is much higher for those under 18. It is no accident that the movement is most popular among the young. Most "mixed-race" organizations are a mixture of people 25 or younger, interracial couples with young children, and parents involved in interracial adoptions. Part of my argument is that the particular notion of mixed race circulated now is a distinctly post-Civil Rights era phenomenon that was coded quite differently before the 1950s.
5. For one of the many lists of mixed-race courses, see http://www.eurasiannation.com/articlespol2003-04mixedstudies.htm, last accessed March 10, 2004. The article, "The Explosion in Mixed Race Studies," interestingly enough, had a banner picture of me as a supposedly representative mixed-race faculty member teaching a course on the subject, although the picture was actually taken several years prior for an unrelated promotional campaign for the University of Puget Sound and then sold to an advertising agency, gettyimages.com.
6. These include the Association of MultiEthnic Americans (www.ameasite.org), Famlee (www.scc.swarthmore.edu/-thompson/famless/home.html), Hapa Issues Forum (www.hapaissuesforum.org), My Shoes (myshoes.com), Interracial Individuals Discussion List (www.geocities.com/Wellesley/6426/ii.html), Project RACE—Reclassify All Children Equally, Inc (projectrace.home.mindspring.com, www.eurasiannation.com, www.swirllinc.org, www.anomalythefilm.com, *Interracial Voice* (www.webcom.com/-intvoice), *MAVIN: The Articulate Journal of the Mixed-Race Experience* (www.mavin.net) and *Metisse Magazine* (www.metisse.com) among many others. Many of the educational reforms just before and after the Census 2000 were advanced by white parents of grade-school children, minors whose identity is claimed on their behalf by their parents as bi- or multiracial; hence the focus on early childhood education.
7. Mary Louis Pratt, *Imperial Eyes: Travel Writing and Transculturation* (Routledge, 1992) and Mary Poovey, *Making a Social Body: British Cultural Formation, 1830–1864* (University of Chicago Press, 1995).
8. I elaborate this argument in "Towards Desegregating Syllabuses: Teaching American Literary Realism and Racial Uplift Fiction" in *Teaching Literature: A Companion*, eds. Tanya Agathocleous and Ann C. Dean (New York: Palgrave Macmillan, 2003), 59.
9. Kenji Yoshino, *Covering: The Hidden Assault on our Civil Rights* (New York: Random House, 2007).
10. Teresa Williams-León and Cynthia L. Nakashimi, eds., *The Sum of Our Parts: Mixed Heritage Asian Americans* (Philadelphia: Temple University Press, 2001); Jonathon Brennan, *Mixed Race Literature* (Stanford: Stanford University Press, 2002); *Mixed: An*

Anthology of Short Fiction on the Multiracial Experience, ed. Chandra Prasad (New York: W. W. Norton, 2006).
11. Root, Maria P. P. and Matt Kelley, eds. *Multiracial Child Resource Handbook: Living Complex Realities*, eds. (Seattle, WA: MAVIN Foundation, 2003); Pearl Fuyo Gaskins, ed. *What Are You? Voices of Mixed-Race Young People* (New York: Henry Holt, 1997); Root, Maria P. P., ed. *Racially Mixed People* (Thousand Oaks, CA: Sage, 1992); Root, Maria P. P., ed., *The Multiracial Experience: Racial Borders as the New Frontier* (Thousand Oaks, CA: Sage, 1995); Loretta I. Winters and Herman L. DeBose, eds., *New Faces in a Changing America: Multiracial Identity in the 21st Century* (Thousand Oaks, CA: Sage, 2002).
12. The mission statements for most of the larger mixed-race organizations and watch-groups—MAVIN, AMEA (including New Demographics and Mixed Race Media Watch, etc.—all subscribe to this assumption that being seen is, to some extend, a political end in itself. For a fuller discussion of how mixed-race organizations understand their relations to civil rights, see Kim M. Williams, *Mark One or More: Civil Rights in Multiracial America* (Ann Arbor: University of Michigan Press, 2006) and *Mixed Race America and the Law*, ed. Kevin R. Johnson (New York: New York University Press, 2003).
13. Toni Morrison, "Unspeakable Things Unspoken: The Afro-American Presence in American Literature *The Norton Anthology of African American Literature* (2nd ed.), Henry Louis Gates, Jr. and Nellie Y. McKay, eds. [New York: W. W. Norton, 2004], 2306.
14. See Roderick Ferguson *Aberrations in Black: Toward a Queer of Color Critique* (University of Minnesota Press, 2003). Quote from "Editorial Review," http://www.amazon.com/Aberrations-Black-Critique-Critical-American/dp/0816641293/ref=pd_sim_b_3/104-3761771-0121516?ie=UTF8. Last accessed June 2, 2006.
15. James Gaines, "From the Managing Editor," *Time*, Special Issue on "The New Face of America: How Immigrants are Shaping the World's First Multicultural Society," 142.21: 2. Much has been written on this *Time* issue, from Donna Haraway to Joane Nagel. See also Evelynn M. Hammonds's excellent discussion of this image in "New Technologies of Race" in *Processed Lives: Gender and Technology in Everyday Life*, eds. Jennifer Terry and Melodie Calvert (New York: Routledge, 1997): 107–122.

One of the earliest but still, I find, most exhaustive and nuanced readings of it is by Michael Rogin, in the opening to *Blackface, White Noise: Jewish Immigrants in the Hollywood Melting Pot* (Berkeley: University of California Press, 1998).
16. William Byrd, *Histories of the Dividing Line Betwixt Virginia and North Carolina* (New York: Dover, 1967), 4.
17. Kimberly McClain DaCosta, *Making Multiracials: State, Family, and Market in the Redrawing of the Color Line* (Stanford: Stanford University Press, 2007), 156.
18. Fazier, Sundee, *Check All That Apply: Finding Wholeness as a Multiracial Person* (NY: InterVarsity Press, 2002).
19. Quoted in Kawash, who similarly argues that "…hybridity is a challenge, not only to the question of human "being," but to the status of knowledge itself, the question of how and if we can *know* identity or hybridity. To rest with the conclusion that identity is really always hybridity deflects the real challenge of hybridity itself, a challenge posed to the very conditions of modern epistemology and subjectivity" (Samira Kawash, *Dislocating the Color-Line: Identity, Hybridity, and Singularity in African-American Literature*, Stanford University Press, 1997). On hybridity see also Robert C. Young, *Colonial Desire: Hybridity in Theory, Culture and Race* (New York: Routledge, 1995).
20. See discussion of the racial politics of genre in Howell's patronage of Dunbar in Chapter 2 of my *Race, Work and Desire in American Literature, 1860–1930* (Cambridge: Cambridge University Press, 2003), 58–60.
21. Werner Sollors, *Neither Black nor White yet Both: Thematic Explorations of Interracial Literature* (New York: Oxford University Press, 1997), 10.
22. Prashad, Chandra, ed., *Mixed: An Anthology of Short Fiction on the Multiracial Experience*, Introduction by Rebecca Walker (New York: W. W. Norton, 2006).

23. Graham is one of the earliest advocates of the multiracial census category, testifying with her child before the Office of Management and Budget. For a scathing criticism of her involvement, see Jon Michael Spencer.
24. Paul R. Spickard, *Mixed Blood: Intermarriage and Ethnic Identity in Twentieth-Century America* (University of Wisconsin Press, 1991), 93. As tempting as it is to include Cuban writers like Nicolas Guillén, Aimé Cesaire, Eduoard Glissant, and many others who write about mestizaje, it is beyond the scope of what I can discuss here.
25. Noel Ignatiev and John Garvey, *Race Traitor* (New York: Routledge, 1998).
26. Harry J. Elam. Jr., "Change Clothes and Go: A Post-Scrip to Post-Blackness" in *Black Cultural Traffic: Crossroads in Global Performance and Black Popular Culture*, eds. Harry J. Elam Jr. and Kennell Jackson (Ann Arbor: University of Michigan Press, 2005): 379–388.
27. Elizabeth McHenry, *Forgotten Readers: Recovering the Lost History of African American Literary Salons* (Durham, NC: Duke University Press, 2002), 261.
28. Harold Cruse, "The Integrationist Ethic as a Basis for Scholarly Endeavors" in *The Essential Harold Cruse: A Reader*, Introduction by Stanley Crouch (New York: Palgrave Macmillan, 2002): 117–124.
29. Paula M. L. Moya ("Introduction") and Linda Martín Alcoff ("Who's Afraid of Identity Politics") in *Reclaiming Identity: Realist Theory and the Predicament of Postmodernism* (Berkeley: University of California Press, 2000).
30. Michael Hames-García, "Which America Is Ours? Martí's 'Truth' and the Foundations of 'American Literature,'" *Modern Fiction Studies*, 49.1 (Spring 2003): 19–53. Special Issue: Fictions of the Trans-American Imaginary, eds. Paula M. L. Moya and Ramón Saldívar.
31. Paula M. L. Moya, *Learning from Experience: Minority Identities, Multicultural Struggles* (Berkeley: University of California Press, 2002), 156–158. Quoted in Hames-García, "Which America Is Ours?" 30.
32. Related to this notion of classroom conflict is the role of educators and the need for them to be self-critical as well. By this I am not suggesting the self-flagellating gestures some white educators and critics in the early 1990s seem compelled to make in prefaces to their articles or in opening their lectures with apologia (for being white). These have the inadvertent effect of waiving critical engagement with their own race with the justification that they do not share the same experience as minority students. But this does not advance analysis of experience; it quarantines it and the teacher, who has in effect, removed him- or herself from applying the critical discussion of racial experience except when it applies to "others."
33. Kimberlé Crenshaw, "Mapping the Margins: Intersectionality, Identity Politics, and Violence Against Women of Color," *Stanford Law Review*, 43 (July 1991): 1241–1265.
34. Hortense J. Spillers, "'The Tragic Mulatta': Neither/Nor—Toward An Alternative Model," in *The Difference Within: Feminism and Critical Theory*, ed. Elizabeth Meese and Alice Parker (Philadelphia: J. Benjamins, 1989), 168.
35. This is an argument commonly raised in debates about reparations—see for instance, Saidya Hartman and Stephen Best, "The Redress Project."
36. See Danzy Senna, "Mulatto Millenium" in *Half and Half: Writers Growing Up Biracial and Bicultural* (New York: Pantheon, 1998).
37. This hardly needs elaboration but for a reminder, one need go no further than the tragic case of Phillippa Schuyler. See Kathryn Talalay's excellent biography of the "mixed-race" genius who ended up repudiating any association with blackness.
38. An excellent and thoroughly documented glimpse into the complexity of two of these related histories, in the revolutionary period in the United States and during the same colonial period in South Africa can be found in George Fredrickson's *White Supremacy* (New York: Oxford University Press, 1983), Chapter 3.
39. See Calvin Hernton's early work, first published in the late 1960s, *Sex and Racism in America* (New York: Anchor Books, 1992) on interracial love, one of the earliest studies of the way racial stereotyping and race/power dynamics inform intimate interpersonal relations even if the couple sees themselves as a sanctuary from the world of race politics.

REFERENCES

Alcoff, Linda Martín. "Who's Afraid of Identity Politics," in *Reclaiming Identity: Realist Theory and the Predicament of Postmodernism*. Berkeley: University of California Press, 2000.

Best, Saidiya Hartman, and Stephen Best. "Fugitive Justice." *Representations: Special Issue on "Redress,"* 92.1 (2005): 1–15.

Brennan, Jonathon, ed. *Mixed Race Literature*. Stanford: Stanford University Press, 2002.

Byrd, William. *Histories of the Dividing Line betwixt Virginia and North Carolina*. New York: Dover 1967.

Crenshaw, Kimberlé. "Mapping the Margins: Intersectionality, Identity Politics, and Violence against Women of Color." *Stanford Law Review*, 43 (1991): 1241–1265.

Cruse, Harold. "The Integrationist Ethic as a Basis for Scholarly Endeavors," in *The Essential Harold Cruse: A Reader*, ed. William Jelani Cobb. New York: Palgrave Macmillan, 2002, 117–124.

DaCosta, Kimberly McClain. *Making Multiracials: State, Family, and Market in the Redrawing of the Color Line*. Stanford: Stanford University Press, 2007.

Elam Jr., Harry J. "Change Clothes and Go: A Post-Scrip to Post-Blackness," in *Black Cultural Traffic: Crossroads in Global Performance and Black Popular Culture*, eds. Harry J. Elam Jr. and Kennell Jackson. Ann Arbor: University of Michigan Press, 2005, 379–388.

Elam, Michele. *Race, Work and Desire in American Literature, 1860–1930*. Cambridge: Cambridge University Press, 2003.

———. "Towards Desegregating Syllabuses: Teaching American Literary Realism and Racial Uplift Fiction," in *Teaching Literature: A Companion*, eds. Tanya Agathocleous and Ann C. Dean. New York: Palgrave Macmillan, 2003. 59.

———. *Mixtries: Mixed Race in the New Millennium*. Stanford: Stanford University Press, forthcoming.

Fazier, Sundee. *Check All that Apply: Finding Wholeness as a Multiracial Person*. New York: InterVarsity Press, 2002.

Ferguson, Roderick. *Aberrations in Black: Toward a Queer of Color Critique*. Minneapolis: University of Minnesota Press, 2003.

Fredrickson, George. *White Supremacy*. New York: Oxford University Press, 1983.

Gaines, James. "From the Managing Editor" on Special issue "The New Face of America: How Immigrants Are Shaping the World's First Multicultural Society." *Time* (1993): 2.

Gaskins, Pearl Fuyo ed. *What Are You? Voices of Mixed-Race Young People*. New York: Henry Holt, 1997

Hames-García, Michael. "Which America Is Ours? Martí's 'Truth' and the Foundations of 'American Literature.'" *Modern Fiction Studies*, 49.1 (2003): 19–53.

Hammonds, Evelynn M. "New Technologies of Race," in *Processed Lives: Gender and Technology in Everyday Life*, eds. Jennifer Terry and Melodie Calvert. New York: Routledge, 1997, 107–122.

Hernton, Calvin. *Sex and Racism in America*. New York: Anchor Books, 1992.

Ignatiev, Noel, and John Garvey. *Race Traitor* New York: Routledge, 1998.

Johnson, Kevin R., ed. *Mixed Race America and the Law*. New York: New York University Press, 2003.

Kawash, Samira. *Dislocating the Color-Line: Identity, Hybridity, and Singularity in African-American Literature*. Stanford: Stanford University Press, 1997.

Kelley, Matt. *Chasing Daybreak: A Film about Mixed Race in America*. 2006. Directed by Justin Leroy and Produced by Matt Kelley. Mavin Foundation, Seattle, WA.

Lopez, Alejandra. *The Population of Two or More Races in California*. Stanford: Stanford University, November, 2001.

McHenry, Elizabeth. *Forgotten Readers: Recovering the Lost History of African American Literary Salons*. Durham, NC: Duke University Press, 2002.

Morrison, Toni. "Unspeakable Things Unspoken: The Afro-American Presence in American Literature," in *The Norton Anthology of African American Literature*, eds. Henry Louis Gates, Jr., and Nellie Y. McKay. New York: W. W. Norton 2004,

Moya, Paula M. L. "Introduction," in *Reclaiming Identity: Realist Theory and the Predicament of Postmodernism*. Berkeley: University of California Press, 2000.
———. *Learning from Experience: Minority Identities, Multicultural Struggles*. Berkeley: University of California Press, 2002.
Moya, Paula M. L., and Ramón Saldívar. "Fictions of the Trans-American. Imaginary." *Modern Fiction Studies*, 49 (2003): 1–18.
Poovey, Mary. *Making a Social Body: British Cultural Formation 1830–1864*. Chicago: University of Chicago Press, 1995.
Prasad, Chandra, ed. *Mixed: An Anthology of Short Fiction on the Multiracial Experience*. New York: W. W. Norton, 2006.
Pratt, Mary Louis. *Imperial Eyes: Travel Writing and Transculturation*. New York: Routledge, 1992.
Rogin, Michael. *Blackface, White Noise: Jewish Immigrants in the Hollywood Melting Pot*. Berkeley: University of California Press, 1998.
Root, Maria P. P, ed. *Racially Mixed People*. Thousands Oaks, CA: Sage, 1992.
———, ed. *The Multiracial Experience: Racial Borders as the New Frontier*. Thousand Oaks, CA: Sage, 1995.
Root, Maria P. P., and Matt Kelley, eds. *Multiracial Child Resource Handbook: Living Complex Realities*. Seattle, WA: MAVIN Foundations, 2003.
Schlaikjer, Erica. "The Explosion in Mixed Race Studies." *Eurasian Nation* (2003). March 4, 2004 <http://www.eurasiannation.com/articlespol2003-04mixedstudies.htm>.
Senna, Danzy. "Mulatto Millenium." *Half and Half: Writers Growing up Biracial and Bicultural*. New York: Pantheon, 1998. 12–27.
Sollors, Werner. *Neither Black nor White yet Both: Thematic Explorations of Interracial Literature*. New York: Oxford University Press, 1977.
Spickard, Paul R. *Mixed Blood: Intermarriage and Ethnic Identity in Twentieth-Century America*. Madison: University of Wisconsin Press, 1991.
Spillers, Hortense J. "The Tragic Mulatta': Neither/nor—Toward an Alternative Model." *The Difference Within: Feminism and Critical Theory* Ed. Elizabeth Meese and Alice Parker. Philadelphia: J. Benjamins, 1989. 165–188.
Talalay, Kathryn. *Composition in Black and White: The Life of Philippa Schuyler*. New York: Oxford University Press, 1995.
Williams, Kim M. *Mark One or More: Civil Rights in Multiracial America*. Ann Arbor: University of Michigan 2006.
Williams, Teresa Kay Cynthia L. Nakashima, George Kitahara Kich, and G. Reginald Daniel. "Being Different Together in the University Classroom: Multiracial Identity as Transgressive Education," in *The Multiracial Experience: Racial Borders as the New Frontier*, ed. Maria P. P. Root. Thousand Oaks, CA: Sage, 1995, 359–379.
Williams-León, Teresa, and Cynthia L. Nakashimi, eds. *The Sum of Our Parts: Mixed Heritage Asian Americans*. Philadelphia: Temple University Press, 2001.
Winters, Loretta I., and Herman L. DeBose, eds. *New Faces in a Changing America: Multiracial Identity in the 21st Century*. Thousand Oaks, CA: Sage, 2002.
Yoshino, Kenji. *Covering: The Hidden Assault on Our Civil Rights*. New York: Random House, 2007.
Young, Robert C. *Colonial Desire: Hybridity in Theory, Culture and Race*. New York: Routledge, 1995.

6

ETHNIC STUDIES REQUIREMENTS AND THE PREDOMINANTLY WHITE CLASSROOM

Kay Yandell

As a professor specializing in minority American literatures, I often teach ethnic and gender studies courses with titles such as "Race and Ethnicity in the Lives of United States Women," "Nineteenth-Century Literature by African-American Women," and "Introduction to Native American Women's Literature," in a university where around 90% of students self-identify as "White."[1] Because this lack of diversity can create an ethnocentric climate that, in turn, discourages minority-culture students from attending and graduating, the university has instituted what faculty and students alike generally refer to as the "Ethnic Studies Requirement" (ESR). This requirement was created in 1987, in response to Black Students' protests of certain White fraternity gatherings at which members displayed caricatures of Black "Fiji Islanders."[2] By instituting the ESR, administrators seek to contribute to a more hospitable university environment for minority-culture students, to help increase cultural diversity on campus, and to impart to all students skills that will be helpful in an increasingly diverse and international job market and society. The requirement mandates that undergraduates take at least one course devoted to the study of one or more ethnic groups in the United States, or organized to view a particular subject (literature, in my case), through the lens of United States ethnic relations in general. In this essay, I would like to share and interpret my experiences of teaching this university-wide ESR to a predominantly white student body, to convey some of what I have learned about what the requirement means to students, faculty, administration, and also to the larger community to which it is ultimately directed. I would also like to share some of my conclusions about, and my teaching strategies for the ESR, in hopes that, as universities across the country increasingly adopt diversity requirements, such experiences might be of help to other professors teaching similar courses in American universities. At the turn of the twenty-first century, though U.S. minority populations[3] are growing, institutions of higher learning continue to have predominantly white, and increasingly upper-income student populations. Given this climate, conversations about and strategies for diversity requirements, and for critical access to university education for minority culture students, seem more important now than ever.[4]

Teaching Agency in a Climate of Fracture

Before I enter into an explanation of the workings of the ESR as I have known it, I would like to explain some of the experiences that drew me to a realist approach to gender and ethnic studies in the first place, and the teaching philosophy that these experiences have inspired. I believe experiences like mine are still very common, both inside and outside the American academy, and I believe they show why gender and ethnic studies are still such important components of any, ideally "universal," university education.

When I came to what many describe as a "progressive" consciousness,[5] it was of a very practical sort, taught by university professors, feminist activists, and practical minority people alike, most of whom themselves came to progressive consciousness within the free speech, civil rights, and war protest movements of the 1960s and 1970s. In later years, I was surprised to sometimes hear this sort of perspective described as old fashioned or unsophisticated by the self-described adherents to Continental Philosophy, or Postmodernist, Post-structuralist, or Post-feminist scholars that I began to encounter in the academy in the 1990s. In fact, I developed an interest in teaching literatures that empower women of color, partly in response to some of the less-empowering accounts of minority identity that I at that time encountered among other scholars of ethnic and gender studies; the deconstruction of binary oppositions[6] surrounding identity that scholars sometimes invoked in their ethnic studies courses, often seemed to me to be strategically employed to reinforce the very lack of recognition of power imbalances that I thought progressive cultural studies sought to expose. I found myself in several discussions with scholars who were making what I considered anti-progressive arguments, and rationalizing them *as* progressive by "deconstructing" the categories of oppressor and oppressed, colonizer and colonized, or more powerful and less powerful, in any given history or social situation. I'd like to give a few examples that I think illustrate the ways that deconstruction of identity is often misused to disregard power imbalances between groups.

A very well-educated colleague whose mind I quite respect, for example, recently explained to me that when she was an undergraduate, a professor at her prestigious university had explained to her class that everyone in the class had at times in their lives been the oppressor, and at other times been the oppressed. Astonished at her account of what I regarded as a professor's attempt to dismiss the power imbalances attending individuals' social locations and affecting human actions, I countered that I thought people from her social location, that of a then-eighteen-year-old African American girl from Chicago's poor south side, the daughter of divorced, working-class parents, would probably have a much harder time oppressing another group of Americans than members of more socially powerful groups would have. I soon learned that social analyses of culture that similarly ignored power imbalances between individuals and groups were quite common in the American academy. What many (perhaps mistakenly) call deconstructing, and I sometimes experience as simply disregarding, the power imbalances between individuals from different social locations is itself often viewed as the most progressive element of these arguments. For example I once heard a scholar with a degree in American Studies—who himself taught minority texts—laud Thomas Jefferson's sexual relationship with his slave Sally Hemmings as an example of Jefferson's ability to transcend the racial prejudices of his time in order to "fall in love with" a woman from such a lower social status. I answered that I thought this scholar should reexamine the few facts we have of the relationship: white masters' sexual relations with the women of African descent whom they kept enslaved was common throughout the antebellum United States. Jefferson, one of the most powerful men in the United States, kept Hemmings—30 years his junior, 16 years old at the time of her first pregnancy—enslaved, illiterate, and ignorant of her rights abroad throughout his lifetime,[7] then allowed members of her (his?) family to be separated and sold at auction upon his death. So, whether her sex with Jefferson involved fondness, fear, or more likely both, the fact that an enslaved woman bore her owners' subsequently enslaved children constitutes a fairly typical result, rather than transcendence, of the power of a master's social location over a slave's social location. As an enslaved woman, after all, Hemmings never had the right to say "no."[8] Another example is a scholar of color who explained that she was practicing "cyborg feminism," of the sort Donna Haraway advocates,[9] by having her eyelids surgically altered to look more Caucasian. These sorts of claims ignore even the most obvious power imbalances between social actors, in a way that risks celebrating as progressive or liberatory the oppressive or exploitative actions that result from and reinforce the domination of the less powerful by the more powerful.

And indeed, though these scholars potentially misunderstood or misused the postmodern theorists they invoked to bolster their refusal of social hierarchies, further reading convinced me that there exists a large body of modern social theory designed to relieve the individual of social and historical responsibility by presenting a theory of the individual that is fractured, or divorced, either from her larger social location (by overlooking such social formations as family, class, race, education, gender, region), or fractured even from herself (through the agency of an alien and unknowable subconscious that primarily determines self and action). I'll name a few by now widely recognized examples.

Such influential critics as Walter Ben Michaels, for example, seemingly reduce recognition of one's social location of ethnicity to race consciousness, and ultimately to racism, in order to promote his alternative of an identity-blind society and social analysis. Michaels asks for example in his well-known work *Our America*,[10] what he claims is a liberatory question regarding identity in general:

> What does it matter who we are? The answer can't just be the epistemological truism that our account of the past may be partially determined by our own identity, for, of course, this description of the conditions under which we know the past makes no logical difference to the truth of falsity of what we know. It must be instead the ontological claim that we need to know who we are in order to know which past is ours. The real question, however, is not *which* past should count as ours, but why *any* past should count as ours. Virtually all the events and actions that we study did not happen to us and were not done by us; it is always the history of people who were in some respects like us and in other respects different. When, however, we claim it as ours, we commit ourselves to the ontology of "the Negro," to the identity of the we and they and the primacy of race.[11]

For Michaels, even constructivist theories of ethnicity smack of essentialism and racism. For example, for an African American woman to realize epistemologically that the history of oppression of her family contributes to her own class, regional and educational circumstances, under Michaels's theory, is nonetheless to give into the essentialist "ontology of 'the Negro,' to the identity of the we and they and the primacy of race." I would counter that the example of self-knowledge I postulate above, of the effect of the past upon the present, and of the present upon the self, conveys much more than "the ontological claim that we need to know who we are in order to know which past is ours." Rather, self and social knowledge seem more often to work in the opposite direction, and much more powerfully that Michaels supposes. In the case of the woman I imagine above, the history of oppression that affected others of her constructed social location informs her of what types of oppression she might face in her own life, and likely conveys her family's experiential knowledge of what kinds of resistance work best in specific situations. From this she learns, for example, which behaviors she should probably teach her own children to empower themselves personally against the oppression they might face as members of the same constructed social group. And, her experientially gained, subjective knowledge also has something to teach the larger world. As people in her subject location (African American women, in this example) organize to form what Sánchez-Casal and Macdonald in this anthology describe as communities of meaning, and make the social knowledge gleaned from similar social experiences known among themselves and to a larger populace, the society as a whole can see how it needs to reform to make itself a more just place for all of its citizens. It can learn, as it did, for example, in the 1960s, to abolish discriminatory Jim Crow laws. This conscious reconstruction of social justice ideally benefits members of all social groups. For Michaels, however, "...race in general...is not...a social construction; it is instead...a mistake."[12] Michaels describes all cultural identification as nothing more than a matter of individual choice (e.g., he compares the seeming "choice" to adopt a homosexual identity to the decision to join the Elks' Club). His idea that racial, cultural, regional, or gender identities are nothing more than individual choices, whether right or wrong in itself, seems to assume a world of equally autonomous individuals able to act without regard to the social location they inhabit. Michaels's solution to what he imagines as the racism inherent to any cultural identification—ignoring the existence of ethnicity entirely—"to choose what one is," assumes complete, perfect, and equal individual autonomy, free of cultural or historic factors. At the same time, it ignores the larger and more immediate issues facing those who inhabit African American (or homosexual, or rural, or elderly, or any) identities—the fact that social agency and power are distributed very unequally among these various groups, often severely limiting their members' abilities to make social choices. Michaels's refusal to recognize the power imbalances between groups reinforces these imbalances by allowing them to go unexamined and unchallenged.

Another common theory of selfhood that, it seems to me, removes social responsibility from individual action by ignoring the relative social power individuals employ as members of various social locations, is the distinctly postmodern, psychoanalytic version of self in which the individual is so internally fractured that he cannot recognize or understand even a coherent self, much less an identifiable social location, or an ethical and empowering social action. Regarding gay identity, for example, Judith Butler postulates:

> Is sexuality of any kind even possible without that opacity designated by the unconscious, which means simply that the conscious "I" who would reveal its sexuality is perhaps the last to know the meaning of what it says…? For being "out" always depends to some extent on being "in;" it gains its meaning only within that polarity. Hence, being "out" must produce the closet again and again in order to maintain itself as "out." In this sense, *outness* can only produce a new opacity; and *the closet* produces the promise of a disclosure that can, by definition, never come.[13]

The quotation indicates that people's conscious minds are so fractured from their subconscious minds that they cannot know what their social location (here, sexual orientation) even is. What's more, because being "out of" is the opposite of being "in" the closet, consciously disclosing one's identity for some reason "must" confound one's attempts to know one's identity. The self here is so fractured and self-contradictory that coming "out" always shoves one back "in." The fallacious nature of this reasoning is only eclipsed by its disastrous political consequences for the poor soul unable to discern, with this theory of identity, whether she is gay or straight, out or in. In the cycle Butler here imagines, attempting to know one's identity is exactly the action that most confuses such knowledge. This very common, postmodern notion of a self so fractured from herself that she cannot even know, much less communicate, analyze, or act in recognition of her social location, ultimately serves to ignore the power imbalances between various social groups and, like Michaels's theory of the self fractured from social location, finally disables any personal reflection upon or action based on analysis of one's social location; if I cannot even manage to decide whether I am, say, gay or straight, African American or Asian American, I certainly cannot discern the ways historical and current social constructions of these categories might affect me, nor how I might act by considering these social constructions' effects on my life, nor how the social knowledge gained from my socially located experiences might be used to make a more just society for all social groups.

I mention these prevalent theoretical orientations against identity constructions that have definitions, or opposites, because by reconceiving social location as irreconcilably self-contradictory, unknowable, or fictitious, such notions of selfhood potentially deny power imbalances between social groups. Because socially constructed categories in fact have very real effects on the lives of those who inhabit them, I feel an ethical responsibility not to teach within these common methodologies which, it seems to me, too often unhelpfully seek to interpret social location as unknowable, ambiguous, self-contradictory, incommunicable, or non-existent. In teaching within the ESR, while I consider it very important to address the socially constructed nature of minority and all cultures, I consider it as important to empower students to see the social consequences these constructions impart, and the ways these can be changed to produce more equitable social theory and practice. This is largely because theories like the preceding, which can work to ignore the power imbalances between groups, closely resemble current political and legal attempts to do the same. As Carl Gutiérrez-Jones explains of Walter Benn Michaels's theories, for example:

> Michaels's recent writings also seem very much in line with conservative gestures that are now focusing on Justice Harlan's dissent in *Plessy v. Fergusson*.... Among other things, Michaels's political commitment is to a vision of society made up of ultimately transparent, individual (trans)actions. At the same time, attempts to posit and interpret group, and especially minority-majority conflicts, are discounted as fundamentally mistaken because the links supposedly binding such groups are always presumed to be artificial....[For Michaels,] historical and cultural concerns need to be severely limited in terms of application to decision making. This concept is of course crucial when Michaels disconnects people from any past they have not immediately experienced....[For Michaels,] one's relation to culture and history is posed solely as a matter of choice. Everyone becomes a potential perpetrator; by the same token, questions produced in a results-oriented analysis, or victim-oriented analysis—that is, those results so critical to the development of affirmative action and harassment laws—are radically devalued.[14]

We live within social and legal systems where our individual experiences, opportunities, and perceptions are heavily influenced by our social locations (our gender, first language, ethnicity, region, etc.). Identity blindness keeps the power imbalances that attend various social locations from being questioned, and ultimately reinscribes ethnocentrism among dominant groups, who, under an identity blind ethic, are not taught that there are different cultural perspectives, much less that these perspectives might better their understanding of themselves and the larger world.

With much academic theory and current pedagogical and public policy alike promoting identity-blind social relations, it is little surprise that students who identify as white, at predominantly white institutions, often arrive in a required ESR course on multiculturalism with very ethnocentric social assumptions, which they assume to be value-neutral, unbiased, and objective, rather than the product of their particular social location. Nothing illustrated this problem to me better than my experience of teaching the same ethnic and gender studies course at a predominantly white institution, first as an elective to a class of women of color, and then to fulfill the ESR.

From Identity Blindness toward Multicultural Understanding

Since I started teaching nearly a decade ago, I have taught mainly at Cornell University and the University of Wisconsin, two universities which, overall, have a lot in common. Both are large, prestigious institutions located in predominantly "white" towns.[15] Though the UW is twice the size of Cornell, both are located within a few hours' drive of two of the largest and most cosmopolitan cities in the United States, and both universities attract students from these cities. The subjects of the courses I have taught at both universities have also been similar—undergraduate survey courses designed to introduce students to the art, literature, cultural practices, gender roles, and philosophical worldviews of diverse ethnic groups, and designed within course numbers, titles, and descriptions assigned by the university. I mention the demographic, institutional, and pedagogical similarities of these two universities because my experience teaching gender and ethnic studies courses at two similar institutions has actually been quite different. This is primarily because, while Cornell had no undergraduate diversity requirement that I know of when I taught there, at the UW the gender and ethnic studies courses I teach fulfill the ESR.

When I taught a course called "Introduction to American Minority Women's Literatures" at Cornell, the course was listed as an elective, and filled entirely with undergraduate women of color. On the first day of the course I always ask students to tell the class where they are from, what they plan to study, why they chose this course, and what literature they most enjoy. I learned that, among these students, some were recent immigrants, many were from working-class or single-parent families, some were the first in their families to attend college, and all had grown up in the nearby city. Almost all said they had had few to no literature courses dedicated to the study of gender and ethnicity, and many said they took the course to read literature by women of color because they thought it would more likely engage historical, cultural, and experiential worlds similar to their own. My goal, when I designed the course, had been to introduce young women of color to the literary tradition by women of color that I had first discovered as an undergraduate. I felt at the end of the semester that I had succeeded in this goal. This course was a joy to teach, and I learned a tremendous amount, both about the literature itself and about how to teach these subjects, from the varying backgrounds and perspectives these young women brought to the course. My student evaluations confirmed that many of the students had learned what I hoped the course had to teach. "This course has truly been a blessing," said one of the most heartening. "I didn't know until now that African American women were publishing novels so long ago. It's inspiring to discover this history of intellectual achievement."

Encouraged by this, my first experience teaching a course organized explicitly around issues of gender and ethnicity, I proposed to teach the same course at my next university job. To my surprise, on the first day of this course, in a class of 26 students, there were only one or two people who identified as anything other than "white Americans" (and one of these disappeared without dropping the course by semester's end). The students were predominantly senior, non-literature majors, and, when I asked why they had chosen an introductory course in non-canonical literatures in their senior year, almost all explained that they were taking the course to fulfill the university's ESR, which I had never heard of before this moment. A couple of students elaborated that, in fact, because they had no interest in literature or ethnic and gender studies, they had delayed fulfilling this requirement as long as possible, and were consequently taking it their last semester in college.

Needless to say, my experience teaching this course was radically different from teaching the same course to a group of women of color who had elected to study the subject. A few students were overtly hostile to the study of what one called "less rich, less complex" literatures and cultures. Many more, however, came to the subject with fairly open minds. Though their perceptions were sometimes fairly ethnocentric, in the course of the semester I came to understand that this ethnocentrism was often the genuinely well-intentioned product of relative inexperience with cultural perspectives and experiences significantly different from their own. In the course of this first semester teaching the ESR, I realized I needed to change the goals and methods I had set for the course. My course evaluations convinced me that I was right in this assertion; I had not been prepared for some students' perception that the views presented in course materials were all my personal views. "You need to realize," said one such conflation of professor and text, "that this is a very angry woman." I had underestimated the resulting hostility some students felt toward the subject and toward me as the source of the subject, because they did not express this hostility openly in class: "She sucks," said another evaluation.

Having an ESR at a 90% white campus will undoubtedly produce classes of predominantly white students, and, in the case of students as unaware of cultural difference as mine sometimes are, the goal of such courses becomes very different than in a more diverse classroom. I now dedicate more course time, for example, to the examination of dominant and sometimes well-intentioned ethnocentric assessments of minority U.S. cultures.[16] I review historical facts disproving dominant myths of American exceptionalism,[17] and most importantly, I emphasize the ways that the relative power held by people from a given social location affects how members of that group can, or often do, act in the social world. I'll use a few recent in-class statements from students as examples of the sort of well-intentioned ethnocentrism which I routinely encounter.

"Black people," one student offered recently, "speak black English just because they haven't been well educated. If they had better schools, they could speak proper English just as well as anyone else." This student, a young man of German Catholic ancestry, I think considered himself to be speaking against racism when he said this, because he was claiming that African Americans' "bad" grammar was a product of environment rather than of innate inferiority. I opened this comment up to discussion by the group in a way that, in a more ethnically diverse class, might have exposed the ethnocentrism underlying divisions of "proper" (white) and "improper" (black) speech habits. In this particular largely white class, however, no one had anything to add. The communities of meaning that Sánchez-Casal and Macdonald identify in this anthology as important in-class alliances through which to explore minority experience and social analyses, can be particularly difficult to organize in courses where few or no students self-identify as minorities of any sort. So, I played the role into which the ESR in a largely White classroom often places me, and undoubtedly one that helps create student hostility—that of social foil. I explained that people often consciously choose to speak in ways that identify them with their group, even (or especially) when they know there is widespread prejudice against this group. I used the example of my own Southern speech patterns, and pointed out that lots of people use a dominant way of speaking when they need to, then speak what seems to them a more natural English when not in a professional environment: I, for example, make a conscious effort not to use such terms as "y'all," "fixin' to," and "might could" when I present at conferences, because I know there is widespread prejudice against such terms. At the same time, I would never relinquish them in my vernacular speech, first, because as far as I'm concerned, they are real and good terms that convey shades of meaning tragically absent from more dominant American speech patterns, and secondly because they are the terms of the people with whom I identify most. This, I explained, is often the case for Southerners, for example, and perhaps for African Americans as well. It's often not that they don't "know how" to speak white English (therein lies another discussion to suggest that there is no proper English—that what we think of as standard English is determined by the dominant, largely white culture in this country) but that they don't choose to speak white English in their otherwise largely black lived social realities. I hope that bringing my own identity into play and giving examples from my own experience encourages students to do the same, creating more equitable classroom discussion and avoiding the preception that I am "talking down" to students; it also shows how we can theorize the experiences attending differing social locations without descending into stereotypes. My colleagues have rightly warned, however, that especially in a classroom setting that is likely to create hostility toward the professor, exposing the socially located nature of my own subjective experience and knowledge also heightens the risk that students will dismiss it, and me as a professor, as irrelevant to their own lives. Some other examples: "Whereas female circumcision is torture, male circumcision is necessary for health. It's disgusting *not* to be circumcised as a man." This young woman assumed that she was speaking the unbiased truth of science, rather than expressing a culturally specific belief. She was genuinely surprised to learn that many groups in the United States and the "first world" do not circumcise their boys, who nonetheless enjoy perfect health, and that there might even be uncircumcised men in this course.

After a few semesters of teaching this introductory course on literature by women of color to this new, largely white student body, and of reading other professors' insights on similar experiences, I had identified several problems common to the particular classroom dynamic of a majority of white students who are required to study the perceptions of the minorities who have been oppressed by whites. By recognizing the potential for these problems, I now work to anticipate and correct for them from the beginning of each semester. Among the pedagogical challenges presented by teaching the ESR to a predominantly white student body, the following three seem to present themselves most often.

First, it helps to realize from the outset and openly to address in class the fact that a course requiring dominant culture students to study minority opinions can affect students' perceptions of not only the professor and quality of the course itself, but also of the texts and views that such a course introduces. Though she here describes elective African-American history courses, Allison Dorsey expresses a conclusion similar to my own in her "Reflections on a Decade of Teaching Black History at Predominantly White Institutions (PWIs)":

> I anticipated neither the ways in which the essence of the storyteller or the complexion, status, and/or class of the listeners would condition their understanding of the stories nor the ways I would be conflated with my subject matter.[18]

This problem seems even more pronounced in a mostly-white class required to study minority authors, and often taught by minority professors, since the potential for student hostility and a desired target for that hostility is heightened by the requirement. In my experience, in this classroom environment, students are much more likely to assume that assigned feminist texts, for example, represent the views of their professor if that professor is female, and to dismiss her as "biased" or "irrational" or "angry" because they are hostile to the views presented in the feminist essays they read. Many, I think, would not associate opinions expressed in the same assigned texts as directly with the opinions of, say, an older white male professor, and would not react with as much hostility toward a professor who enjoys white privilege and male privilege, because even though a woman's subject location likely grants her knowledge of women's issues that a man's might not, white American men, because of their dominant subject location, are often still seen to occupy a more objective, unbiased, rational perspective on minority literatures.

Second, white students' hostility often results from the challenge brought by minority viewpoints to students' notions of themselves. As Frances Maher and Mary Kay Thompson Tetreault describe, white students are so accustomed to thinking of themselves and their beliefs as "neutral" that they often enter such courses unaware that they read from a social location at all:

> [They] tended to perceive themselves only as individuals, missing their own connection to the economic and political structures that privilege them in relation to other women and men.... The white students had trouble seeing themselves as white, in seeing racism as part of the social structure in which their own lives were embedded...[19]

ESR courses present minority social analyses that in part work to expose the equally subjective nature of majority cultural assumptions from which students often read. Course readings designed to show white students that their opinions proceed from their own subjective, "biased" social locations as surely as does that of, say, a Black Lesbian Disabled War Veteran (a category one student proposed as the most likely to win university scholarships), can disturb students' notions of self and world in ways that professors should anticipate from the course's beginning and actively address throughout.

Third, white students required to study minority perceptions often exoticize the very different subjectivity they encounter there. Dorsey, a professor of African American history, describes this phenomenon in her own teaching. "For most privileged white students, people of African descent are exotic objects of study whose lives serve as a link to an oppositional culture and styles of expression.... This reasoning is unconscious, which is to say that students operate from racial innocence."[20] Dorsey here describes the phenomenon as it arises in elective courses that students themselves have chosen to attend. *Requiring* all white students to engage, for example, what bell hooks calls "radical black subjectivity" through the ESR, on the other hand, insures that those students who overtly consider minority cultures to be inferior will be present in class as well. Overall, however, I agree that the discomfort that students often feel in courses of this sort, and their frequent romanticizing or exoticizing of course authors, usually stem from a lack of experience with different subject locations, rather than from conscious malice. One preliminary goal, of course, is to humanize the exoticized Other, to show that her perspective is as likely valid as one's own, to learn empathy through that validation, and to re-imagine a world in which different social locations contribute toward, rather than inhabit, the "objective" center of social practice.

And, of course, such innocence and romanticism are exactly the sorts of perceptions that the ESR hopes to teach students to transcend. It is one of the most important, and most difficult jobs that we as professors can undertake with and for our students. As bell hooks reminds us in *Teaching to Transgress*, "courses that work to shift paradigms, to change consciousness, cannot necessarily be experienced immediately as fun or positive or safe and this is not a worthwhile criteria to use in evaluation."[21]

A New Outline for Teaching Resistance Literatures in the ESR

When I first started teaching literatures by and about women of color, I designed a course *for* women of color. I had hoped to inspire and to embolden, to allow young women from historically oppressed backgrounds to explore self models and social strategies offered by women from similar social locations and experiences. I think that's very important work. Having gained some experience teaching these same marginalized selves, experiences, and social strategies to students from the dominant culture, however, I see that this work is just as important. I've also seen that it has to be done in a very different way than my original project, because predominantly white classes often proceed from an entirely different set of assumptions, entirely different worldviews, than do more diverse or predominantly, say, African American, Latin@, or Native American student populations. With this fact in mind, I have established a set of rules for myself, as follows, to accomplish the paradigm shift that the university employs me to effect for these students. The following practices constitute the best solutions I have found for overcoming the pedagogical challenges posed by teaching the ESR to a predominantly white student body.

First, I discuss with students the history that inspired—and the University's goals for—the ESR, stressing the fact that this history and these goals affect even white students, personally. I explain that the university has developed this requirement to present suppressed minority texts, views, literatures, and representations, because its leaders know such resistant views might otherwise never reach students' consideration. I explain that consideration of diverse viewpoints is vital to the functioning of any democratic society, and that if they consider the views presented in this course to be "one-sided" or "biased" (as they sometimes claim), they should consider that this requirement is *designed* to give opposing viewpoints on many social issues. This design decision I defend with realist theoretical claims: different social locations, different experiences, produce different social theories in different cultures; we must consider these differing social theories to create our best society. Beginning the course with these explanations of our purpose somewhat alleviates the hostility some students feel for being required to read resistance literatures, and I hope it better allows them to see that I personally am not the object of their hostility.

Second, I find it vital to show predominantly white classes that the knowledge taught in this course affects them personally, in very real ways, even though most of them consider themselves part of the majority culture. For example, in the early 2000s, several major U.S. corporations, among them Alcoa Aluminum, Cargill Agriculture, General Motors, Procter and Gamble Pharmaceuticals,

Kimberly-Clark Paper, Ford Motors, and Hewlett Packard Technologies, announced plans to stop recruiting at our campus, because of the lack of diversity, and because they find that our graduates are not "culturally competent" to work in an increasingly multicultural and international workplace.[22] In other words, not knowing what this course has to teach has had a direct and adverse financial affect on students' lives. My goal as a professor of the ESR, I explain, works as part of the university's 10-year plan to increase ethnic diversity on campus and to give students the cultural competence they need to succeed in an increasingly diverse workforce.

Third, while addressing the ideological topics around which the course is organized, I find it very helpful to list, define, and actually have students learn by name the assumptions on which this alleged cultural incompetence, and on which general U.S. ethnocentrism, are based; these are the assumptions this course works to overturn. These assumptions include, for example, the founding, and still very common notion that the United States has a God-given Manifest Destiny to destroy, control, or assimilate other cultures, both on this continent and around the world. I address the related notion of Social Darwinism, which still lies unstated beneath common U.S. assumptions that there are primitive and sophisticated races, peoples, ethnicities, or cultures in the United States and the world, and that cultures "evolve" along a predetermined track toward more sophisticated modes of expression, employment, and belief—toward such modes as alphabetic writing (and away from, say, ideographic writing), toward industrialization (away from farming, herding, or hunting), and toward capitalism (away from collective land ownership or government regulated industry), toward individualism (away from, say, tribalism), toward the nuclear family (away from extended families living together)—in short, toward dominant U.S. culture. These notions of social evolution assume that it is natural and inevitable for sophisticated cultures to destroy others (to "survive") in the imagined cultural competition toward the Survival of the Fittest. We begin and sustain the course by naming and engaging these and other myths supporting U.S. ethnocentrism and imperialism, myths such as the Bootstraps Narrative, the myth of the Individual, and the American Dream, three related notions of U.S. meritocracy positing that all U.S. citizens begin life with the same opportunities, that if anyone works hard enough, she can be and attain whatever she wants, and that a person's position in life is a result of her own talent or hard work, uninfluenced by the social position and network of contacts into which she chanced to be born. Along with these, over the years I have learned to engage another unspoken myth that thoroughly permeates U.S. culture: the claim that the way best to create equality is to ignore inequality. This myth, which I have referred to as Identity Blindness, posits, for example, that even mentioning the lower socioeconomic status often held by people of color or women, in itself blames these groups for the problem, and itself dashes hopes within these groups by publicly humiliating them for their lower social status. This myth is one of the most important to engage from the course's beginning, since many of my students have been taught that it is rude or vulgar to mention social inequalities between groups, and that the most ethical, and certainly the most polite, thing to do is not to mention the problem. This chosen Identity Blindness manifests itself in class in ways it took me years to understand: in my ESR courses, students sometimes claim that being a victim—of anything—is an entirely optional, psychological state; a victim is one who chooses to acknowledge social inequalities or discrimination (and not, as I define it, one to whom something bad has happened). I occasionally hear that it is feminists who disempower women, by stating the fact, say, that on the whole women still earn less than men. The success of ESR courses depends on students seeing through such myths and assumptions of polite silence, apathy, or superiority, to recognize and analyze the history that actually caused the imbalances, and the ways that these myths perpetuate power imbalances in U.S. society.

Fourth, I work to expose the underlying, virtually always economic, motives behind the creation of these myths. Without explicit acknowledgement that specific groups with specific motives cause and perpetuate social imbalances, many students cannot understand why these social power imbalances exist, or why they can't easily be fixed. For example, students often have difficulty believing that the history of racism in this country is really as codified, legalized, systematic, and long-standing as it actually is. If students do not learn how racial differences were used to serve imperialism and capitalism by, for example, creating a visual distinction between most antebellum slaves and free people, and that there were very real economic interests served by instituting racial hierarchies, they tend to be left with no answers to such questions as "why would anyone think light skin is better than dark skin?" A question like this one shows that a student has not moved beyond an understanding of racism to its underlying cause: a history of imperialism that financially benefits specific groups. A professor needs to have ready, or preferably already have established in class discussion, larger theoretical explanations of the underlying economic systems that created and perpetuate social myths and the social inequalities they justify.

Fifth, I've learned to design reading and discussion questions in advance to combat the realities of ethnocentrism as they arise around our subjects of study. For readings, I mix fictional literature (I'm assuming a literature course, since that's what I teach) with history, personal essay, and social theory, to convey ideas more clearly. Students sometimes claim, for example, that Sethe, the protagonist of Toni Morrison's *Beloved*,[23] is simply a child abuser for killing her baby to spare her the life of slavery, but few can deny the horrors of being a woman enslaved when *Beloved* is clustered with non-fictional readings from Harriet Jacobs's *Incidents in the Life of a Slave Girl*, in which the autobiographer hides in a low attic literally for years to avoid being raped by her owner. Students sometimes claim sexism is a thing of the past in the United States after reading Kate Chopin's *The Awakening*,[24] but they better understand the epistemic level at which sexism still works today when *The Awakening* is clustered with, say, Gloria Steinem's essay "If Men Could Menstruate,"[25] which lists examples of common ethnocentric and sexist American perceptions, and humorously exposes the absurdity of these beliefs. In discussions, I find it helps to state for students that the rules of Academic Freedom apply to them as well as to their professors. If students don't feel they can express controversial opinions, theorize or think aloud about previously unexamined beliefs, assumptions, or attitudes, they can't come to new understandings that better explain the social facts of ethnic experience and interaction as they have known them. I also find it helpful, often through past examples of my own ethnocentrism, to assure students that ethnocentrism is inevitable and understandable (after all, what standards do we have to measure the world, other than those our own culture has given us?), but also largely surmountable through increased knowledge of and empathy for the life of the Other. The paradigm shift required to affect this empathy is one this course's most important goals.

Sixth, I try to emphasize the epistemological, rather than the ameliorative, reasons for creating an ethnically democratic, multicultural society—the realist notion that we will produce better social knowledge in a multicultural society, rather than simply minimize the effects of past wrongs. Despite the goals of the ESR, I also try not to lose sight of the fascinating, eye-opening discoveries about the breadth of possibility for human culture, that accompany the study of the Other.

ESR Discussions

The existence and implementation style of the ESR draws scrutiny from all sides. Within the academy, many professors would rather teach their ethnic studies courses as electives than under the requirement, in order to draw more motivated students, and decrease or eliminate the student hostility they sometimes face in ESR courses. This is especially true for professors who are themselves members of the minority groups they present in class, since in this case, student expressions of hostility and ethnocentrism can come to seem like personal attacks. Since, so far as I know, there is no official institutional acknowledgement of or correction for the lower student evaluations that ESR courses can generate, some professors (especially some untenured professors) avoid teaching within the ESR to avoid having their teaching evaluations brought down. Some professors add that there is no structured training or preparation for teaching the ESR. Administrators, similarly, are pressed to discern which courses should fulfill the ESR (it was recently decided, for instance, that courses concerning ethnicity outside the United States would not be included as ESR courses), and must negotiate with departments who do not feel that Ethnic Studies are relevant to the educations of their students; students of the Engineering Department, for example, are allowed to fulfill the ESR on a pass/fail basis, rather than for the usual letter grade. Students complain both that they have to fulfill a requirement they sometimes feel is irrelevant to their course of study, and, among minority students and parents, that the ESR allows several courses on groups who are U.S. minorities in terms of number, but not in terms of social power or representation in the university.

In April 2004, when the state newspaper announced that several major corporations would stop recruiting on campus, this debate widened to include the surrounding community who would be affected by the decision. State newspaper accounts reported the event as the result of a lack of diversity that the university acknowledges is long-standing, and has already started working to improve, with increasing success. The Law School, for example, was featured in legal journals as a school "where diversity works"[26] for its 27% minority student body in an 88% white state. The state's college preparatory course for minority middle and high school students began only in 1999, and has already increased its enrollment over 10 times in the past six years. The newspapers also mention the work of the ESR as a successful step to increase awareness and diversity on campus.

Community responses to these articles to my mind show both how often the general public conflate Identity Blindness with its near opposite, multiculturalism, and how desperately the much-maligned Ethnic Studies Requirement really is needed in universities across the United States. The responses in the local newspaper's opinion section to news of the recruiters' withdrawal sound very much like opinions some students express in ESR courses. An opinion piece titled "Multiculturalism Fails in Its Mission: White Students Demonstrate So Little Cultural Competency that They Can't Get Jobs"[27] demonstrates misunderstanding of the problem and potential solutions alike, similar to that I see in some students. This article unwittingly demonstrates exactly the sorts of misunderstanding of cultural interaction that the ESR seeks to correct. For example, very much like many of my students and many of the cultural theorists and political decision makers I mention earlier, this author assumes Identity Blindness (i.e., pretending that racial and cultural differences do not exist, thereby reinstating ethnocentric world views) is the same as a multicultural education:

> The irony of today's...graduates being labeled "culturally incompetent" is that today's grads have been steeped in multiculturalism all their lives. Since their first day in kindergarten, they...have been told that skin color is irrelevant and that to judge someone on the basis of race, ethnicity, gender, sexual orientation, religion etc., is just plain wrong. They've discovered they can say the "F" word with impunity, but never the "N" word. They've been pounded with lectures on the virtue of sensitivity.

This author's graduates have apparently been through very different schools than my students, who claim to have had virtually no "cultural sensitivity training." More disturbing, however, this quotation equates the mere teaching of students not to acknowledge or express racism (not to say the "N" word) with a multicultural education, the goal of which is not the refusal of, but the acknowledgement and genuine appreciation for other practical and attendant epistemological differences of other cultures. This report assumes that it is enough to teach students not to call their coworkers derogatory names, and seems not even to imagine the existence of the much larger and more rewarding goals, means, and benefits of a truly multicultural education, to maintain and support the distinctive identities of the cultural groups within a society.

Finally, the report goes on to conclude—just as I've heard a shocking number of students, professors, and administrators conclude over the years—that the "cultural incompetency" of majority students is the responsibility of minority students:

> It's possible the proliferation of student groups designed to promote cultural awareness actually enforce cultural isolation: If there's a club just for black students, or Hispanic students, or Armenian students, they'll be less likely than ever to mix with members of other groups. This denies other students the opportunity to improve their "cultural competency."

This assertion is a common one among educators and students alike, who assume that part of the job of coming to university as a minority person is to work for members of the cultural majority that surrounds you, to teach them the "cultural competence" they need to succeed. In reality, the main goals of culturally oriented, minority student groups are not "to promote cultural awareness" of minority cultures for majority-culture students. These clubs, rather, seek to provide a more hospitable and comfortable environment for minority students, exactly so that they are not as pressured constantly to explain their minority selves to the majority around them.[28] It is very doubtful that the two or three hours a week, at most, that these clubs meet, interferes in any way with students making cross-cultural friendships from the classes, dormitories, eating halls, intramural sports teams, and so forth, that form the larger part of university life.

This opinion article reiterates its ethnocentric perspective when it names its alternative to the multicultural education that it claims has failed (and that I would counter has not even been fully implemented or fully begun its work). The article offers the following as a more adequate solution: "[T]he university can't change society alone. Parents of black students must disabuse their kids of the notion that doing well in school is "acting white" if they want to expand university graduation rates.... The multiculturalist strategies we've been so doggedly pursuing for the past three decades haven't produced any real results." So, while this anonymous author wants black students to give up their cultural clubs (the places they can most "act black" without feeling out of place or self-conscious about it), she doesn't want to hear the university represented as a place where African American have to "act white." In other words, she wants African American students to do all the adjusting to a predominantly white university, and not to complain about having to do so.

Other published community reactions differ from the above assumptions. A senior from the nursing school remarks that she has never been trained on the relation between cultural difference and health care availability or belief, and that she thinks she needs to be: "We are not educating nurses to care for people of different backgrounds. How can we decrease the scarcity of health care if we are not exposing student nurses, most of whom are female and white, to [people of] different...races or abilities...?"[29] The majority of respondents, however, agree with the solicited opinion section article, and more explicitly blame minority students for not changing enough to fit into a country that was, at least according to the following article titled "Minorities Are the Real Klutzes," built from a blank land, entirely by and for the white middle-class:

> The minorities are the cultural klutzes. Remember that they are trying to fit into our world, our language, our academics and so on which our parents, grandparents and great-grandparents created long ago. Acceptance of minorities by students is an individual choice. Most students are intelligent and can see when diversity is being crammed down their throats. The modern American Black culture is worlds away from anything remotely civil. If the Hispanic culture takes the time to learn and teaches their young at an early age, they have the most hope of learning and working side by side with our youth.[30]

Another, titled "Smells like Political Correctness" goes on to suggest that the Law School is practicing "reverse discrimination" and that majority students' rights to make racial slurs to minority students is protected by First Amendment rights. Teaching students that offensive comments about minority students "is inappropriate and detrimental to campus climate" "reeks with PC,"[31] the article concludes.

I sample here the local opinions and problems surrounding the ESR and the university's larger attempts to increase diversity, to provide some sense of how new, unknown, misunderstood, and occasionally just hated, the project of a multicultural education still is to community members and students taking the ESR. University projects for diversity, perhaps not unlike this essay, still work at the level of explaining, again and again, why these projects are necessary, what they do, how they will benefit minority and majority social groups alike. When I teach in the ESR, I dedicate the first week of the course, and then reiterate throughout, the history, necessity, goals, benefits, and hopes for the required study of minority cultures. This work facilitates the eventual cultural discoveries I hope I bring to my students, of how different cultures organize their worlds, and how we can all learn from their insights. With that in mind, I would like to end by answering the question that Walter Benn Michaels asks in the first section of this essay. Michaels asks:

> What does it matter who we are? The answer can't just be...that our account of the past may be partially determined by our own identity....It must be instead the ontological claim that we need to know who we are in order to know which past is ours. When [we claim a past] as ours, we commit ourselves to the ontology of "the Negro," to the identity of the we and they and the primacy of race.[32]

In my experience, people virtually never claim an identity ("African American," in Michaels's example) because they want to claim the history most associated with that identity, or want to determine the past by that identity, as Michaels claims. Nor are identity claims necessarily essentialist, as Michaels posits. Rather, as realism posits, it matters who we are because identities are the best ways to explain our lived experiences. If I am an African American woman, it is not because I want to claim slavery and jazz as my very own. Rather, this title, "African American woman" matters because it is the title that makes the most sense of, say, my speech patterns, where I live, what my parents do for a living, why I fear racism in a way others don't, why getting my hair done as a child burned my scalp, why I can't trace my family tree back before 1865, what music or literature I know best, and so on. This explanation—this identity—therefore, brings with it real social insights, types of interpretation, and modes of enjoyment that people from other identity categories can learn from and share. This is why it matters who we are, and this is why it is important to create a truly multicultural education for our students. This, despite the obstacles, is why I continue to teach within the ESR.

Notes

1. "In fall 2006 UW-Madison enrolled a total of 802 minority [undergraduate] students, 650 of whom were Wisconsin residents and most of whom had graduated from a public school in the previous spring. Total undergraduate minority enrollment was 12%, which is a little less than the minority representation in the Wisconsin population (13.8%) and a lot more than the minority representation among high school graduates who took the ACT the previous spring (8.3%)." "Wisconsin's Minority Population and the Race/Ethnic Diversity in

the Recruiting Pool for UW-Madison Undergraduate Admissions," http://apa.wisc.edu/Diversity/RecruitingPool_Notes_July2007.pdf (accessed July 28, 2008). These statistics do not represent my own experience teaching within English Department and ESR courses, which generally seem to have many fewer than 12% minority students.
2. "Report of the Ethnic Studies Requirement Review Committee," July 18, 2002: 7, http://mendota.english.wisc.edu/~danky/esr.pdf (accessed July 28, 2008).
3. By "minorities," I mean what universities sometimes call targeted minorities or underrepresented minorities, that is, groups whose representation in higher education and other measures of social power (such as elected government positions, socioeconomic class, top private sector jobs, et cetera) is lower than their percentage in the general population. For this reason, women are considered minorities in many ways: though they comprise a slight majority of the general U.S. population, they hold only a small fraction of elected positions in the U.S. government, for example.
4. David Leonhardt, "As Wealthy Fill Top Colleges, Concerns Grow Over Fairness," *New York Times*, Thursday, April 22, 2004, A1. "Students from upper-income families are edging out those from [the] middle class at prestigious universities around [the] country, from flagship state colleges to [the] Ivy League; [the] change is fast becoming one of [the] biggest issues in higher education."
5. By "progressive conscious" I mean a political orientation that acknowledges the power imbalances this country's imperialist history has created, to oppose discrimination, and to promote social equity. This orientation often assumes that historically oppressed peoples best overcome resulting social power imbalances through (1) organizing to identify similar experiences of oppression, their underlying causes, and potential solutions (2) raising consciousness so that other people, both those who do and who don't identify as members of that group, come to know these social problems and potential solutions, and (3) activism, through which government representatives and legal codes are changed to solve social problems.
6. Jacques Derrida, the inventor of the term "deconstruction," famously did not specify its meaning. A 1993 paper he presented at the Benjamin N. Cardozo School of Law in New York defined it thus: ". . . deconstruction, if there is such a thing, takes place as the experience of the impossible." Asked to define the term in a 1998 interview, he explained, "It is impossible to respond. I can only do something which will leave me unsatisfied." Jonathan Kandell, "Jacques Derrida, Abstruse Theorist, Dies at 74," *The New York Times*, October 10, 2004, http://www.nytimes.com/2004/10/10/obituaries/10derrida.html (accessed July 30, 2008). Scholars of Derrida have defined deconstruction as the demonstration "through comparisons of a work's arguments and its metaphors, that writers contradict themselves—not just occasionally, but invariably—and that these contradictions reflect deep fissures in the very foundations of Western culture." Mitchell Stephens, "Deconstructing Jacques Derrida: The Most Reviled Professor in the World Defends His Diabolically Difficult Theory" *Los Angeles Times Magazine*, July 21, 1991, Sunday, http://www.nyu.edu/classes/stephens/Jacques%20Derrida%20-%20LAT%20page.htm (accessed July 30, 2008). While most scholars readily agree to the facts of social construction (the notion that our words, thought, and beliefs are determined by the society that surrounds us), many feel it is important to stress that there can be human agency to perceive and alter our social constructions, rather than positing, as strict deconstructionists sometimes do, that social constructions are fictions that unite what is in reality an inherently fractured, self-contradictory, unknowable self, language, society, or history. The problem deepens when scholars overlook the fact that social construction of categories and wholes are vitally necessary for social action and regulation. We cannot define human rights, for example, if we cannot construct anything coherent enough to be called "human." These constructed realities are also often overlooked when scholars employ the Foucaultian notion that social constructions are so completely structured by power hierarchies that there can be no political realization, organization, protest, or revolution, that is, no social action, that is not itself a part of a dominance-saturated, hierarchical social construction. Extensions of this view sometimes posit that all political or social action, therefore, merely reinscribes power imbalance; the best we can do is simply to parody these constructions with ironic gestures

that expose them, such as wearing a necktie as a woman, e.g., to show the constructed masculinity attendant upon wearing a necktie. It bears noting that not even Michel Foucault himself could live with such a cynical theory, protesting near the end of his life for AIDS awareness.
7. Jefferson wrote to a friend from Paris explaining that it's safe to take slaves to France because, though they have the right to emancipate themselves there, they will never know of this right.
8. For an example of an enslaved woman's perspective on her owner's demands for sex, see Harriet Jacobs, *Incidents in the Life of a Slave Girl* (Cambridge, MA: Harvard University Press, 1987).
9. See Donna Haraway, "A Cyborg Manifesto: Science, Technology, and Socialist-Feminism in the Late Twentieth Century," in *Contemporary Literary Criticism: Literary and Cultural Studies*, ed. Robert Davis and Ronald Schleifer (New York: Longman, 1998), 696–727.
10. Walter Benn Michaels, *Our America: Nativism, Modernism, and Pluralism* (*Post-Contemporary Interventions*) (Durham, NC: Duke University Press, 1995).
11. Michaels, *America*, 128
12. Walter Benn Michaels, "Autobiographies of the Ex-White Men: Why Race Is Not a Social Construction." Paper presented at The Futures of American Studies Conference, Dartmouth College, August 15, 1997.
13. Judith Butler, "Imitation and Gender Insubordination," in *Inside/Out*, ed. Diana Fuss (New York: Routledge, 1991), 13–31, 15–16.
14. Carl Gutiérrez-Jones, "Color Blindness and Acting Out," in *The Futures of American Studies*, ed. Donald E. Pease and Robyn Wiegman (Durham, NC: Duke University Press, 2002), 248–265, 249, 256.
15. At the time of the 2000 census, Ithaca New York, the first town in which I taught, had 29,287 residents, who identified themselves as 73.97% White, 13.65% Asian, 6.71% Black, 5.31% Hispanic, 1.86% Other, 0.39% American Indian, and 0.05% Pacific Islander. Cornell University had 20,000 graduate and undergraduate students. Madison Wisconsin, the second town in which I taught, in the 2000 census counted 218,432 inhabitants, who identified as 83.96% White, 5.84% Black, 5.80% Asian, 4.09% Hispanic, 1.67% Other, 0.36% American Indian, and 0.04% Pacific Islander. The University of Wisconsin had 41,552 undergraduate and graduate students.
16. E.g., I sometimes list for students some prevalent beliefs that members of dominant cultures often learn about differences between their own and minority cultures. These lists change by location, historic era, and group, of course, but such generalizations can nonetheless allow students to see some double standards for individual behavior that even positive stereotypes about cultural difference can create. Here is one such list:

European American Cultures	American Indian Cultures
Modern, Present	Past, Gone
Social Change typifies, improves nation	Social change renders nation corrupt or inauthentic
Literate	Illiterate
Technological	Non-technological
Civilized, Sophisticated	Savage, Primitive
Owns land	Does not own land
Farms	Does not farm. Hunts and gathers
Stationary	Nomadic
Rational	Spiritual
Mature	Developing
Destined to Dominate	Destined to Disappear
Not influenced by Native Culture	Influenced by Euro-American Culture

Lists like this one of differences that United States citizens often assume about European American and Native American cultures help students understand why they might tend to speak of Native people in the past tense, why they sometimes express concern that gaming industries corrupt Native American culture (but not European American culture), or why some of them might believe individual American Indian people have a larger obligation than they do to learn ancient cultural traditions.

17. Many students assume that virtually every industrial technology or social innovation was invented in the United States, or that the newest, biggest, fastest, or best of everything can be found in the United States. As relevant to any particular reading assignment, culture, or history studied, it helps to review other peoples' cultural and technological innovations, for example.
18. Allison Dorsey, "'white girls' and 'Strong Black Women': Reflections of a Decade of Teaching Black History at Predominantly White Institutions (PWIs)," in *Twenty-first-Century Feminist Classrooms*, ed. Amie Macdonald and Susan Sánchez-Casal (New York: Palgrave Macmillan, 2002), 203–231, 206.
19. Frances A. Maher and Mary Kay Thompson Tetreault, *The Feminist Classroom: An Inside Look at How Professors and Students Are Transforming Higher Education for a Diverse Society* (New York: Basic Books, 1994), 220–221.
20. Dorsey, "White Girls," 203.
21. bell hooks, *Teaching to Transgress: Education as the Practice of Freedom* (New York: Routledge, 1994), 53.
22. Though I have seen no official University statements on this issue recently, students tell me some of these recruiters have returned. Perhaps this fact suggests that perceptions of UW graduates' "cultural incompetence" are abating among recruiters.
23. Toni Morrison, *Beloved* (New York: Knopf, 1987).
24. Kate Chopin, *The Awakening* (New York: W. W. Norton, 1976).
25. Gloria Steinem, "If Men Could Menstruate," http://www.haverford.edu/psych/ddavis/p109g/steinem.menstruate.html (accessed July 30, 2008).
26. *The National Jurist*, March 2004.
27. "Multiculturalism Fails in Its Mission: White Students Demonstrate So Little Cultural Competency that They Can't Get Jobs," *Wisconsin State Journal*, April 12, 2004: E1.
28. On the importance of cultural clubs and program houses on university campuses, see Amie A. Macdonald, "Racial Authenticity and White Separatism: The Future of Racial Program Housing on College Campuses," in *Reclaiming Identity: Realist Theory and the Predicament of Postmodernism*, ed. Paula M. L. Moya and Michael R. Hames-García (Berkeley, CA: University of California Press, 2000), 205–225.
29. Tamaria Parks, "Where Is the Education?" *Wisconsin State Journal*, April 18, 2004: E1.
30. Marc David Wilson, "Minorities are the Real Klutzes," *Wisconsin State Journal*, April 18, 2004: E1.
31. Dan Bach, "Smells Like Political Correctness," *Wisconsin State Journal*, April 18, 2004: E1. See also Ted Koehler and Carlos Miranda, "Are Today's UW-Madison Graduates Cultural Klutzes?: No Incentive Encourages Awareness," *Wisconsin State Journal*, April 11, 2004: B1. *The Oxford English Dictionary* defines "politically correct" as a term "from the early 1970s, specifying conforming to a body of liberal or radical opinion, especially on social matters, characterized by the advocacy of approved causes or views, and often by the rejection of language, behavior, et cetera, considered discriminatory or offensive." The dictionary notes that the term is often used dismissively.
32. Michaels, *America*, 128.

7

HISTORICIZING DIFFERENCE IN *THE ENGLISH PATIENT*: TEACHING KIP ALONGSIDE HIS SOURCES

Paulo Lemos Horta

Michael Ondaatje's *The English Patient* (1992) ostensibly invites a postcolonial reading when it describes how an English officer nicknames the Sikh sapper Kip after viewing his first bomb disposal report: "the officer had exclaimed, 'What's this? Kipper grease?' and laughter surrounded him. He had no idea what a kipper was, but the young Sikh had thereby translated into a salty English fish. Within a week his real name, Kirpal Singh, had been forgotten."[1] Kip will only reemerge as Kirpal Singh at the close of the novel,[2] when he blames England for the American bombing of Japan and decides to return to India and reclaim his identity. Critics have obliged this suggested line of interpretation, sounding the appropriate notes on the subjects of naming and the emergence of the postcolonial identity from the imperial. And yet in the classroom my students and I have found it pertinent to ask: what sources provide Kip with his 'real' name and identity? Whose identity and experience does Ondaatje seek to rescue from erasure and forgetfulness? Ondaatje's own acknowledgments in *The English Patient*, which credit *The Tiger Strikes, The Tiger Kills, A Roll of Honour,* and *Martial India*[3] as his sources for Kip,[4] point to a prior and more determinant act of naming. *Martial India* singles out the bravery of a Kirpal Singh who was decorated for capturing with a handful of men a large village held in strength by the Germans, prompting author Yeats-Brown to gush, "the cavalry spirit survives, and hearts beat as high as they ever did, amongst these stalwart yeomen."[5] The four works provide Ondaatje with a composite portrait of the Sikh holy warrior after which he models Kip, and with a positive articulation of the martial mode of multiculturalism that Kip embodies in his interaction with the Canadian nurse Hana and the English patient (Almasy).[6] This chapter recounts a pedagogical experiment of teaching the novel alongside its sources to obtain a more objective gauge of Ondaatje's success in rendering minority identity and experience.

Realist theory, as articulated by Satya P. Mohanty, Paula Moya, Linda Martín Alcoff and others, highlights the pivotal role of identity in generating experience and knowledge of ourselves, the social relations we inhabit, and the aesthetic ideals we express.[7] Drawing on this theory, Amie A. Macdonald and Susan Sánchez-Casal have developed a pedagogy that aims to democratize the classroom for all social identities and enable "critical access" for members of historically oppressed and excluded groups.[8] This pedagogy envisions the classroom as an epistemological laboratory where all students can revise their theories about the world by becoming aware of the "mediated connection between identity and knowledge-making."[9] In this enterprise of democratizing the classroom the diversification of curricula proves a necessary but not sufficient condition, as Michael Hames-García and Paula M. L. Moya have demonstrated. Michael Hames-García cautions in particular against the notion that social change can be effected merely through the positive representation of diversity in course syllabi, pressing the need for a critical assessment of what is taught in the name of diversity and how it is taught. His case study of anthologies of American literature notes both the omission of writings that are skeptical of the American self-image of meritocracy and the glossing over of the radical and contentious dimension of the minority writing that is included. For Hames-García it is imperative that curricula not merely present "a smorgasbord of diverse experiences" but rather situate texts in relation to one another "historically, rhetorically, and ideologically," for only then will students "feel something personally at stake" in literary texts and the social conflicts they represent and be empowered by their education to redress oppression and inequality.[10] In this light the challenge we face as pedagogues might be stated simply as: how to contextualize? How can we translate the call for the historical and political contextualization of minority literatures into effective pedagogical strategies? What experiments might transform the classroom into the epistemological laboratory envisioned by Macdonald and Sánchez-Casal? As Paula Moya has argued persuasively, the challenge ahead entails not only the inclusion of radical minority voices neglected so far, but also the radical recontextualization of authors and texts already canonized as representative of diversity and minority experience.[11]

It is this second imperative, the need to radically recontextualize works from the new canons of multicultural literature, that this chapter addresses and tests with reference to the experiment of teaching Michael Ondaatje's *The English Patient* along with its acknowledged historical sources. In Canada Ondaatje's perceived positive representation of non-Western cultures and minority experience and his inclusion in the main canon of Canadian letters has been credited with "mainstreaming" multiculturalism in Canadian academia and society.[12] *The English Patient* proves an ideal test case due to both the manner in which it claims the ideals of multiculturalism and cosmopolitanism,[13] and the wide extent to which scholars have obliged this line of interpretation. The novel's acknowledgments detail the author's efforts to ground his fictional portrayal of the Sikh sapper in the British Army, Kirpal Singh, with reference to accounts of Indian, and in particular Sikh, soldiers under British command in World War II. Winner of the Booker Prize in 1992 and the basis for the 1996 film that won nine Academy Awards including best picture, Ondaatje's novel has become a staple of canons of multicultural literature in diverse academic environments in Britain, Germany, Canada, Australia, and the United States. Ondaatje's fiction of diversity has been celebrated for enabling—if not already constituting—positive social change in precisely the superficial manner Hames-García cautions against.

In *The English Patient* the character of the Sikh sapper in the European theater of war, Kirpal Singh, possesses a preternatural poise, stoicism, and self-sufficiency, and his unparalleled mastery of technology is confirmed when he is able to defuse the detonator that had defeated his mentor and leading world expert on mines, Lord Suffolk. The conspicuous absence of any suggestion of a flaw or vulnerability in Kip's character has not attracted critical attention from scholars. On the contrary, the only complaint frequently voiced by scholars with regard to Kip is that there is not enough of him in the film version of the novel (written and directed by Anthony Minghella, 1996), which privileges the narrative of the explorer Lazlo de Almasy and his affair with Katharine. These scholars, who are in the majority opinion concerning the merits of the novel's adaptation to the screen, share an assumption common within the broader scholarship on the novel.[14] The key to its postcolonial sensibility and progressive politics is seen as lying in the narrative of Kip, from his apprenticeship under Lord Suffolk in an experimental bomb disposal unit in England to his friendship with the other members of the quartet in the Villa San Girolamo near the war's end—Hana, Caravaggio, and the English patient (Almasy). It is principally in this storyline that critics locate the novel's engagement with discourses of alterity and belonging, its critique of British racism, imperialism and Orientalism, and its affirmation of minority experience and multiculturalism. According to this common interpretation, the film's most damning oversight is the omission of the novel's conclusion in which Kip becomes disillusioned with the Allied campaign upon hearing of the bombing of Hiroshima and Nagasaki.[15]

To teach *The English Patient* in light of its engagement with its sources is to question the standard reading of Kip as emblematic of the progressive multiculturalism in Ondaatje's novel. When I have taught Ondaatje's novel in Canada, Kip's invulnerability sometimes leads to an impasse in class discussion. Is Kip to be embraced as a sympathetic portrayal of a Sikh by a Sri Lankan born writer, or is he rather wooden and unpersuasive? While immigrants and other minority students tended to be the most disaffected with Kip, non-immigrant students often expressed difficulty understanding objections to such a positive portrayal of a minority character. Was that not the very purpose of multiculturalism? While the pattern of disagreement was clear, there appeared to be no objective manner to move forward. To move beyond this stalemate, in subsequent iterations of the course I proposed to students that we examine Ondaatje's acknowledged historical sources for Kip. Ondaatje's British and 'Indian' sources alike date to the 1940s and are without exception works of war propaganda that articulate a distinctly martial conception of multiculturalism as the comradeship in arms of the Commonwealth's many nations. In this light a more objective assessment of Ondaatje's characterization of Kip seemed possible. If the character did not diverge significantly from the cited sources, he would constitute a positive stereotype inherited from the British literature on martial India. If however Ondaatje could be said to revisit these sources with irony, or contradict them with recourse to the accounts of serving under the British by the Indian and Sikh soldiers that are available in modern scholarship, then Kip might represent a postcolonial answer to the stereotypes of war propaganda. Is Ondaatje complicit with the mystification of Sikh identity found in his sources, or does his choice of sources suggest a conscious project of identifying and remedying error? These underlying questions motivated the experiment of recontextualizing *The English Patient*, with the goal of making all students more conscious of the mediated connection between identity and knowledge-making. Both my original disciplinary formation as a political scientist and my experience as an immigrant inclined me to be skeptical of superficial discourses of multiculturalism: I had internalized Charles Taylor's imperative that meaningful cultural exchange presupposed "a fused horizon of standards"[16] and from my student days felt this necessitated comparative attention of substance to other cultures.[17]

In accordance with their status as military propaganda, Ondaatje's acknowledged sources for Kip, *Martial India, The Tiger Strikes, The Tiger Kills*, and *A Roll of Honour* function first and foremost as repositories of the received ideas of British imperialism with respect to the division of the peoples of the Indian subcontinent into "martial" and "servile" races. This official British writing on India, penned by military commanders and printed by a government press in London and Directorates of Public Relations in the colonial capitals, was defined by the necessities of waging imperial war and upholding British power in the subcontinent and casts selected minorities as surrogates and proxies for this imperial power. The division of Indian peoples into servile and martial races was disseminated in the late nineteenth century by Roper Lethbridge's history of India that, shaped by the experience of the Mutiny of 1857, cast as servile the Hindus who had been instrumental in the expansion of British India, and privileged as warriors the recently conquered peoples who were perceived to have been loyal during the revolt.[18] In this writing on India, Sikhs are defined by their role in the wars of imperial expansion and preservation, first as the worthiest of adversaries and then as the most loyal of British subjects. The wartime texts consulted by Ondaatje, produced to assure India of the value of its contribution of a volunteer army of two million and to reassure a beleaguered English public of the sufficiency of this contribution, do not shy away from privileging Gurkhas and Sikhs above others as martial peoples.

Sikhs in particular emerge in this literature as holy warriors for whom ceremonies of faith are intimately interwoven with feats in battle, in passages that anticipate the lyrical and mythologizing connection Ondaatje draws in *The English Patient* between the Sikh sapper's recollection of the hymns sung at the Golden Temple and his "mystical" affinity for the technologies of warfare.[19] "A remarkable people, the Sikhs, with their Ten Prophets, five distinguishing marks, and their baptismal rite of water stirred with steel," proclaims *Martial India*, "a people who have made history, and will make it again."[20] This work lyrically recounts how Sikh armies would take the "book of hymns," the Granth Sahib, into battle "as the Ark accompanied the Israelites."[21] *Martial India* details how the Sikhs proved worthy of the title of Singh (Lion) even in the view of their Arab enemies who recognized them in Mesopotamia as "the Black Lions," and asserts that Sikhs are never despondent in the frontlines for they fight to "the last breath" and "die laughing at the thoughts of Paradise."[22] The Sikh holy warriors are prized for earning the friendship of the British by fighting valiantly against them in the 1840s and then siding alongside them during the mutiny: "a friendship baptized and confirmed in blood...for the fiercer the fighting, the warmer the subsequent friendship."[23] The British author marvels at how the Sikhs inflicted upon the British losses greater than those suffered in the Napoleonic wars in six battles in 1845 and 1849 and at how "only eight years later," during the Mutiny of 1857, the Sikhs and the British were fighting side by side.[24] A troop of 50 Sikhs is eulogized for having been "prepared to die to a man" to defend 12 Englishmen against 2,000 mutineers armed with artillery when they "might have easily saved themselves by surrendering."[25] The wartime literature traces a genealogy of the Sikhs' singular valor through a "hundred conflicts from Poperinghe to Pekin, and from the Halfaya to the Ngakyedauk Passes" to World War II where the relatively small community served under British command at "the highest proportion of any race in India," contributing 300,000 men.[26] Thus a composite portrait emerges of the Sikh holy warrior, and the sapper in particular,[27] as possessing all the virtues of a martial race—discipline, self-sufficiency, loyalty, unrivalled courage, and comradeship—and virtually none of the vices.

Ondaatje's chosen sources for Kip are distinguished by their emphasis on the shared experience of World War II as enabling what might be termed a martial mode of multiculturalism. "These Indians are aggravating devils at times, because they don't know English, but by God they can fight," proclaims a representative passage from *Martial India*, "[t]here is sometimes a gulf between our ways of thinking, but never between our ways of fighting."[28] Subsequent accounts of the role of cultural diversity in the British war effort did not always prove as celebratory: for instance Alan Moorehead's account of the North African campaign, *Desert War*, presents multinationalism as an obstacle the Allied army had to overcome in contrast to the Axis powers who had "just two languages to cope with, two temperaments to consider."[29] Ondaatje privileges sources that present a uniformly positive assessment of the martial ends of multiculturalism. *The Tiger Kills* admires the comradeship between British commander and Indian soldier and the strengthening of the "wonderful spirit of the war."[30] *A Roll of Honour* describes the British Army as a true League of Nations bound by a "special friendship" forged by "twenty months of comradeship on the field of battle."[31] "Today millions of them and millions of us are engaged in a common task," chimed in *Martial India*, stressing that though "the languages of the lower deck are indeed Babel—Urdu, Tamil, Telugu, Malayalam, Kanarese, and Punjabi," through all these "runs the tongue of Shakespeare and the terms of endearment of the British sailor."[32] Ondaatje's depiction of the comradeship among the desert explorers and the inhabitants of the Villa San Girolamo inherits from these sources a recognizable mode of multiculturalism in which ethnic and religious lines are transcended by the common experience of war and a common pleasure in the high culture of English literature and the popular culture of the artifacts of war. Ondaatje's Egyptian, Austro-Hungarian, and English explorers bond over Milton's *Paradise Lost* and G.C. McCauley's 1890 translation of Herodotus' *Histories*. At the Villa San Girolamo Hana and Caravaggio, the Sikh sapper, and the English patient bond over morphine, condensed milk sandwiches, and a shared delight in Kipling.

In this fashion *Martial India, The Tiger Strikes, The Tiger Kills*, and *A Roll of Honour* provide not only the composite traits that make up Kip's character, but also the basis for this character's cross-cultural interaction with Hana, Caravaggio, and Almasy via courage, mastery of the technologies of war, and loyalty to mentors. Ondaatje's Kip corresponds to the composite portrait of the Sikh holy warrior in his sources: he is a "knight, a warrior saint," "fully comfortable in this world" in "his self-sufficiency."[33] In acts of war, he exhibits the cool courage attributed to the Sikhs, never speaks of the dangers he encounters, and refuses morphine for his pain. In times of peace, he remains apart yet respectful of others' dancing, revelry and smoking. *Martial India* recounts the first time a Sikh sapper laughed,[34] and *The English Patient* relates Kip's first lie and attributes to him an impeccable motivation. Kip is further indebted to Ondaatje's sources in his strictness,[35] immaculate grooming,[36] reluctance to admit his homesickness, shyness about his accomplishments,[37] proficiency in bomb-disposal and marksmanship, and predilection for motorcycles[38] and what Ondaatje terms the charms of Western technological invention.[39] In Ondaatje's source texts, the Indian soldier earns his English officers' friendship and admiration through his mastery of "the technicalities of the modern weapons with which he would have to fight."[40] Likewise in *The English Patient* Kip boasts that Sikhs are "brilliant at technology" and earns the admiration of mentors and friends due to his "mystical closeness" with and "affinity" for "machines."[41]

A consideration of *The English Patient* in light of its acknowledged historical sources allows for a recontextualization of the narrative of Kip's adoption and rejection of an English persona that has hitherto been deemed the chief evidence of the novel's postcolonial and progressive sensibility. War propaganda explains the cooperation between Indians and the British as a matter of friendship and character. Indian and Sikh loyalty during the mutiny, *Martial India* ventures, should not be interpreted as "fidelity to the British cause, which is after all an abstraction, but to *individuals*."[42] *Martial India* illustrates the nature of this loyalty with the anecdote of an English judge who was rescued from a mob during the mutiny by a man whom he had ruled against, but who recognized his just character. In *The English Patient*, Kip finds it difficult to pledge his loyalty to what he will term "voices of abstract order,"[43] be they for empire or independence. He enters the war convinced "that there was a greater chance of choice and life alongside a personality or an *individual*,"[44] an inclination that leads him to sign up for service as a sapper under Lord Suffolk. The novel explicitly contrasts Kip's aversion to abstract causes such as nationalism, which seduces his brother in India,[45] and his own attachment to individuals "who had the abstract madness of autodidacts, like his mentor, Lord Suffolk, like the English patient."[46] Suffolk does not disappoint: he is brilliant, noble, and eccentric, boasting "strange bits of information" such as where to procure the best tea during a bombardment. Kip regards him as "the first real gentleman he had met in England," and in his person he comes to love "the best of the English."[47] By the time Lord Suffolk is killed by a mine, something of a transfusion of identity (defined both in terms of character and culture) seems to have taken place between the Englishman and the Indian—Kip is said to contain the "knowledge of Lord Suffolk."[48] If Kip loves Lord Suffolk for his eccentricity, in turn Lord Suffolk and his secretary inform him they had decided on his selection prior to the admissions process based on a first impression of his "brilliance and character."[49]

This notion of loyalty defined by personal affinity inherited from *Martial India* shapes key moments in Kip's character arc, from the idealized portrait of his meteoric rise in a British Army presented as a veritable multicultural meritocracy to his disillusionment with the British war effort following the American bombing of Hiroshima and Nagasaki. Ondaatje invites a postcolonial reading of Kip by having him reject the newfound community of the quartet of friends at the Villa San Girolamo and decide to return to India after hearing of the nuclear bombing of Japan. Doubtless the author intended the Sikh sapper's didactic reprimand of his friends, in which he questions "all those speeches of civilizations from kings and queens and presidents,"[50] as a moment of rupture and revelation for Kip's character. Yet in light of the novel's debt to a notion of loyalty articulated in its wartime sources, the terms of Kip's disenchantment appear less persuasive and less radical. From the outset Kip had pledged his loyalty to an individual and not to the abstract order of colonial or anti-colonial ideology. Given Kip's youthful skepticism of ideology, the presentation of his questioning of the rhetoric of queens and statesmen as a sudden revelation seems out of character.[51] The forced quality of Kip's political awakening after hearing of the bombing of Japan over the radio[52] can be attributed to the nature of Ondaatje's reliance on his sources: it is difficult to deploy a character that is a composite of propaganda in the service of a critique of this same propaganda. Ondaatje's Kirpal Singh does not differ in essence from the composite Kirpal Singh of British wartime propaganda: he is bereft of interiority and remains a type—the positive and romanticized stereotype of the Sikh holy warrior. There is little evidence for the common assumption that Ondaatje must have invested his own minority and immigrant experience in his novelistic portrait of Kip. Reminiscent of 1940s propaganda, the mode of multicultural comradeship Kip inhabits in his cross-cultural friendships and romance at the Villa San Girolamo is recognizably martial.

Drawing on this research I sought to devise an experiment that seeks to transcend the mere celebration of diversity in the classroom and stresses the mediation of identity as enabling of both error and knowledge for writer, teacher, and student alike. The assignment asked senior undergraduate students to consider in a short research paper the extent to which the notion of multiculturalism embodied in the character of Kip was indebted to Ondaatje's historical sources. Students interested in the cross-cultural bond between Kip and Lord Suffolk, his fellow sapper Hardy, Caravaggio, and the English patient, friendships in which experience and knowledge of the technologies of war was key, consulted one or more of *Martial India, The Tiger Kills, The Tiger Strikes,* and *A Roll of Honour*. Those more concerned with Kip's relationship with Hana were encouraged to read one of Ondaatje's acknowledged sources for her character, G. W. L. Nicholson's laudatory popular histories *The Canadians in Italy, 1943–1945* and *Canada's Nursing Sisters*.[53] The assignment was designed to afford students greater objectivity in gauging the relative veracity and success of representations of minority identity through examining Ondaatje's choice and handling of historical sources. Students were to consider the following questions: What authority does Ondaatje ascribe to knowledge from experience, and which experiences does he privilege in fashioning minority identities in the text? And how might we become more aware of our own epistemic status as a function of our engagement with the nature of these texts and the borrowings from them? The experiment of teaching Ondaatje's novel alongside its sources would demonstrate the benefit of exercises that make all students in the classroom more attentive to the role of identity for author, scholar, and reader as constitutive of both knowledge and error.

Teacher and student alike were intrigued by the extent to which a consideration of the novel's non-literary sources—the war propaganda and other texts including Major A. B. Hartley's *Unexploded Bomb: A History of Bomb Disposal*[54]— illuminated Ondaatje's largely inherited notions of diversity. Students registered their surprise at the naïveté with which Ondaatje accepted the propagandistic proclamation of the equality between British and Indian soldiers as an actual meritocracy, which academic English sources (let alone the Indian) make clear never existed. They noted the unlikelihood that—in the manner of Ondaatje's Kip—Indian soldiers would be shipped 12,000 miles to train in England, captain experimental bomb squads, command (rather than serve under) British and Allied forces, and be addressed by British officers as "sir" rather than (as even in the propaganda) "sepoy." Students speculated that many of the instances of praise addressed to the Sikhs in Ondaatje's sources and in his novel constitute cases of making virtue of necessity or expediency. Sappers entered enemy territory to clear mines and erect bridges for the army advancing behind them, often cutoff from other forces (hence "self-sufficient") and enduring disproportionate losses (and hence "prone to self-sacrifice"). Some students noted that the sources at Ondaatje's disposal, read for their statistics as well as the rhetoric, betrayed a conception of Indian troops as disposable. One student found that sappers in the arena were denied the benefit of the cutting-edge research and experiments that Kip enjoys in Ondaatje's novel, and that for many months sappers of the British Army stationed in Italy died trying to defuse British bombs, having been refused sensitive information that might have assisted them.[55] Students were frustrated: they had expected Ondaatje's portrayal of the Sikh experience of serving under the English in World War II to more substantially interrogate and differ from the texts of British war propaganda.

An initial resistance to the notion of deploying source study as a window into the formative role of identity for both knowledge and error came in the form of what Bat-Ami Bar On terms the boredom of received expectations.[56] Many students inherited from previous classes an assumption concerning multiculturalism that "all identity-based knowledge is automatically accurate and reliable," which Macdonald and Sánchez-Casal rightly hold untenable.[57] In past iterations of my own course, students often displayed the assumption that Ondaatje, as a writer from Sri Lanka, must be investing his experience and sympathy in the character of Kip, and that this affinity between author and character lent the novel its perceived postcolonial sensibility and progressive politics. The assumption reflects the terms of the celebratory early reception of Ondaatje as a multicultural author in literary scholarship, rather than subsequent analysis of Ondaatje's affiliation with the Burger class traditionally aligned with the Dutch and British and not with Tamil or Sinhalese identity.[58] In keeping with the broader assumption that Ondaatje must be presenting an accurate representation of minority experience via Kip, students often articulate the related expectations that multiculturalism entails the affirmation of a liberal notion of meritocracy, and that the mere inclusion of minority experience, and cross-cultural friendship and romance in particular, suffices to make a novel progressive and empowering. The recontextualization of *The English Patient* with respect to its sources led students to become aware of these expectations, not as self-evident truths, but rather as hypotheses that needed to be tested against literary and historical evidence.

For students the principal hypothesis to be tested against the source material was that the progressive politics of Ondaatje's novel resided in the cross-cultural friendship and romance Kip forges in the abandoned monastery at the Villa San Girolamo at the close of the war. Indeed, the appeal of cross-cultural friendship to students makes sense both intuitively and at the level of theory. Paula Moya persuasively points to the positive epistemic value of friendship that crosses ethnic lines, in particular among members of different minority communities.[59] The assignment demonstrated that the difficulty in this respect was that the *The English Patient* follows its propagandistic sources too closely. Even with Almasy, the Hungarian mistaken for an English patient, the bond is established (as in Ondaatje's sources) through a shared delight in military technology:

> Kip, hearing from Caravaggio that the patient knew about guns, had begun to discuss the search for bombs with the Englishman [Almasy]. He had come up to the room and found him a reservoir of information about Allied and enemy weaponry.... Soon they were drawing outlines of bombs for each other and talking about the theory of each specific circuit. "The Italian fuzes seem to be put in vertically. And not always at the tail." "Well, that depends. The ones made in Naples are that way, but the factories in Rome follow the Germany system. Of course, Naples, going back to the fifteenth century..."[60]

Ondaatje's description of Kip's first meeting with Almasy follows a familiar pattern identified by students, according to which Kip, like Kirpal Singh in *Martial India*, earns the respect and admiration of the English and other Allied troops through his martial knowledge and stoicism in acts of war. Students noted that Kip seems to be content to be defined by the approval he receives in British eyes, and earnestly appreciates this acceptance by the English officers. In accordance with the loyalty of Indian soldiers to brilliant individuals recorded in war propaganda, Kip simply transfers his loyalty from one eccentric mentor to another, from Lord Suffolk to Almasy. The semblance of equality portrayed in the novel between Kip and Almasy turns out to be indebted to the hierarchical relationships of Ondaatje's sources and to the vocabulary of war that defines them. Students found that no real epistemic breakthroughs were made possible through this friendship: in coming to admire Kip as a Sikh holy warrior, like Suffolk, Caravaggio and Hana before him, the English patient only learns what he already knows.

Kip's cross-cultural romance with Hana, too, would turn out to be overdetermined by Ondaatje's sources. Students noted with interest that Hana's storyline echoed hints at the sexual availability of nurses in the laudatory works on Canadian nurses serving in World War II, as evident in Ondaatje's reference to the "last dances" nurses granted soldiers with the knowledge of their imminent mortality.[61] Kip's romance with Hana would also prove reminiscent of British propaganda on the contribution of India's "martial" races to World War II. *Martial India* singles out the delight of an injured Indian soldier in the "prettiness" of the white nurse who tends to him, "the first white woman (he told me) that he had seen for seven months."[62] A student also noted the peculiar nature of Kip's sexual appeal to Hana in the novel, where references to his strong build are balanced throughout by an attention to his small stature and the delicacy of his wrists.[63] The image of the Indian soldier as small yet powerful also proved to have antecedents in the source material. "There are more soldiers than students in India, and they are better representatives of the real spirit of the nation," pleads the author of *Martial India*, "I wish my English and American friends who judge Indians by the clerkly classes they meet in London or New York could have seen these young men with their powerful chests and muscles like whipcord."[64] In this manner in *Martial India* the occasional reference to the small stature and fine limbs of the Indian soldier is countered through the vivid and sexualized description of his prowess at wrestling: "the wrestlers were locked together, straining and gasping, then a brown and glistening body would flash through the air, landing with a thump on the matting. Quicker than the eye could follow, his adversary was astride him, trying to force both shoulders down simultaneously, but the other writhed free, and in an instant attacked again."[65] Hana likewise desires in Kip his small wrists, his "shirtless brown body," his "chest with its sweat," and "dark, tough arms."[66] Ondaatje would later claim he sought to portray as relatively innocent the romance between Kip and Hana in contrast to the violent sexual bond that connects Katherine and Almasy.[67] Indeed, the debt the students traced of this romance to Ondaatje's propagandistic sources cautions against privileging the subplot of Kip and Hana as evidence of the novel's progressive politics.

The assignment on source material enabled students to draw on untapped reservoirs of knowledge pertaining to their experience as immigrants and members of minority communities. Many students felt emboldened to speak and write openly about their identities in a way that they did not seem as comfortable doing at the outset of the course. For instance a female Sikh student whose grandfather had fought under British command in World War II at first did not feel comfortable talking about this in class. She ultimately wrote a research paper on Ondaatje's use of historical sources in the portrayal of the Sikhs, demonstrating how this literature was contradicted by Indian accounts that Ondaatje neglected. Only in the conclusion to her paper did she feel comfortable lending credence to the experience of her grandfather as constitutive of objective knowledge, in questioning the ease with which Kip slips into a happy professional and family life upon his return from the war at the close of Ondaatje's novel:

> Personally, having a grandfather who once served for the military, it can be asserted that many soldiers did not return to a healthy state of living after returning from the military. After serving in the infantry, my grandfather did not return to his homeland to lead a wonderful life. Instead, he suffered multiple strokes which were worsened by the drinking habit he had acquired while in the military. The military itself was a place of harsh conditions, where the men were hardened to be the "tough warrior-like" Sikhs they had to become in order to fight. Thus, to see Kip illustrated in so positive a light and returning to India to lead such a normal life appears to be highly unlikely and instead highly romanticized.[68]

Recognizing the value of evidence she introduces under the rubric of personal knowledge, the student corrects what she perceived to be the error and mystification in Ondaatje's characterization of the Sikh experience of war and its aftermath.

The experiment of reading *The English Patient* alongside its sources further enabled the development of communities of meaning in the classroom. Classroom communities of meaning, as defined by Macdonald and Sánchez-Casal, refer to groups of students whose shared social location and experience enable them to arrive at common truths about the social world.[69] This descriptive category acknowledges the collective aspect of knowledge-making that can group students according to a shared epistemic affinity, and that may transcend ethnic lines as in the cited example of new immigrants who share a common stance toward figures of authority.[70] The assignment on Ondaatje's sources afforded students the chance to probe the epistemic affinities available via a shared identity defined either in terms of community, as in the case of Sikh students interested in the historical experience of Sikhs during the war, or a comparable social location, as in the case of minority students interested in the history and literary (mis)representation of other minorities. A significant number of students felt empowered by the opportunity to research the experience of someone with a background similar to their own—as a Sikh, or a second-generation female immigrant to Canada (such as Hana). As with the student whose grandfather fought under British command in World War II, these students often had not had a chance to substantially draw upon their own experience as potentially constitutive of objective knowledge.

Given the choice, many minority students chose to research sources pertaining to characters whose experiences and backgrounds differed from their own. It later emerged in discussion that some of these students were reacting to previous exposure to essentialist and celebratory multiculturalist pedagogies that called upon them only as authorities on their identities as (mis)recognized by instructors. A male Sikh student, for instance, recalled being discouraged in another class from pursuing a research interest in early-twentieth-century race riots against Chinese and Japanese immigrants and being encouraged to speak and write instead on Indian issues. Such students found appealing the opportunity to research a minority social location comparable but not identical to their own, so as to be able to examine their own experiences and identities in a comparative framework. Overall, students of all backgrounds tended to display less resistance to pedagogical experiments that emphasized the epistemic value of identity when (drawing from Realist theory on identity) I clarified that these were not intended as exercises in essentialism, but rather as windows into the possibility of connecting to other minority students and to majority students sympathetic to issues pertaining to minority experience and its representation.

An unintended but welcome consequence of the exercise in source study was that many students began to wonder how less erroneous and mystified portraits of minority experience might have been woven from the same fabric that Ondaatje had at his disposal. Some students became so invested in their chosen minority experience, whether that of Sikh sappers in the war or of second-generation Canadian immigrant women who volunteered as nurses, that they needed to believe Ondaatje did more justice to it than first suggested by direct comparison of source and novel. They would sometimes embroider on the characters available in the text with recourse to elaborate back stories of their own research and invention that might reconcile the details of Ondaatje's fiction with the facts of history pertaining to Sikh sappers or Canadian nurses serving in the British military. Ondaatje's narrator, for instance, makes evident with a didactic anti-utilitarianism that Hana abandons her unit to look after the English patient in the Villa San Girolamo because she had lost her belief, not only in the cause of war, but also in any appeal on behalf of the greater good.[71] One student argued that Hana might fear instead, as thousands of nurses reportedly did, the loss of professional and personal purpose that the imminent end of war in the European arena would bring with it. In this view, Hana flees to the Villa to tend to a single dying man to escape not the war, but the war's end. There may not be enough corroborating evidence for this reading in the novel, but it resonates with accounts of nurses' experiences during the war.[72] Other students followed the lead of director Anthony Minghella, who attributed a wit and irony to Kip on the subject of India's relationship to Britain that Ondaatje recently complained he found too "political."[73] These students sought to read irony and sarcasm into the most perfunctory of Kip's dialogue in the novel, and demonstrated wonderfully the extent to which students are willing to project their own experiences, identities, and epistemic affinities onto the novel to accommodate Ondaatje's choices as a writer. For as students read, research and write, they in effect supplement the text with glosses shaped by their own experience, knowledge, and sense of possibilities and politics.

Most students were conscious of where Ondaatje's mystification of Kip's Sikh identity ended and the corrective work of their research and rewriting began, often hinting in their papers at their fascinating alternative versions of Ondaatje's novel in notes or parenthetical asides. In response I encouraged students to hand in a one-page sketch of the novel they would have written with Ondaatje's sources at their disposal. The result persuaded me of the value of integrating this new component in future assignments dealing with the literary construction of minority identity. This may turn out to be the most empowering element of this pedagogical experiment for students. Consider the scenarios suggested in three representative answers. One student picked up on a passing reference in Yeats-Brown's *Martial India* to Indian women's fascination with modern machinery and imagined a plot in which an Indian woman would volunteer for and then become disenchanted by the war. Another, noting the tendency in the novel to define Kip by the approval he earns in British eyes, imagined an alternative fiction in which the Sikh sapper would be ironic in his assimilation to English ideals from the start. A third student ventured that the novel's vision of multiculturalism would be more meaningful had Kip been allowed to feel anger toward British colonialism, noting that multiculturalism must require work and constitute more than Anglophilia. These responses indicate the students' own engagement with attaining greater objectivity in the presentation of minority experience, and their frustration with unsubstantial articulations of multiculturalism.

The establishment of critical access for minority students and the formation of communities of meaning among and beyond minority groups necessitate institutional changes that transcend changes in curricula and pedagogical practice in the classroom.[74] Nevertheless, in accordance with the pedagogy of Macdonald and Sánchez-Casal, it is possible to experiment with pedagogical strategies that may aid in making students more conscious of identity as productive of both knowledge and error about the social world. These strategies seek to facilitate the identification of error and mystification in the presentation of minority experience and to empower students to tap into family and communal histories as possibly constitutive of objective knowledge about the social and political world they inhabit. A first step, following the lead of Hames-García, is to ask why and in what fashion works of minority writers are already included in the canon. There is a tendency to include works by minority writers that are flattering to a nation's self-conception as a multicultural meritocracy and to gloss over textual elements of contentiousness and rebellion. This tendency reflects contentment with a superficial notion of multiculturalism. Chandra Mohanty warns against a conception of multiculturalism as a "benign variation" that "bypasses power as well as history to suggest a harmonious, empty pluralism." Susan Sánchez-Casal notes pointedly that "'Celebrate diversity!', multiculturalism's favorite slogan, has taught students that the mention of multiculturalism should be followed by a party, exotic foods and dances, colorful costumes—anything but a political critique."[75] To heed Chandra Mohanty's imperative to historicize difference,[76] it is useful to read literary texts in relation to the historical and ideological texts they comment on.

It is with this objective in mind that I experimented with teaching Kip alongside his sources, prompting students to historicize the mode of multiculturalism and the portraits of cross-cultural friendship and romance that the novel presents. In contrast to previous classroom discussions where only minority students were likely to interrogate the progressiveness of Ondaatje's portrayal of Kip, now minority and majority students alike collaborated in historicizing and questioning the martial mode of multiculturalism that Ondaatje inherits from his chosen sources. Perhaps most productively, they became imaginatively invested in recreating alternative histories and fictions that better represented both their own experiences and research into sources used or neglected by Ondaatje. In completing the assignment I designed for them, they taught me the value of a new question that might do further justice to the claims of minority experience and identity laid upon them as readers, students, and producers of literature—namely, how might they have deployed comparable sources and experiences to different ends? Exposed to a pedagogy informed by Realist theory on identity, students exhibited a great curiosity toward multiculturalism conceived not merely as a celebration of diversity but as a comparative and cross-cultural mode of inquiry.

Notes

I owe more than the customary thanks to the editors and those who gave feedback on this chapter when it was first presented as a paper at a conference of the Future of Minority Studies Research Project, and to the readers of the chapter. Thanks to Satya P. Mohanty for first inviting me to take part in this think tank. I would like to acknowledge conversations on Ondaatje with Chelva Kanaganayakam, Kanishka Goonewardena, and Vaithees Ravindiran. Above all thanks to my students at the University of Toronto and Simon Fraser University, co-conspirators in these experiments and the source of constant inspiration.

1. Michael Ondaatje, *The English Patient* (Toronto: McClelland & Stewart, 1992), 87.

2. "His name is Kirpal Singh and he does not know what he is doing here [in Europe.]" Ondaatje. *The English Patient*, 287. Note the narrator refers to him as Kirpal upon his return to India, 302.
3. Walter George Hingston, *The Tiger Strikes: The Tiger Kills: The Story of the Indian Divisions in the East African Campaign* (Calcutta: Directorate of Public Relations, 1942); Hingston, *The Tiger Kills: The Story of the Indian Divisions in the North African Campaign* (Great Britain: His Majesty's Stationery Office, 1944); F. Yeats-Brown, *Martial India* (London: Eyre and Spottiswoode, 1945); Major General J.G. Elliott, *A Roll of Honour: The Story of the Indian Army 1939–1945* (London: Cassell, 1965).
4. "Acknowledgments," Ondaatje, *The English Patient*, 304–306.
5. Yeats-Brown, *Martial India*, 193.
6. It is a truly composite portrait of a martial mode of multiculturalism: *Martial India* begins by recommending "official publications about the Indian Army" including *The Tiger Strikes* and *The Tiger Kills* (v), *The Tiger Kills* itself cites *The Tiger Strikes*, and Field Marshal Sir Claude Auchinleck writes the forward for both *The Tiger Kills* and *A Roll of Honour*, which, though published in 1965, relies uncritically on the war propaganda.
7. See in particular Satya P. Mohanty, *Literary Theory and the Claims of History: Postmodernism, Objectivity, Multicultural Politics* (Ithaca, NY: Cornell University Press, 1997); Paula Moya, "Introduction," and Linda Martín Alcoff, "Who's Afraid of Identity Politics?," in *Reclaiming Identity: Realist Theory and the Predicament of Postmodernism* (Berkeley: University of California Press, 2000): 1–26, 312–344; Moya, *Learning from Experience: Minority Identities, Multicultural Struggles* (Berkeley: University of California Press, 2002); and Mohanty, Moya, Martín Alcoff and Hames-García, eds., *Identity Politics Reconsidered* (New York: Palgrave Macmillan, 2005).
8. See Susan Sánchez-Casal, "Unleashing the Demons of History: White Resistance in the U.S. Latino Studies Classroom," Amie Macdonald "Feminist Pedagogy and the Appeal to Epistemic Privilege," and their introduction to their co-edited volume *Twenty-first-Century Feminist Classrooms: Pedagogies of Identity and Difference* (New York: Palgrave Macmillan, 2002): 59–85, 111–133, 1–8.
9. **Chapter 1 (this volume).**
10. Michael Hames-García, "Which America Is Ours? Martí's 'Truth' and the Foundations of 'American Literature.'" *Modern Fiction Studies*, 49.1 (Spring 2003): 33.
11. See Hames-García, "Which America Is Ours?" and Moya's discussion of the privileging of Richard Rodriguez as representative Hispanic author in "Cultural Particularity vs. Universal Humanity: The Value of Being Asimilao," in *Learning from Experience*, 100–135.
12. Laura Moss, "Mainstreaming Multiculturalism: Canada Reads *In the Skin of a Lion*," *Canadian Association for Commonwealth Literature and Language Studies*, 2003. See also Geert Lernout, "Multicultural Canada: The Case of Michael Ondaatje," in *"Union in Partition": Essays in Honour of Jeanne Delbaere*, eds. Gilbert Debusscher and Marc Maufort (Liège, Belgium: L3-Liège Language and Literature, 1997). For a dissenting view, see Stephen Henighan's polemic *When Words Deny the World: The Reshaping of Canadian Writing* (Erin, Ontario: Porcupine's Quill, 2002).
13. See Paulo Lemos Horta, "Ondaatje and the Cosmopolitan Desert Explorers: Landscape, Space and Community in *The English Patient*," in *Moveable Margins: The Shifting Spaces of Canadian Literature*, ed., Chelva Kanaganayakam (Toronto: Toronto South Asian Review Publications, 2005), 65–84.
14. See Susan Hawkins, "The Patients of Empire." *LIT: Literature Interpretation Theory*, 13.2 (April–June 2002): 139–154; Gillian Roberts, "'Sins of Omission': The *English Patient*, The *ENGLISH PATIENT*, and the Critics." *Essays on Canadian Writing*, 76 (2002 Spring): 195–215; Subhash Jaireth, "Anthony Minghella's *The English Patient*: Monoscopic Seeing of Novelistic Heteroglossia." *UTS Review: Cultural Studies and New Writing*, 4.2 (November 1998): 57–79; Maggie Morgan, "The English Patient: From Fiction to Reel." *Alif*, 18 (1998): 159–173; Jaqui Sadashige, "Sweeping the Sands: Geographies of Desire in *The English Patient*." *Literature/Film Quarterly*, 26.4 (1998): 242–254.

15. Saul Zaentz, the film's producer, found Kip's reaction implausible and incompatible with his own experience of the end of the war in the European arena (Commentary, 2004 Collector's Series DVD).
16. Charles Taylor, *Multiculturalism, or, the Politics of Recognition*, 70. Taylor cautions: "a favorable judgment made prematurely would not only be condescending but ethnocentric. It would praise the other for being like us (70)."
17. My introduction to the chapbook *Exiles Write Back* cautioned against the "complacency" of official multicultural policy in Canada in the realm of publishing and translation (Toronto: Massey College, 2001): 5. A year earlier my contribution to a manifesto for the Witness Portraits Project in Toronto, which documented the experience of exiled authors such as Reza Baraheni, called for the teaching of world literature in translation and ventured "in literary matters multiculturalism should mean something other than rewarding those who flatter and reinforce our prejudices regarding Portugal, Italy, China, India and Latin America."
18. Roper Lethbridge, *A Short Manual of the History of India, with an Account of India as It Is, the Soils, Climate and Productions, the People, Their Races, Religions, Public Works and Industries; Civil Services, and Systems of Administration, with Maps* (London: Macmillan, 1881).
19. Ondaatje, *The English Patient*, 271.
20. Yeats-Brown, *Martial India*, 31.
21. Ibid., 30.
22. Ibid., 31.
23. Ibid., 52.
24. Ibid.
25. Ibid., 68.
26. Ibid., 52, 29.
27. "Like a gold thread through the history of the Indian army runs the story of the Sappers," ibid., 184.
28. Ibid., 150, 200.
29. Moorehead, *The Desert War*, 191. Ondaatje may have been familiar with this work, as he acknowledges another work by Moorehead in *The English Patient* (*The Villa Diana* [London: Hamish Hamilton, 1951]) and he wrote his novel in the private library of his brother Christopher Ondaatje, which includes Moorhead's work on Sir Richard Burton and the Nile exploration (Moorehead, *The White Nile* [New York: Harper, 1960]).
30. Hingston, *The Tiger Kills*, 7.
31. Elliott, *Roll of Honour*, 246, 266, and 267.
32. Yeats-Brown, *Martial India*, 200, 114.
33. Ondaatje, *The English Patient*, 73, 90, 209. See also page 217, where Kip "speaks of warrior saints" and Hana "feels he is one."
34. Yeats-Brown, *Martial India*, 193.
35. Note "He had always been dutifully in line at the crack of dawn, holding out his cup for the English tea he loved," and "He continues his strictness," Ondaatje, *The English Patient*, 86, 126. Cf. The praise for the discipline of Indian troops throughout *Martial India*.
36. "He [Kip] is the only one of them [in the Villa] who has remained in uniform. Immaculate, buckles shined, the sapper appears out of his tent, his turban symmetrically layered, the boots clean and banging into the wood or stone floors of the house," Ondaatje, *The English Patient*, 74. Compare to Yeats-Brown, *Martial India*, "Every man at this centre wears a clean bush-shirt," 41.
37. Note "He was someone who felt uncomfortable in celebrations, in victories," and "'He dismantled a large bomb, a difficult one. Let him tell you about it.' The sapper shrugged, not modestly, but as if it was too complicated to explain," Ondaatje, *The English Patient*, 112, 107. Cf. Yeats-Brown, *Martial India*, 185. "We'll ask him, but he's shy."
38. Compare Ondaatje, *The English Patient*, 71, with Yeats-Brown, *Martial India*, 85, 197.
39. Ondaatje, *The English Patient*, 270.

40. Elliott, *A Roll of Honour*, 225.
41. Ondaatje, *The English Patient*, 272.
42. Yeats-Brown, *Martial India*, 68. My emphasis.
43. Ondaatje, *The English Patient*, 285.
44. Ibid., 187. My emphasis.
45. Kip faults his brother's abstract idealization of Asia for neglecting certain facts: "Japan is a part of Asia, I say, and the Sikhs have been brutalized by the Japanese in Malaya." Ondaatje, *The English Patient*, 217.
46. Ibid., 111.
47. Ibid., 186, 187, 195.
48. Ibid., 196.
49. Ibid., 189.
50. Ibid., 285.
51. The novel's concluding images conform to its overall pattern of privileging personal over abstract affiliations, affirming the claims of belonging that will come to bind Kip to India, to be familial rather than political. With its final image—which connects Hana and Kip across time and space—the novel insists on the force of personal connection and affinity, not of abstract causes.
52. The extraneous nature of this scene resonates with the inclusion in Ondaatje's *Divisadero* (2007) of a similar scene in which cardsharps and their companions in small town Nevada are shocked by a NPR report of "America bombing a civilian city [Baghdad]" into the realization (in an echo of Kip's accusation in the earlier novel) that no one present at their gathering is innocent of the bombardment (*Divisadero* [Toronto: McClelland & Stewart, 2007], 161–162). In each case the moralizing seems strained—labeled a distraction in the apt summation by the *New York Times* reviewer of *Divisadero* (Erica Wagner, "Picking up the Pieces," *The New York Times*, June 17, 2007).
53. G. W. L. Nicholson, *The Canadians in Italy 1943–1945* (Ottawa: Queen's Printer, 1956) and *Canada's Nursing Sisters* (Toronto: A.M. Hakkert, 1975).
54. Major A. B. Hartley, *Unexploded Bomb: A History of Bomb Disposal* (London: Cassell, 1958).
55. Ondaatje has Kip encounter only German bombs and mines. Cf. Hartley, *Unexploded Bomb*, 175–176. The R. A. F. would not share the details of their bombs outside of a small circle of authorized people, and "even the Bomb Disposal Directorate staff were denied complete information" (176).
56. Bat-Ami Bar On, "Teaching (about) Genocide," in Macdonald and Sánchez-Casal, *Twenty-first-Century Feminist Classrooms*, 235.
57. Chapter 1, this volume.
58. See for instance Kanishka Goonewardena, "*Anil's Ghost*: History/ Politics/ Ideology." Paper presented at Canadian Association of Geographers Annual Conference, McGill University, Montreal, May 29-June 3, 2001.
59. "'Racism is not Intellectual': The Epistemic Significance of Inter-Racial Friendship." Paper presented at Reading Identity: Literature, Pedagogy, and Social Thought, University of Wisconsin-Madison, October 9–11, 2003.
60. Ondaatje, *The English Patient*, 88–89.
61. Ibid., 85.
62. Yeats-Brown, *Martial India*, 84.
63. For instance, Hana jots down in her diary, "[h]e is small, not much taller than I am," adding almost immediately, "a toughness to his nature doesn't show" (Ondaatje, *The English Patient*, 209, emphasis in the original).
64. Yeats-Brown, *Martial India*, 36.
65. Ibid.
66. Ondaatje, *The English Patient*, 72, 127.
67. In his commentary on the 2004 Collector's Series DVD edition of the 1996 film version of his novel.
68. Reena Gill, "The Characterization of Kip in Michael Ondaatje's *The English Patient*" (term paper, Simon Fraser University, Vancouver, Canada, n.d.).

69. Chapter 1 (this volume), 33.
70. Ibid., 33–34.
71. "She would not be ordered again or carry out duties for the greater good," Ondaatje, *The English Patient*, 14.
72. See G. W. L. Nicholson, *Canada's Nursing Sisters* (Toronto: A.M. Hakkert, 1975): 158.
73. Listen to the commentary with Minghella and Ondaatje on the 2004 Collector's Series edition DVD of the film.
74. See chapter 1 in particular the first section, "Redefining Democratic Access to Education," 1–10.
75. Respectively, Mohanty quoted in Macdonald and Sánchez-Casal, and then Sánchez-Casal, "Unleashing the Demons of History," in *Twenty-first-Century Feminist Classrooms*, 65.
76. Mohanty quoted in Macdonald and Sánchez-Casal, "Feminist Reflections on the Pedagogical Relevance of Identity," in *Twenty-first-Century Feminist Classrooms*, 9.

3
REALIST PEDAGOGICAL STRATEGIES

8

TEACHING DISCLOSURE: OVERCOMING THE INVISIBILITY OF WHITENESS IN THE AMERICAN INDIAN STUDIES CLASSROOM

Sean Kicummah Teuton

In a course on identity in American Indian literature, in our discussion of Maria Campell's *Halfbreed*, Kevin, a "white" student, insists that cultural identities are relentlessly bound by racial features, "I can't just say I'm black and that means I'm black," he scoffs. In the fall, I hear my name called from across the campus square; it's Kevin visiting with several other students, all African American men. The moment stays doggedly with me as I consider the successes, mistakes—and missed opportunities—of the course. I never talk with Kevin after that, and he never discloses to me his identity as African American, but looking back on his uncharacteristically savvy comments on race, his interest in black culture, and his frustration with students' views on cultural identity, I see that Kevin very likely is African American. Perhaps, for several reasons surrounding undisclosed identity politics in the classroom and on campus, he never gave full voice to that identity so vital to the subject of the course. Kevin's experience with my course poses a particular challenge to professors who teach in seemingly less-diverse classrooms.[1] In this chapter, I explore the hidden role of suppressed social, racial, and sexual identities, and present the risks, benefits, and ways of enabling such identities to emerge more fully in the classroom.

Before turning to the delicate question of revealing students' hidden social identities, we might first consider the role of identity in the classroom and its broader relationships to social knowledge on college campuses and in the United States. According to a growing number of scholars who have been developing a "realist"[2] approach to social knowledge and concepts such as cultural identity and personal experience, our conclusions that knowledge is unavoidably mediated need not preclude the pursuit of more comprehensive accounts of a world all of us, despite our differing perspectives, inhabit. This condition of viewing the world from our different social locations might explain why students often produce divergent readings of the same passage in a text, and of that text's social meanings in the real world. Though personality or idiosyncrasy no doubt play a role in such interpretations, social identity often mediates how students approach and value the world represented in a work of fiction different from their own experiences of the world. In fact, I imagine this experience of the culturally or socially unfamiliar is what attracts yet frustrates non–American Indian students to American Indian literature, and conversely, what attracts Native students who want to learn more about their tribal backgrounds: "Like most American college students, Native American students are typically not fond of the novels and poems that make up the reading lists of literature courses. Instead, they want both to understand their tribal histories and to develop ways to understand their place in their tribal cultures," writes Robert Warrior.[3] For example, in my course on identity, the class approaches *Halfbreed* from positions that are unavoidably influenced by their identities. While most of the European American students at first find Maria Campbell to be sadly "caught between two worlds," Steven, one of the few American Indian students in the classroom, eventually provides an alternative reading, suggesting that Maria is not trapped, but participates in two cultures, and, because of her ties to her Cree grandmother Cheechum, identifies more strongly with her Indian side. As Roger Dunsmore observes, "It is important to point out...that Indian students are often confronted with intimidating situations in classrooms where they are expected to know and speak publicly about their traditions."[4] This said, in an almost entirely European American class, the understandably reluctant presence of a single non-white-identified student can surprisingly alter a class of students who are otherwise non-reflexive and even self-assured in their readings of minority literatures and cultures.

From a realist perspective, the mediated nature of categories such as identity should indeed be investigated and disclosed. But this approach advances beyond the critique of the constructed quality of identity to consider not only how this often ignored social mediation influences the values we come to hold and the conclusions we make about books and the world, but also the ways some forms of mediation might actually enable different, more productive readings. Such disclosures can enable us both to discover our own locations of reading and to produce better "translations" of American Indian literature, as Greg Sarris explains.[5] Such a position on identity in the classroom, however, is built on a view of knowledge as a socially mediated, collective process in which a diversity of viewpoints contributes to a more comprehensive understanding of literary texts and differently situated people in the world. As Paula Moya puts it: "If we believe that one of the purposes of education is to teach us the truth about the world we live in, we will argue for a postpositivist realist examination of a plurality of perspectives in both curricular offerings and pedagogical strategies. Only by remaining open to the habits of interaction and ways of relating to the world that other cultures offer can any of us fairly evaluate 'our' way of being as one worth preserving and perpetuating."[6] On a smaller scale, the classroom comprises just such a world in which all of us work to come to terms with a text that challenges our known worlds and asks that we expand that world into a broader one. The world is certainly stratified by hierarchies of power, but because hegemony is not random but systemic, relations of power in that world can be located and disclosed. In fact, this project must also take place at the higher institutional levels of education, in the way Michael Apple assesses the relation of "ideology and curriculum." Here, he not only exposes the institutional bias that can misguide curriculum, but, most important, even suggests that this same institutional mediation may reform curriculum:

> It is not the case that a critical perspective is "merely" important for illuminating the stagnation of the curriculum field. What is even more crucial is the fact that means must be found to illuminate the concrete ways in which the curriculum field supports the widespread interests in technical control of human activity, in rationalizing, manipulating, "incorporating," and bureaucratizing individual and collective action, and in eliminating personal style and political diversity.[7]

The disclosure of unexamined interests can also be undertaken in the classroom, so that what once functioned as a suppressed form of ideology and, as such, is further enabled to reproduce, is now externalized, demystified, and accounted for. We make visible the lives of the powerful and the powerless in this first step toward social justice.[8] In investigating the form of the institutions that reproduce privilege, we come to learn what it would take to change them. From this marxist insight,[9] we imagine what kind of social conditions and innovative institutions would best enable human flourishing. In any search for social change, whether we innovate a new model of democracy or discover an alternative family, scholars and community leaders first act to expose the workings of power to control even the very names of their institutions. In my life as a teacher, it is during this exposure that I and my students together learn what kind of classroom community restricts human expression, reflection, and development. But like any creative process, we proceed on a sense that something more is possible. On seeing Kevin again and sensing that I let him down, I was reminded that enabling the expression of minority voices in the classroom was not enough. I would have to rethink a classroom that carefully engages the invisible power of whiteness to silence those like Kevin.

White Invisibility

Like John Dewey, who famously writes of bringing democracy to education,[10] we can ramify this view of social change in society to the classroom, thus building a community from the ground up that will support the robust production of diverse social identities to enable a more complex understanding of minority literature and lives. Of course, like intellectual workers the world over, we first must confront and describe the function of ideology in U.S. culture. While capitalism has always served the United States to maintain wealth among an elite few, out of which grows cultural groups from the Bostonian Brahmans to the most impoverished share croppers and factory workers, historically, the racial category of "white" preserves its power in excluding other groups from attaining its privilege. As we know, racial groups that we today commonly understand to be white, such as Italian Americans, Jewish Americans, or Irish Americans, at one time, were not white. Instead, whiteness had to be bought and learned, and usually by raising one's class status through the generations. Today, so-called white Americans tend to identify first as white, and second as Norwegian or German; while one's ancestors might have declared, "I'm a Swede," one now claims not a noun-based, but an adjectival ethnicity, "I'm Swedish." The invention of whiteness, in fact, shares a surprisingly similar past with the invention of "ethnicity." The history of whiteness is also a history of cultural surrender. To become white then and today one often had to give up one's mother tongue, cultural practices, and religious beliefs. It is this exchange, whether willingly pursued for a higher class and economic opportunity, or reluctantly accepted for an unfettered daily life or for the safety of one's children, that leads often to either the denial of a cultural past or the regret for its loss. In the United States even today, there is a forceful incentive to suppress one's deeper identifications and become a member of the mainstream, a white, straight, middle class, capitalist Christian.

As whiteness reigns in the nation, so it pervades mainstream university campuses and classrooms. For many of us, as for today's students, it is in college, when we arrive from very different nations, regions, religions, and cultures that we begin our assimilation to whiteness. Those students who bear the physical features to allow them entry, and possess the economic status to "wear" whiteness often do so, while those who can look white but do not have the resources to act white often do all they can to obtain them, working a part-time job to afford the consumer lifestyle. In its strangely protean quality, whiteness expands to absorb previously excluded racial groups and yet makes no clear behavioral demands. Whiteness is ever-present in a consumer aesthetic, but nowhere does it explicitly define its origins and beliefs, values and practices, "Whiteness is everywhere in U.S.A. culture, but it is very hard to see.... As the unmarked category against which difference is constructed, whiteness never has to speak its name, never has to acknowledge its role as an organizing principle in social and cultural relations," writes George Lipsitz.[11] White American is thus an inessential identity. Amie Macdonald, who develops a realist pedagogy, explains that while scholars insist minority identities must rely on foundational beliefs, whiteness never does: "Ironically, then, the American racial group 'white' is one of the best examples of inessential conceptions of racial identity, composed as it is of diverse nationalities, ethnicities, races, religions, and skin colors."[12] In the cultural trade of a rich cultural identity for an abstract consumer identity, whiteness offers surprisingly little in exchange for the surrender of one's cultural and social self-conception. Whiteness does not offer a community, a history, or a set of values—it rather offers the very absence of these. Freed from the moral and social demands of culture, whiteness promises radical individualism, and the power to command a self-evident, rational, objective view of U.S. history and society: "By refusing to question the partiality and exclusivity of hegemonic national history," writes Susan Sánchez-Casal, "white students are able to retain an unspoken cultural identity that conflates whiteness with righteousness."[13] Macdonald extends her claims regarding whiteness: "In fact, one could argue that the U.S.A. racial category 'white' is also a political identity, which masks itself as natural and performs a central role in the maintenance of white hegemony."[14] Whiteness thus delivers a kind of empowered invisibility.[15] The category of white American is less a cultural identity than a default category students assume, often regardless of their skin tone, class, or religion. In fact, we are popularly taught that if we have fair skin, we are necessarily white, that to refuse one's whiteness is to deny one's racial privilege in a racist world. In its overwhelming presence, promise of power, and consumer availability, whiteness in the classroom can only be confronted by focusing on what we often trade for its status. Asking self-identifying white students to consider when their families became white and what they gave up to do so is one way to begin interrogating whiteness. To meet this task in my course, I turn to texts such as *Halfbreed*.

Halfbreed (1973) is the autobiography of Maria Campbell, a woman from Canada who, as the title suggests, identifies as Métis or "Halfbreed." Reared in extreme poverty during the mid-twentieth century, Campbell's people, rather unlike other mixed groups in the United States, are a nearly syncretic amalgam of races, cultures, and practices: "The Isbisters, Campbells, and Vandals were our family and were a real mixture of Scottish, French, Cree, English, and Irish. We spoke a language completely different from the others. We were a combination of everything: hunters, trappers, and *ak-ee-top* farmers. Our people bragged that they produced the best and most fearless fighting men—and the best looking women" (25). Perhaps because of their racial category that defies both Cree and white Canadian social and legal definitions, Campbell's Métis people, commonly called "Road Allowance People," have no land base and are literally marginalized as squatters on borrowed government land along the roadside. Like many biologically mixed people, the Métis display a variety of racial features from light to dark shades of skin, hair, and eyes, for example. Campbell herself has unusually dark skin, green eyes, and black curly hair, while one sister, Delores, has red hair and blue eyes and another, Peggie, has brown hair and hazel eyes. Her brothers, Edward and Geordie, have brown curly hair and hazel eyes, while Danny has "black hair as straight as an Indian's" and "as dark as [Maria's], with huge eyes" (93–94). Despite their range of racial features, Campbell's people share the characteristic of a pervasive, deplorable poverty that readers often cannot believe they are encountering in a twentieth-century text. The Campbells travel by wagon, eat trapped and roasted gophers, and suffer the constant threat of the Canadian welfare system that wishes to disband the family.

Representing a range of identity factors such as race, culture, language, history, region, class, and nation, *Halfbreed* challenges students to understand just what holds the Métis together, what keeps them from becoming either white or Indian. Further, because some of Campbell's siblings can pass for white, students ask why they do not become white. During our discussion, some students come to understand how whiteness is a restrictive category, whose gatekeepers refuse entry to those who might corrupt the purity of the "race." Since to many white-identified townspeople, the Métis are the object of scorn, the opposite of upstanding citizens, they by definition cannot become white. In fact, it is largely by contrasting themselves to the hated "halfbreed" that European Canadians maintain their whiteness. Students are often willing to engage this reading of whiteness in Canadian history, as we gently move the discussion to explain whiteness in the United States. In one approach, we review the tacit use of the term "the white man" among most students as a paradigmatic abstraction of Europeans from their colonial history. Throughout, students are asked to distinguish "negative" identities—those defined by what they are not—from positive, more realist ones: those that grow from and describe the world. In our readings, we often focus on such textual instances of cultural interaction between the two groups, as here, when the Métis go to town:

> The day would come when we had enough seneca roots and berries to sell, so we would all get bathed, load the wagons and go. The townspeople would stand on the sidewalks

and hurl insults at us. Some would say, "Halfbreeds are in town, hide your valuables." If we walked into stores the white women and their children would leave and the storekeepers' wives, sons, and daughters would watch that we didn't steal anything. I noticed a change in my parents' and other adults' attitudes. They were happy and proud until we drove into town, then everyone became quiet and looked different. The men walked in front, looking straight ahead, their wives behind, and, I can never forget this, they had their heads down and never looked up. We kids trailed behind with our grannies in much the same manner. (36)

In discussion, mainstream students sometimes consider how physical appearance is not the only mark of entry to whiteness, but middle class wealth and behavior as well. In the passage above, the Métis people appear as twentieth-century peasants, riding horse-drawn wagons to town to sell gathered roots. The European Canadian townspeople assume that the Halfbreeds steal, the adults reproducing their hatred in their children, asking them to surveille the Campbell family when they enter the stores. But perhaps more an identifying feature than race, poverty marks the Métis as other, as not-white. Their roots, their wagons, their clannish entry into the town—their impoverished consumer identity—utterly disallows them whiteness. Ironically, however, Campbell's people do not attempt to assume a more mainstream behavior, to blend, as it were, into the European Canadian town, but instead dare not look white people in the eye, keeping their heads down. Most important, the Métis, like the European Canadians, reproduce this behavior in their children, who "trail behind with [their] Grannies in much the same manner." Focusing on poverty and class identity, in discussion I encourage students to step outside racialized notions of identity, if only for a moment, to consider how being poor asserts its own "glass ceiling," in which even visibly fair-skinned people are told they can and should achieve middle-class whiteness. Often, students identifying as white, but from low-income homes, find themselves beginning to ally with Campbell. Though light skin is certainly a privilege in Canadian and U.S. societies, it is often not enough to grant full access to resources such as safe neighborhoods, higher education, and respectful employment. In discussions of this passage, students from poor backgrounds are often understandably reticent to admit their economic status. To broach this discussion, I ask students to share their experiences of coming to college from either first- or multi-generation college-educated families. In basing the discussion on personal experience, students indirectly discover the economic opportunities or barriers that influenced their own arrivals on campus. Ironically, middle class pressure in the classroom works to silence the democratic right to speak freely about class, as bell hooks argues: "Most students are not comfortable exercising this right—especially if it means they must give voice to thoughts, ideas, feelings that go against the grain, that are unpopular. This censoring process is only one way bourgeois values overdetermine social behavior in the classroom and undermine the democratic exchange of ideas."[16] We discuss what it would mean to identify as one of "the poor" in the United States, and some lower-class students quickly reject the notion of poor as an identity for the reason that poverty, in itself, is only to be eradicated. Others, however, piece together some of the productive values nonetheless honored in a low-income heritage, values that inform their identities today. Returning to the image of the Métis coming to town, students slowly disclose the tendency both to romanticize and demonize poverty in the service of middle-class white society.

In discussions of the autobiography, I am careful to consider the culture of whiteness, and whether the Campbells identify as white or Scottish, Irish, and so on. Ironically, Campbell calls the European Canadian descendents white, suggesting that they are less attached to their European heritage than the Métis people. Campbell and her people explicitly recognize and value their European heritage, incorporating it into their identities, the treatment of the townspeople notwithstanding. She declares that her people, for example, despite their depression and rage for being excluded from the resources of the mainstream, would never side against the Catholic Church: "Our people talked against the government, their white neighbors and each other, but never against the church or the priest regardless of how bad they were" (32). Instead, they attempt to combine Christianity with traditional Cree religious views:

> In our community lived an old, old man called Ha-shoo, meaning Crow. He was a Cree medicine man. Ha-shoo loved to chant and play the drum. When Saint-Denys arrived he asked some young men to go about the settlement and tell people about the church services. When the messenger arrived at Ha-shoo's house, the old man asked, "What do they do?" The boy said, "Oh, Grandfather, they talk and sing." The old man answered, "I'll be there and I'll bring my drum."
>
> So to the service he went. The minister conducted it in Cree with lots of hollering and stamping. Finally he said, "Now we will sing." Old Ha-shoo, who was sitting on the floor, took up his drum and began to chant. The minister yelled, "Ha-shoo, you son-of-a-bitch! Get the hell out of here!" The old man got up and left, and so did the rest of the congregation. (29)

The students usually find the story humorous yet poignant. Though the Evangelical minister goes so far toward cultural interaction as to give his sermon in Cree, he draws the line at Ha-shoo's Cree sermon of drumming and singing. The scene grows sadly comical, however, in knowing that Ha-shoo was willing to enter another religious house, expecting to "talk and sing," even though the religious leaders come from different places. When the people, in solidarity with Ha-shoo, refuse the minister's disrespect for their Métis priest, we are reminded that the Métis people, despite their respect for Catholic beliefs, put their community first. Native students tend to understand the passage in the context of Indian education, in which tribal knowledge has historically been destroyed in European American schools.[17] To prevent the class from polarizing down colonial lines on this issue, I work throughout the semester to integrate the classroom. This often requires just a seemingly playful gesture of "musical chairs" before a discussion, in which students are asked to move their seats randomly across the room. To depersonalize colonial relations, I also initiate debates in which three or more groups of students must present a particular side of an issue. Such organizations not only intervene to disturb entrenched colonial relations between Natives and U.S. citizens but also model alternative political identities for students to "try on." In the classroom, the passage also provides a moment to consider religious identity. To appreciate Campbell's willingness to incorporate non-Native religions into her world, I ask the class whether anyone has investigated or embraced a religion other than that of their parents. Often, students realize they accept (or consciously choose) the religious identity of their parents. In so doing, the students come to realize that not only do we often passively accept a religious identity given to us, but often refuse to consider others. Because many students come from formal religious backgrounds, such as Judaism or Catholicism, which have often been suppressed in their move toward whiteness, this discussion provides an opportunity to consider the suppression of religions in the United States. From such extended discussions, students often call home and return to explain how their grandparents, for example, had a different faith. In rare moments, students themselves imagine what religious practices and their attendant values might have been forgotten within their own families.

Unlike many of us, Campbell does not assume the Christian religion of her mother, though it dominates the region. Instead, she struggles with social practices and meanings that Christian or Indian religions provide, and, like her father and her grandmother, consciously decides to accept a Cree religious identity, less for metaphysical than for epistemological reasons. Campbell finds the religious views of her Cheechum and the Cree people more meaningful than the Catholic religion of her mother, which Campbell feels wrongly judges and shames her for her early marriage: "Her philosophy was much more practical, soothing, and exciting, and in her way I found comfort. She told me not to worry about the Devil, or where God lived, or what would happen after death. She said that regardless of how hard I might pray or how many hours I spent on my knees, I had no choice in what would happen to me or when I would die. She said it was a pure waste of time that could be better used more constructively" (72). Confronting and discussing such literary moments in the contexts of students' lives, we build an understanding of how we inherit identities, on the one hand, and consciously adopt identities, on the other. While we are given racial, historical, and national identities, we can nonetheless interrogate these and also choose social and political identities that, in turn, transform our previous historical identities.

In her story, Campbell's family identifies more closely with their Cree side, partly because the Cree people appear to be a little more accepting than the European side. After considerable discussion, in which we piece out the differences among inherited, imposed, and chosen identities, students begin to understand that our identifications with nations, lands, religions, or the past, as well as with ethnic groups, can be diminished or improved—based on our freedom to interrogate, challenge, and relate to them. From this view, students begin to understand that our challenge in the classroom as well as society is to set the conditions to more openly identify with others and the world. For it is through this engagement with social others—others who might be similar to or different from ourselves—that we evaluate our principles and actions and develop as human beings. In class, one student, Charlene, who views herself as bi-racially African American and Irish American, still insists that Maria, the mixed Cree and white narrator, could identify exclusively with neither racial category, but must accept the indeterminate condition of her racially mixed cultural identity. As students soon discover, however, the category Halfbreed is, to some extent, a default category resulting from limited choices, restrictive identity groups that refuse entry.[18] Surprisingly, Halfbreedness, like whiteness, operates less consciously than other self-examined identities. The Campbells come to be seen not as tragic breeds unavoidably caught between two worlds, but inhabitants of a world of our own making, which, if it is so, can be changed. For example, Justin, a student who is American Indian and visibly European, understands why Maria identified so strongly with her Cree grandmother Cheechum, and less with her more socially distant European grandparents. Native students who are identified as racially European often participate in, and thus understand, two distinct cultures: "The reality of a culture experienced by a student may be a collage of values and perceptions that does not resemble very closely the statements in the literature [about Indian education]. The student's reality does not negate traditional realities of the culture but exists beside or intertwined with these realities," writes Gregory Cajete.[19] Of course, this is not to say that the Métis people desire to become either wholly European or Indian, but, instead, that all of us must work to provide forcibly indeterminate groups with more identity choices. Most important, students must understand that Campbell's people do not necessarily desire, above all, to be accepted as European Canadian, but rather expect the same opportunities available to European Canadians. Given the opportunity to retain their culture and traditions—all the while reserving the chance to attain middle class privileges—many students discover other routes to economic security than pursuing whiteness. But, as students figure out, this opportunity is often blocked by a social world that insists racial features and class appearance are the neutral, inherent, and immutable markers of identity.

Reluctant Passing

Each fall, I lead an introductory lecture in American Indian literature. Though most all of the three hundred or more students appear to be white, a few of these students invariably come to my office to talk about their confused feelings regarding their cultural identities. The student will often confess to me that she or he has been told by a parent that the family has "Indian blood."[20] Because the students are not enrolled tribal citizens, almost never know the tribe to which they belong, or the land from which the tribe comes, it's hard to tell whether they have Native ancestry, or merely speak of an erroneous family legend. These are individualist pronouncements that Western education encourages, and rather contrary to the way Native people are identified in terms of a community who claims them, as Vine Deloria explains: "The final ingredient of traditional tribal education is that accomplishments are regarded as the accomplishments of the family and not the individual."[21] After thinking over this phenomenon for a few years, I have made a few conclusions. First, the rise of whiteness is only part of the story of race in the United States. While whiteness grants invisibility, many sense that it somehow comes at a cultural cost. However, rather than embrace one's European heritage, for example, U.S. citizens who otherwise identify as white often desire the indigenousness promised in a trace of Native heritage. Indeed, for many ethnic groups, the prevalent myth of American Indian heritage offers one the chance to "belong" to the land. The process of "going native," however, bears a long history in the colonial imagination, and is utterly natural, considering the human attraction of people to lands. Second, U.S. citizens are so thoroughly race conscious that not only skin tone, but, indeed, blood itself is held to be the primary door to either close or open one's access to the world. In such a race-based society, students feel they should ignore race to prevent racism, on the one hand, and embrace their own racial blood to promote culture, on the other. Their eyes glazed with longing for an Indian racial past, the above students rarely realize that, as with any dominated identity, being Indian is not merely a matter of embracing one's distant heritage, but of difficult cultural work. Such work invites fascinating classroom exercises; for example, I ask students, when home on semester break, to research the ancestral title to their neighborhood lands. On their returns, students report on their family's land tenure, as sometimes documented in treaties and homesteads, or recorded in storefronts, sharecropping, migratory farming, or plantations. The fact that white-identified students with possible Native heritage come to my office as if to receive my private acceptance underscores my third and final point: many students feel they don't have a *right* to exchange their assumed identities for others, whether based on blood descent or cultural heritage.

This is because, in U.S. popular culture, those who don't appear, but identify as, Native are often ridiculed as members of the "wannabe" tribe. Whether one is European or African in physical appearance, to call oneself "Native American" is, to many U.S. citizens, a romantic attempt to escape one's obvious and "true" racial identity. Those with European features and skin tone who identify as American Indian are especially dismissed, for these persons, it is believed, wish to have it both ways, the social privilege of whiteness and the cultural roots of Indianness. From this view, those Indians who are also middle class, heterosexual, and male possess the most privilege of all. Race is certainly a factor in defining Indianness. But less like U.S. citizens, for whom race is central to identity, tribal people tend to determine Indian identity through kinship and nationhood. Through heritage, enrollment, and adoption, tribal nations claim their citizens, and this, the affiliation with a tribal community, is the primary source of Indian identity for Native people. In my classroom, I foreground white racial privilege by encouraging discussions about how certain physical appearances grant social and economic access to the world, but also how the term wannabe (which is actually in the dictionary) might represent a deeper longing that the class should take seriously. We begin by considering when Campbell, as a child, desired her hair straightened, with her great-grandmother declaring, "Just wait, my girl, your Cheechum will make your hair straight yet" (49). Having internalized the colonial image of the Indian, Campbell works to make her hair long, straight, and dark, as she believes authentic Native women's hair appears. Students, especially non-white students, understand this desire on entering the university, when, for example, Indian students either cut their long hair or grow it out, attempting to meet either the traditional or progressive image of the American Indian. In discussing what it means to "want to be" other than one is often made to be, students come to understand the restrictions placed on all of us in a socially managed world.

From such discussions, students discover that the restricted ability to choose one's identity limits not only one's group membership, but also one's personal, moral, cultural, and political growth. The class soon concludes that all of us either passively accept the identities imposed on us or actively select and develop those identities that better suit our vision of ourselves and our experiences in the world. To illustrate the challenge, we consider less determinate and more malleable forms of chosen identity, in political affiliations with progressive causes such as feminism, to which male students are often surprised to learn they can actually belong, and, in our discussion of *Halfbreed*, the cause of the Métis people, led by Louis Riel, to establish a nation in Canada in 1869. Perhaps more than any other identity marker, political identity shapes Maria Campbell's narrative. She begins her first chapter thus: "In the 1860's [*sic*] Saskatchewan was part of what was then called the Northwest Territories and was a land free of towns, barbed-wire fences and farm-houses. The Halfbreeds came here from Ontario and Manitoba to escape the prejudice and hate that comes with the opening of a new land" (9). The Red River Rebellion of 1869 was, of course, a failure, and their leader Riel was hanged for treason by the Canadian government in 1885. Such historical background to a First Nation culture in Canada, two borders removed from the United States, gives students enough distance from their own nation to begin to examine national identity, patriotism, citizenship, and belonging, and U.S. colonial privilege, similar to the way Michico Hase critiques Japanese nationalism to begin discussions of U.S. nationalism from a safe distance.[22] In the passage above, U.S. students, recalling the mythologized national narrative, identify with the Métis in their flight west to escape religious prejudice and work the soil in a seemingly uninhabited land, thus interrogating our national "process of collective remembrance," as Roger Simon calls it.[23] The class soon learns that, depending on one's perspective, the Métis people are either merely rebels or refugees with a right to a homeland and a life free of persecution. In such debates, students learn that national identity and its attendant political claims regarding the right to lands and self-governance must be defended on ethical grounds, and that citizens must be accountable for the history and actions of their nation in the world.

Their brief national hopes crushed, Campbell's people now and again experience a revival of cultural pride, as when Campbell's father begins to attend the CCF party meetings to demand rights for the Métis. Maria Campbell feels a growing sense of hope, and among all citizens, Métis and white alike: "For the first time I saw whites inviting Halfbreeds into their homes" (66). In such passages, students discover that a common cause to overcome poverty and to protect the rights of all citizens need not divide along racial lines; though gender and racial difference should be recognized, political identities intersect across such differences. Soon after, however, the Canadian government infiltrates the meetings, divides the white and Indian members, and crushes the "rebellion" again. Campbell recalls the return of despair to her family; when remembering her father's depression, she asks, "Have you ever watched a man die inside?" (68). The phrase is often repeated in the classroom as students work to understand the growing sense of hope in the blossoming of an alternative, enlightening worldview, and its new identity. Campbell's father still looks the same on the outside, but his humanity has been trampled, and is thus made to accept the politically neutralized identity formerly ascribed to him by the dominant Canadian government.

So not to simplify the issue of choosing an identity, students consider the great complexity regarding the comparative power not to accept but to choose an identity. Broaching this issue, Kelly, who shyly sits at the back wall of the room, explains to the class that she is Winnebago. With her fair skin, blond hair, and blue eyes, students dismiss her, rolling their eyes. Janet, an African American student, chuckles at the contention that Kelly is an American Indian.

I feel I must intervene, and remind the class that we view our lives in extraordinarily complex ways that defy a simple logic of skin tone. This said, the class also understands the racial complexity of Janet's response; white skin privilege allows Kelly to embrace and announce her Indian blood in a way that Janet cannot. Though all of us have, from the first day, introduced ourselves and our home towns, our cultural backgrounds, and so on, I am committed to disclosing my own national, cultural, gender, and racial identifications throughout my teaching so that students, by example, come to account for their identities and imagine new ones. In our discussion of freedom to choose identities, I feel I must further elaborate my own. I explain that it is perhaps universal that all of us, despite our particular backgrounds, undergo a process of identity development, as we transform from children to adults, for example. Since my parents divorced while I was still quite young, I was reared less by my father, who is European American, than by my mother, who is a dark-skinned Cherokee woman. Having been born in a low-income neighborhood, in Compton, California, and having grown up in a mixed Latina/o community, I never really imagined whiteness until I went to college. There, at a largely white institution, with my light skin, I became marked as white. Though I explained that I was a Cherokee Indian, European American students never really took me seriously as Native and instead attempted to claim me as white. Awed by the apparent privilege of whiteness, I attempted to be accepted among my white-identified classmates. In the end, however, I could not overcome the economic, and hence social, barriers. Despite my "secret" part time job in the dormitory cafeteria, my revoked credit card to purchase preppy clothes, and my mother's disapproval, I could not afford whiteness.

As I explain to students, in mainstream culture, no matter what I call myself, my light skin, brown hair and eyes, and middle class behavior often lead people to designate me as white. Despite how I view myself, as the son of a strongly Cherokee woman, from a traditionally matrilineal culture in which non-Cherokee men marry into Cherokee culture and become Cherokees, white culture attempts to claim me, while Indian people tend to perceive the cues that identify me as Indian. So living with this complicated racial identity has trained me to practice a kind of double consciousness, in which I am always aware of the disjuncture between how I see myself and how others see me, when, for example, European American men befriend me at a Native gathering, assuming I'm not a participant, but a fellow observer. This racial contradiction grows when one considers that my tribal enrollment is regulated by race through the U.S. Department of the Interior, which lists my "certified degree of Indian blood," in deviation from a supposed pristine age of Cherokee racial purity. Such explicitly racist systems of blood purity to reduce the Native population and U.S. commitments[24] have only been used in the American South to enslave Africans, in Nazi Germany to control Jews, and in Apartheid South Africa to segregate indigenous people. In my disclosure, I end by discussing that race operates so centrally in U.S. society that non-Native citizens assume the right to determine the racial authenticity of Indian people by physical features. Such race-based evaluations often surprise Native people, however, who identify primarily not as ethnic minorities, but as citizens of tribal nations. When Indian students explain themselves not so much as members of racial groups but rather Native nations, with legal, treaty-based relationships with the United States, white-identified students often grow frustrated with a group who cannot be easily racialized and absorbed into the white middle class, while minority students tend to marginalize Indian students from antiracist movements, even though anticolonial struggles work hand-in-hand.

Similar to my experience of being "misidentified" as white, despite how Campbell understands her own cultural identity, others do not know how to place her, viewing her as African American, Aboriginal, or simply other. For this discussion of disjuncture between how we view ourselves and how others view us, I work to bring complexity to what we commonly call racial "passing." Michael Hames-García's revised concepts of "multiplicity" and "restriction" provide a model for complex group membership. He carefully explains the challenge presented to those who belong to a cultural group that often marginalizes those who also belong to another social group. He discusses the concern in Michael Nava's novel, *The Hidden Law*, of being a gay Chicano protagonist and bearing strong allegiances to a Chicano community even though that community is reluctant to include him. Of course, many people from a variety of groups experience a similar tension: "Black women, gay Chicanos, and Asian American lesbians are examples of people who have memberships in multiple politically subordinated groups in the United States."[25] Hames-García advances a crucial point for understanding the challenge of multiple group membership, that it will not do to explain social-cultural identity as a contributive formulation, as a mathematical equation: "[O]ne cannot understand a self as the sum of so many discrete parts, that is, femaleness + blackness + motherhood."[26] He argues that to do so would be to essentialize one aspect of one's identity to stabilize the construction of another. Imagine, for example, that an American Indian woman recovers from a marriage in which she was battered, and then becomes a feminist. It would be wrong to suppose she developed a "new" feminist identity as an addition to a genderless American Indian cultural identity; conversely, she decides to become a feminist not from a pre-racialized cultural position, but as a Native woman, a cultural location that likely has something to do with the radicalization of her femaleness. According to Hames-García: "The whole self is constituted by the mutual interaction and relation of its parts to one another. Politically salient aspects of the self, such as race, ethnicity, sexuality, gender, and class, link and imbricate themselves in fundamental ways. These various categories of social identity do not, therefore, comprise essentially separate 'axes' that occasionally 'intersect.' They do not simply intersect but blend, constantly and differently...."[27] Hames-García's redefinition of multiplicity helps explain the political transformation of the woman I mention above, avoiding simplistic or reified notions of how people participate in a variety of communities, changing and developing in fact as a consequence of complex allegiances. As Hames-García explains, social location is not immobile, but moves and resituates, the self often reconstituting in response to one's surroundings or moment in life.

Imbalances in social power, however, affect the potential benefit of this kind of fluidity of the self, through what Hames-García calls "restriction." This is a useful term for a situation in which many of us often find ourselves, a modern problem in a world that often seeks to reduce the complexity of social representation and recognition, to simplify the challenge of knowing social others, but also to serve political interests. Restriction occurs when one is viewed in terms of only one social identity at the expense of all others: "According to the fracturing logic of domination inhering in capitalist cultures, this multiplicity of the self becomes restricted so that any one person's 'identity' is reduced to and understood exclusively in terms of that aspect of her or his self with the most political salience."[28] A society often decides for us exactly which aspect of our complex self is to be recognized as the defining feature of our social existence. When one's own view of oneself is in agreement with that of the dominant cultural construction, one's interests are "transparent," explains Hames-García. But for those who do not assimilate or conform to the dominant white American construction of the citizen, this restriction is oppressive and reductive, limiting the possibility for fruitful social expression. According to Hames-García, those who experience this restriction of their social multiplicity have suppressed, "opaque" concerns.[29] Clearly, the challenge facing social theorists in Indian Country is to create conditions that allow a robust multiplicity of social identities to emerge. Hames-García explains that those who have transparent interests occupy a privileged position in a cultural or social group in which their desires for social expression are not questioned and thus take for granted their ability to be fully expressive. But in an unequal world, this feeling of transparency is often an illusion produced by the very dominant social network the transparent support, a false transparency in which desires for multiplicity can be unwittingly suppressed. In this regard, always recognizing and overcoming opacity is a vital process of a just society watchful of blockages to social expression and invention. Hames-García advocates a realist theory of social selfhood that allows for a whole yet multiple self without fragmenting the self so that one is always partly outside at least one social community. This view of multiplicity, he argues, best serves anticolonial and antiracist struggles because it understands the necessity of including particular interests in order to always reconsider and revise the goals for which we, all of us, collectively work.

Enabling Disclosure

One day early in the semester, James, a student with European features, shakes his head during a discussion on white racial dominance, and tells the class, "But I'm not white, I'm German and Irish." Every now and then, a European American student will not identify as white, but often only to evade discussing how we are racialized in the United States to dominate or empower specific groups. In my courses, however, James's refusal to recognize his own racialization as white by other people identifying as white presents an early opportunity to introduce issues for the days ahead, such as the difference between race and racialization, imposed identities and chosen identities, and, as I have been presenting here, white skin and white dominance. In this last instance, students are challenged to make the philosophical discovery that race is indeed an invention—but one that affects our very lives. Though we can be racialized within a dominant category, whiteness is often a matter of self-racialization:

> Race is a cultural construct, but one with sinister structural causes and consequences. Conscious and deliberate actions have institutionalized group identity in the United States, not just through the dissemination of cultural stories, but also through systematic efforts from colonial times to the present to create economic advantages through a possessive investment in whiteness for European Americans.[30]

This is a critical moment in the classroom; explaining the privilege of European features, such as light skin tone, to James and other students who share his resistance requires careful discussion so that white identified students don't become defensive and shut down the opportunity for self-reflection. If I were, for example, to demand that James, before his gazing classmates, deal then and there with his white dominance, he would only grow silent and complain later that I crushed dissent in the democratic classroom. Instead, as I have been explaining, European American students who identify as white must be given an alternative to whiteness if they are to renounce whiteness. One alternative is to provide students who identify with white dominance a political identity such as "antiracist." bell hooks recalls this transformation among white Americans in the apartheid South as a profound moral choice: "No one is born a racist. Everyone makes a choice. Many of us made the choice in childhood. A white child taught that hurting others is wrong, who then witnesses racial assaults on black people, who questions that and then is told by adults that hurting is acceptable because of their skin color, then makes a moral choice to collude or to oppose."[31] We might list the steps toward enabling the disclosure and dismantling of whiteness:

1. Create an environment of trust and respect so that students can disclose their suppressed identities.
2. Encourage students to identify with characters in literary texts as models through which to share experiences.
3. Situate the power of race in relation to "chosen" identities such as religions or political movements.
4. Provide students with alternative cultural identities that enable richer views of the world.

One way to set the context for disclosing whiteness is to provide alternatives in American Indian literature, such as in *Halfbreed*, where students discover other models for identity. Most important, white-identified students must be shown that some identities can be consciously chosen, such as religious or political identity. From this understanding, they realize that whiteness can be disclosed as a dominant, self-perpetuating category that many of us continue to benefit from, and discarded for more enriching cultural identities. Of course, the task of having students with dominant identities recognize and surrender their power can only be appealed on ethical grounds that students should, but often do not, accept. Paulo Freire famously describes the challenge: "Discovering himself to be an oppressor may cause considerable anguish, but it does not necessarily lead to solidarity with the oppressed. Rationalizing his guilt through paternalistic treatment of the oppressed, all the while holding them fast in a position of dependence, will not do. Solidarity requires that one enter into the situation of those with whom one is solidary; it is a radical posture."[32] In my view as an American Indian teacher, this is our greatest challenge, to nourish empathy in students from dominant locations so that they recognize and hand over their white dominance to pursue just relations. As Maria Campbell puts it, "I believe that one day, very soon, people will set aside their differences and come together as one. Maybe not because we love one another, but because we will need each other to survive" (156–157).

Notes

1. At my institution, the University of Wisconsin-Madison, non-European American enrollments are quite low. In fall 2003: men: 19,876 or 47.8%; women: 21,712 or 52.2%; ethnic minority students: 4,108 or 9.9%; African American: 994 or 2.3%; Asian American: 1,838 or 4.4%; Native American: 230 or 0.6%; Hispanic: 1,046 or 2.5%; from Wisconsin: 25,974 or 62.5%; from other U.S. states: 12,077 or 29.0%; international students: 3,571 or 8.6%; U.S. states represented: 50; countries represented: 120 ("UW Student Profile." The Board of Regents of the University of Wisconsin System. http://www.uc.wisc.edu/profile/quickfacts.php?file=qfstudents. January 20, 2004). In the state of Wisconsin, the median income for a four-person family in 2002 is $68,000, above the national average of $64,169, though Wisconsin workers, especially women, tend to work longer hours ("The State of Working Wisconsin, 2004." Center on Wisconsin Strategy. www.cows.org).
2. For scholars developing realist approaches to knowledge, identity, and experience, see Paula M. L. Moya and Michael R. Hames-García, eds., *Reclaiming Identity: Realist Theory and the Predicament of Postmodernism* (Berkeley: University of California Press, 2000). For realist pedagogies, see Amie A. Macdonald and Susan Sánchez-Casal, eds., *Twenty-first-Century Feminist Classrooms: Pedagogies of Identity and Difference* (New York: Palgrave Macmillan, 2002).
3. Robert Allen Warrior, "Literature and Students in the Emergence of Native American Studies," in *Studying Native America: Problems and Prospects*, ed. Russell Thornton (Madison: University of Wisconsin Press, 1998), 111–129, 117.
4. Roger Dunsmore, "Introduction," in *Earth's Mind: Essays in Native Literature* (Albuquerque: University of New Mexico Press, 1997), 1–13, 2.
5. Greg Sarris, "Storytelling in the Classroom: Crossing Vexed Chasms," in *Keeping Slug Woman Alive: A Holistic Approach to American Indian Texts* (Berkeley: University of California Press, 1993), 149–168.
6. Paula M. L. Moya, "Learning How to Learn from Others: Realist Proposals for Multicultural Education," in *Learning from Experience: Minority Identities, Multicultural Struggles* (Berkeley: University of California Press, 2002), 136–174, 174.
7. Michael W. Apple, *Ideology and Curriculum*, 1979 (New York: Routledge, 1990), 128.
8. On subordinated groups naming the powerful to denaturalize and resist them, see Vaclav Havel, "The Power of the Powerless," in *Citizens Against the State in Central Eastern Europe*, ed. John Keane (Armonk, NY: M. E. Sharpe, 1985).
9. Here, I am thinking of Marx's "practical" responses to Feuerbach, especially Theses II and XI. See Karl Marx, "Theses on Feuerbach," in *The Marx-Engels Reader*, ed. Robert C. Tucker (New York: Norton, 1978), 143–145.
10. On the relationship of education to democracy, Dewey writes: "The devotion of democracy to education is a familiar fact....A democracy is more than a form of government; it is primarily a mode of associated living, of conjoint communicated experience. The extension in space of the number of individuals who participate in an interest so that each has to refer his own action to that of others, and to consider the action of others to give point and direction to his own, is equivalent to the breaking down of those barriers of class, race, and national territory which kept men from perceiving the full import of their activity" (*Democracy and Education: An Introduction to the Philosophy of Education*, 1916 [New York: Dover, 2004], 83). Sonia Nieto insists that education include an account of the failings and "contradictory dimensions" of democracy ("From Brown Heroes and Holidays to Assimilationist Agendas: Reconsidering the Critiques of Multicultural Education," in *Multicultural Education, Critical Pedagogy, and the Politics of Difference*, eds. Christine E. Sleeter and Peter L. McLaren [Albany: State University of New York Press, 1995], 191–220, 207).
11. George Lipsitz, *The Possessive Investment in Whiteness: How White People Profit from Identity Politics* (Philadelphia: Temple University Press, 1998), 1.
12. Amie A. Macdonald, "Racial Authenticity and White Separatism: The Future of Racial Program Houses on College Campuses," in *Reclaiming Identity: Realist Theory and the*

Predicament of Postmodernism, eds. Paula M. L. Moya and Michael R. Hames-García (Berkeley: University of California Press, 2000), 205–225, 218–219.
13. Susan Sánchez-Casal, "Unleashing the Demons of History: White Resistance in the U.S.A. Latino Studies Classroom," in *Twenty-first-Century Feminist Classrooms: Pedagogies of Identity and Difference,* eds. Amie A. Macdonald and Susan Sáchez-Casal (New York: Palgrave Macmillan, 2002), 59–85, 63.
14. Macdonald, "Racial Authenticity and White Separatism," 219 n. 28.
15. Here I distinguish invisibility as a protected privilege from invisibility as an imposed form of dehumanization, such as explored in Ralph Ellison, *Invisible Man.* I had been describing the invisibility of whiteness before I discovered the essay on white invisibility by Monica Beatriz Demello Patterson, "America's Racial Unconscious: The Invisibility of Whiteness," in *White Reign: Deploying Whiteness in America,* eds. Joe L. Kincheloe, Shirley R. Steinberg, Nelson M. Rodriguez, and Ronald E. Chennault (New York: St. Martin's, 1991), 103–121.
16. bell hooks, "Confronting Class in the Classroom," *Teaching to Transgress: Education as the Practice of Freedom* (New York: Routledge, 1994), 177–189, 179.
17. On the destructive U.S. education policies, see Jorge Noriega, "American Indian Education in the United State: Indoctrination for Subordination to Colonialism," in *The State of Native America: Genocide, Colonization, and Resistance,* ed. M. Annette Jaimes (Boston: South End, 1992), 371–402; and Donald Fixico, "Indian Minds and White Teachers," in *The American Indian Mind in a Linear World* (New York: Routledge, 2003), 83–103.
18. Since Campbell freely uses the term "Halfbreed" (with a capital "H"), students assume they too can use the term to describe Campbell's people. Of course, the term bears a racist history, and students learn this history during discussion, and especially how group members use terms for their own members that other non-members should not use. I encourage the class to use the term "Métis" instead. Formerly "Halfbreeds," the Métis Nation is now recognized in the Canadian Constitution.
19. Gregory A. Cajete, "The Native American Learner and Bicultural Science Education," in *Next Steps: Research and Practice to Advance Indian Education,* eds. Karen Gayton Swisher and John W. Tippeconnic III (Charleston, WV: ERIC, 1999), 135–160, 152.
20. On European Americans desiring a legacy of Indian blood, see Eva Garroutte, *Real Indians: Identity and the Survival of Native America* (Berkeley: University of California Press, 2003).
21. Vine Deloria, Jr., "Knowing and Understanding: Traditional Education in the Modern World," in *Spirit and Reason: The Vine Deloria, Jr., Reader* (Golden, CO: Fulcrum, 1999), 137–143, 141.
22. Michiko Hase, "Student Resistance and Nationalism in the Classroom: Reflections on Globalizing the Curriculum," in *Twenty-first-Century Feminist Classrooms: Pedagogies of Identity and Difference,* eds. Amie A. Macdonald and Susan Sáchez-Casal (New York: Palgrave Macmillan, 2002), 87–107.
23. Roger I. Simon, "Forms of Insurgency in the Production of Popular Memories: The Columbus Quincentenary and the Pedagogy of Counter-Commemoration," in *Between Borders: Pedagogy and the Politics of Cultural Studies,* eds. Henry A. Giroux and Peter McLaren, (New York: Routledge, 1994), 127–142, 130.
24. On the U.S. policy to regulate Native identity through race and blood, see M. Annette Jaimes, "Federal Indian Identification Policy: A Usurpation of Indigenous Sovereignty in North America," in *The State of Native America: Genocide, Colonization, and Resistance,* (Boston: South End, 1992), 123–138.
25. Michael R. Hames-García, "'Who Are Our Own People?': Challenges for a Theory of Social Identity," in *Reclaiming Identity: Realist Theory and the Predicament of Postmodernism,* eds. Paula M. L. Moya and Michael R. Hames-García (Berkeley: University of California Press, 2000), 102–129, 104.
26. Ibid., 103.
27. Ibid., 103.
28. Ibid.,104.

29. Michael Hames-García adapts his concepts of transparency and opacity from María C. Lugones's work on "transparent" versus "thick" group membership; see "Purity, Impurity, and Separation," *Signs*, 19.21 (1994): 458–479, 474.
30. Lipsitz, *Possessive Investment in Whiteness*, 2.
31. bell hooks, *Teaching Community: A Pedagogy of Hope* (New York: Routledge, 2003), 53.
32. Paulo Freire, *Pedagogy of the Oppressed*, 1968, Trans. Myra Bergman Ramos (New York: Continuum, 1974).

References

Apple, Michael W. *Ideology and Curriculum*. 1979. New York: Routledge, 1990.
Cajete, Gregory A. "The Native American Learner and Bicultural Science Education." *Next Steps: Research and Practice to Advance Indian Education*. Eds. Karen Gayton Swisher and John W. Tippeconnic III. Charleston, WV: ERIC, 1999. 135–160.
Campbell, Maria. *Halfbreed*. Lincoln: University of Nebraska Press, 1973.
Deloria, Vine, Jr. "Knowing and Understanding: Traditional Education in the Modern World." *Spirit and Reason: The Vine Deloria, Jr., Reader*. Golden, Co: Fulcrum, 1999. 137–143.
Dewey, John. *Democracy and Education: An Introduction to the Philosophy of Education*. 1916. New York: Dover, 2004.
Dunsmore, Roger. Introduction. *Earth's Mind: Essays in Native Literature*. Albuquerque: University of New Mexico Press, 1997. 1–13.
Fixico, Donald L. "Indian Minds and White Teachers." *The American Indian Mind in a Linear World*. New York: Routledge, 2003. 83–103.
Freire, Paulo. *Pedagogy of the Oppressed*. 1968. Trans. Myra Bergman Ramos. New York: Continuum, 1974.
Garroutte, Eva Marie. *Real Indians: Identity and the Survival of Native America*. Berkeley: University of California Press, 2003.
Hames-García, Michael R. "'Who Are Our Own People?': Challenges for a Theory of Social Identity." *Reclaiming Identity: Realist Theory and the Predicament of Postmodernism*. Eds. Paula M. L. Moya and Michael R. Hames-García. Berkeley: University of California Press, 2000. 102–129.
Hase, Michiko. "Student Resistance and Nationalism in the Classroom: Reflections on Globalizing the Curriculum." *Twenty-first-Century Feminist Classrooms: Pedagogies of Identity and Difference*. Eds. Amie A. Macdonald and Susan Sáchez-Casal. New York: Palgrave Macmillan, 2002. 87–107.
Havel, Vaclav. "The Power of the Powerless." *Citizens Against the State in Central Eastern Europe*. Ed. John Keane. Armonk, NY: M. E. Sharpe, 1985.
hooks, bell. *Teaching Community: A Pedagogy of Hope*. New York: Routledge, 2003.
———. "Confronting Class in the Classroom." *Teaching to Transgress: Education as the Practice of Freedom*. New York: Routledge, 1994. 177–189.
Jaimes, M. Annette. "Federal Indian Identification Policy: A Usurpation of Indigenous Sovereignty in North America." *The State of Native America: Genocide, Colonization, and Resistance*. Ed. M. Annette Jaimes. Boston: South End, 1992. 123–138.
Lipsitz, George. *The Possessive Investment in Whiteness: How White People Profit from Identity Politics*. Philadelphia: Temple University Press, 1998.
Lugones, María C. "Purity, Impurity, and Separation." *Signs* 19.21 (1994): 458–479.
Macdonald, Amie A., and Susan Sánchez-Casal, eds. *Twenty-first-Century Feminist Classrooms: Pedagogies of Identity and Difference*. New York: Palgrave Macmillan, 2002.
Macdonald, Amie A. "Racial Authenticity and White Separatism: The Future of Racial Program Houses on College Campuses." *Reclaiming Identity: Realist Theory and the Predicament of Postmodernism*. Eds. Paula M. L. Moya and Michael R. Hames-García. Berkeley: University of California Press, 2000. 205–225.
Marx, Karl. "Theses on Feuerbach." *The Marx-Engels Reader*. Ed. Robert C. Tucker. New York: Norton, 1978. 143–145.

Moya, Paula M. L, and Michael R. Hames-García, eds. *Reclaiming Identity: Realist Theory and the Predicament of Postmodernism*. Berkeley: University of California Press, 2000.

Moya, Paula M. L. "Learning How to Learn from Others: Realist Proposals for Multicultural Education." *Learning from Experience: Minority Identities, Multicultural Struggles*. Berkeley: University of California Press, 2002. 136–174.

Nieto, Sonia. "From Brown Heroes and Holidays to Assimilationist Agendas: Reconsidering the Critiques of Multicultural Education." *Multicultural Education, Critical Pedagogy, and the Politics of Difference*. Eds. Christine E. Sleeter and Peter L. McLaren. Albany, NY: State University of New York Press, 1995. 191–220.

Noriega, Jorge. "American Indian Education in the United State: Indoctrination for Subordination to Colonialism." *The State of Native America: Genocide, Colonization, and Resistance*. Ed. M. Annette Jaimes. Boston: South End, 1992. 371–402.

Patterson, Monica Beatriz Demello. "America's Racial Unconscious: The Invisibility of Whiteness." *White Reign: Deploying Whiteness in America*. Eds. Joe L. Kincheloe, Shirley R. Steinberg, Nelson M. Rodriguez, Ronald E. Chennault. New York: St. Martin's, 1991. 103–121.

Sánchez-Casal, Susan. "Unleashing the Demons of History: White Resistance in the U.S.A. Latino Studies Classroom." *Twenty-first-Century Feminist Classrooms: Pedagogies of Identity and Difference*. Eds. Amie A. Macdonald and Susan Sáchez-Casal. New York: Palgrave Macmillan, 2002. 59–85.

Sarris, Greg. "Storytelling in the Classroom: Crossing Vexed Chasms." *Keeping Slug Woman Alive: A Holistic Approach to American Indian Texts*. Berkeley: University of California Press, 1993. 149–168.

Simon, Roger I. "Forms of Insurgency in the Production of Popular Memories: The Columbus Quincentenary and the Pedagogy of Counter-Commemoration." *Between Borders: Pedagogy and the Politics of Cultural Studies*. Ed. Henry A. Giroux and Peter McLaren. New York: Routledge, 1994. 127–142.

The State of Working Wisconsin, 2004. Center on Wisconsin Strategy (COWS). www.cows.org.

UW Student Profile. The Board of Regents of the University of Wisconsin System. http://www.uc.wisc.edu/profile/quickfacts.php?file=qfstudents. January 20, 2004.

Warrior, Robert Allen. "Literature and Students in the Emergence of Native American Studies." *Studying Native America: Problems and Prospects*. Ed. Russell Thornton. Madison: University of Wisconsin Press, 1998. 111–129.

9

Religious Identities and *Communities of Meaning* in the Realist Classroom

William S. Wilkerson

Most of us on the academic left remain agnostic or atheist, despite the existence of a progressive Christian tradition, the oft-noted importance of the black churches in the civil rights movement, and the appealing ideals of liberation theology. There may be a variety of reasons for this: the continued influence of the Enlightenment tradition, the general trend to secularism encouraged by the advance of Capital, the conservative role that the church often plays in places like the United States. Certainly, many out queers (such as myself) have had their fill of conservative and evangelical forms of Christianity and their peculiar homohatred.

When it comes to teaching, however, such religious skepticism produces a dilemma. On the one hand, academic leftists may find it necessary to question, if not outright criticize, certain conservative and dogmatic aspects of students' faith. On the other hand, progressive ideals of education entail critical pedagogy and student-centered teaching, in which the classroom is democratized, the authority of the instructor diminished, and fallibilism and open-ended inquiry stressed. By encouraging us to reach students where they are, to engage their beliefs with respect, and to assume that the instructor may be wrong about important matters, these pedagogical strategies prevent the instructor from strong criticism of religious ideals in the classroom. So what do we do? Do we criticize religion, support their beliefs, or try to find some method for doing both? How can we respect religious identity and yet raise the questions that need to be raised?

Teaching where I do, at a mid-sized public university in the Deep South, these issues confront me daily in my introductory level courses. Informal surveys of my students reveal that a majority of them strongly identify as Christian, typically conservative Baptist, and an even larger number maintain some commitment to a supreme being resembling the Judeo-Christian God. In the cracks of this fairly monolithic culture float a few practitioners of other faiths, often international students who practice Hinduism, Islam, and also a small handful of typically outspoken but outnumbered skeptics and atheists.

Many if not most of my Christian students are biblical literalists; they believe wholeheartedly in creationism and find evolution laughable. In terms of ethics, they typically endorse a fairly unreflective version of divine command theory: particular actions are moral because God commands them. Their Christian religious beliefs do not produce neat political and social lines, however. Most of my students (all races) would claim that racism is wrong based on biblical tenets (which does not preclude them from being racists, of course), and most of them (all races) claim fairly conservative sexual values, rejecting homosexuality and premarital sex and embracing "family values" (although practice and preaching here diverge, not surprisingly). However, I have noted that my white Christian students lean heavily towards conservative political agendas and the Republican party, while my non-white students (mostly black, with a small number of Latinos/as) lean in the other direction.

I want to discuss the challenges and paradoxes that teaching this particular group of students raises with respect to two issues. First, how can we teach ethics to students in such a way that *critically* engages their religious commitments and identities, be they Christian, atheist or something else? In speaking of critical engagement with belief, I have in mind the following ideal: we should avoid two possible, pedagocial mistakes: on the one hand, telling students that their religious viewpoint has no relevance to ethical questions and on the other hand, allowing students' religious view to uncritically decide their ethical beliefs. Religious viewpoints must be *constructively* engaged in a way that yields both knowledge and insight into the production and justification of specifically ethical claims.

Second, I want to discuss the way in which the Bible and Christian teaching can be used to deal with liberatory questions dear to the academic left: how can we show our students that their Christian identity can be a location from which to contest oppression, rather than a foundation for oppression or a way of avoiding new or uncomfortable questions about ethics and the social realities of our time and place?

COMMUNITIES OF MEANING

I would like to begin with the idea of *communities of meaning* as discussed by Amie Macdonald (2002) and Susan Sánchez-Casal, (Macdonald and Sánchez-Casal, 2002). Students enter the classroom sorted into various identity groups that reflect the structural oppression and privileges of society. In my classroom, I have students with differing gender, race, sexuality, and class locations: black or white, male or female, straight and gay, working class or privileged. These identities have a complex reality that emerges as students negotiate their experiences and their lives in the face of social structures that transcend their individuality. A person's racial and gender identity is a real feature of her life, not so much because of biology or other non-social features of personhood, but because the person finds herself located at the nexus of various patterns of social living. She may not ask to be treated as a black woman in society, but she is treated as one, and these experiences and structures produce the reality of her identity.[1]

As Macdonald points out, we often assume that students speak and learn from these identity locations. This idea lies behind using *standpoint epistemology* in the classroom, for standpoint epistemologies generally assume first, that one's identity provides knowledge about society, and second, that those with oppressed identities have an easier time seeing the oppressive structures of society.² Thus, the students with oppressed identities can contribute knowledge that those with privileged identities may not have easy access to. Smart versions of standpoint hold that simply having an oppressed identity is not an automatic guarantee of producing accurate knowledge: the mere fact of being gay does not grant automatic assurance that one's claims about sexuality and the sexual oppression in society will be accurate. However, the person in an oppressed location will at least have the proper standpoint for producing accurate knowledge about society.

Even in this smart version, using standpoint strategies in the classroom is not without problems, as described by Macdonald and others. On the one hand, standpoint strategies tend to put oppressed minorities within the classroom in the position of having to speak for their oppressed location. This is both an unfair burden (why should black folks have to explain racism to racist white folks?), and more troubling, a possible source of inaccurate knowledge, or even knowledge that reproduces prejudice and oppression. Macdonald uses the example of a black Jamaican student who described to her class how homosexuality was wrong from the standpoint of her specific identity. While students in the class were quick to question this student's faulty leap from her identity location to her claim of homohatred, the problem exists that identity never assures that accurate knowledge will be produced. Even if we keep the smart version of standpoint in mind and assume that identity is not a guarantee of truth, the tendency (and it is only a tendency) is to avoid questioning a student's identity claims, because the student is taken to be authoritative about her or his identity.

Furthermore, standpoint tends to put students with privileged identities on the outside and on the defensive: the implicit message seems to be that they do not have knowledge to contribute, and that they cannot trust their own experiences. Again, smart standpoint holds that privilege does not automatically entail false claims about social structure, but the assumption is always made that claims made from privileged positions are subject to more intense scrutiny.

Since standpoint is ultimately based upon the idea that a person's social location grants them experiences from which they can construct knowledge of the world, the idea of a community of meaning cuts directly to the students' experience of the world as informed by identity markers, without starting from the position of their identity. As Madonald and Sánchez-Casal put it, a community of meaning is an epistemic community that is not defined solely by identity markers, but by shared experiences, values, and ways of being and understanding the world. Communities of meaning are shifting and overlapping, and allow for students to speak from the complexities and multiplicities of their identities and social locations. Poor students share many experiences, even if their races are different, racial minorities may share many experiences, even if the peculiarities of their racial and gender identities differ, progressive students may all feel alienated by a conservative environment, even if other features of their identity differ. The important point is to access students' shared experiences and locations, rather than only their identities, as a starting point for creating a community of inquirers and knowledge-seekers.

My next point should be obvious, given my interests: religious beliefs and practices constitute communities of meaning for most of my students, believers or otherwise. In saying this, I want to focus particularly on shared experiences and shared values as a base for creating knowledge. Most of my Christian students share what I will call a Christian experience. While this includes the obvious things like going to church, having a religious family (or alternatively, a conversion experience not shared by family) and a community of fellow believers, in fact the main aspect of this Christian experience are not these obvious social institutions, but instead a specific claim about the power of God in their individual lives. Many of my Christian students believe that God has personally touched their lives, granting them not only peace, happiness, and resilience through hardship, but also structuring their lives and infusing them with values. They have been, in their words, *saved*.

Care must be taken with this very Protestant idea. At this point, I neither want to endorse nor deny the metaphysics behind such a claim. It interests me pedagogically, because in the view of most of my Christian students, this aspect of their lives is as salient and important as any other feature of their identity, and it is the experience base from which they want to approach many traditional philosophical questions as well as current political and social issues. If I want to reach my students where they are, I need to access this community of meaning in my classroom.

There are several complications, however. First, my students are not completely consistent on this score. While some of them genuinely find God and their religious belief and practice to be the central deciding factor in how they live, the majority of them (even Christians) cite a fairly simple notion of happiness as the ultimate value in deciding life plans, career paths, and relationship decisions. I interpret this as proof that they have imbibed U.S. consumer culture and secular values as deeply as they have drank from the cup of religion, and also as a good illustration of the usefulness of communities of meaning in the classroom: there are overlapping values and experiences at work in the lives of my students.

Second, there are non-Christian believers and atheists in my class. Obviously, I cannot simply aim the class at the Christians and succeed. Even among the Christians, there can be many differences with respect to social and political issues. This reveals the need to structure the class around differences among the communities of meaning in the classroom. The idea here, again expressed by Macdonald and Sánchez-Casal, is that differences among students' experiences and social locations should be the origin of knowledge and discovery in class. The different experiences and beliefs that students bring to the classroom can be the starting point for students' own critical assessment of their experiences and beliefs. The fact that some students do not share the same beliefs and experiences; that they are not part of the same community of meaning, opens the possibility that students' interpretation of their experience, and their reasons for holding their beliefs, can be subjected to question, reason, and possibility of rejection or better grounding.

The instructor can facilitate this process, by selecting communities of meaning with some ingenuity. My general strategy, for instance, was not to separate the class into Christians and non-Christians, but monotheists, deist/pagans, agnosticis, and atheists. This had several positive effects. First, it allowed for a wider variety of religious standpoints to come to the fore in the classroom, and put some students in a situation of having to decide exactly what they did believe. Second, it stressed commonalities among Jewish, Christian, and Islamic traditions, and compelled Christian students into dialogue with Muslim students about similar ethical questions. While in many respects these religious traditions are different enough to merit separate communities of meaning, they were, for the purpose of the ethics class, similar in respect to their meta-ethical outlook. Finally, it left nobody in the classroom feeling alone.

I also should note that we spent an entire class period just dividing into groups. I began by handing out a series of questions about religious belief, and then sorted the students based on how they answered these questions. We then discussed, as a class, how different groups answered a specific set of religious and ethical questions, so that students began to see how their ideals and beliefs were and were not informed by religious commitments.

Biblical and Textual Hermeneutics

Once my class was sorted into their smaller communities, my strategy was not to question divine command theory as such (using, for instance, the Euthyphro problem[3]), but rather to raise an epistemological issue: how, if ethics is determined by divine command, would we know a divine command if we came across one?[4] This epistemological issue is, I believe, less threatening to my believing students, for it calls neither the existence of God nor the truth of their religion into question; it only asks what kind of knowledge is possible and what justification we can give for this knowledge. It also works directly with the pedagogical strategy of communities of meaning, requiring students with different experiences and values to engage in dialogue about the justification of their beliefs.

Of course, the majority of believing students hold that divine commands are revealed in sacred texts. For my Christian students, this means the Bible. These students typically enter the classroom with a general and fairly simple view of biblical authority: the Bible's pronouncements concerning morality are authoritative *and sufficient* ethical principles. Behind this version of authority lies the view that biblical pronouncements are self-standing, and require little or no interpretation.

This view, of course, has little hope of succeeding as an ethical principle. To academics in the Humanities, it seems especially hopeless, since most of us have been thoroughly trained to see the extent to which any reading of a text involves fallible interpretation—we are all too aware that we bring our own assumptions, history and experiences to any text we read,[5] and we know, especially since the work of Derrida, that texts can be riddled with nearly irresolvable ambiguities. While students have some knowledge that interpretation may be involved in literature and poetry, they are often not taught that such interpretation is a general part of any reading, and certainly, most of my Christian students have rarely thought to apply this learning to their religious practice. Meanwhile, the outspoken atheists in the class may recognize this principle entirely too well, using it to claim that nothing at all can be learned from the Bible, which also seems to me to be inaccurate and problematic.

This difference among my students' views provide an excellent starting point for discoveries about reasoning, interpretation, and how knowledge claims are offered, justified, and examined, and so class quickly developed into a discussion of hermeneutics. The exercises I used and am about to present neither prove nor disprove that the Bible can be a source of moral knowledge, but rather that it can be a source of knowledge only in the context of reasoning and interpretation about the claims it makes. Accordingly, it neither validates the Christian nor the atheist, but instructs both in the epistemological project of justifying ethical claims. The exercises aim at achieving an understanding of how claims to truth are made through a collective process of reasoning that brings our experience and a variety of other sources into play. Such a project is generally consistent with realism, which holds that knowledge claims can not be granted absolute authority, certainty, or foundation, even if they can be justified and supported to the point of believing them to be true.

In a very helpful essay, Anthony Ellis (1996) lays out some different views on biblical authority, and a number of textual and teaching strategies for examining them. The basic idea behind them is simple: show to students that the Bible contains a great many ethical principles (either in the form of direct moral injunctions or stories) that we do not accept today. Once this point is established, the class can go on to discuss the role of moral reasoning and interpretation in deciding truth. Two examples will suffice:

"Whoever curses his father or his mother shall be put to death" (*Exodus 21:17*)
"You shall not kill" (*Exodus 20:13*)

The first of these seems highly implausible to students. Students are often tempted to ask what is meant by the term curse (according to Ellis, the Hebrew most likely means "belittle" or "insult"), and this discussion should be engaged precisely because it shows students that any moral precept, couched as it is in language, will require interpretation. Hence, the same exercise can and should be engaged with the second injunction which, of course, is the Sixth Commandment. Here the question concerns the meaning of the word "kill." Biblical scholars generally agree that it means something like "wrongful and deliberate killing," but once the word "wrongful" has come into play, we are in the realm of moral reasoning that stands relatively independent of biblical authority, since we must interpret that authority.

The broader question here concerns why we accept one precept, and not the other. There is a quite complex historical and theological story about the status of Jewish Law in relation to Christianity, but I have found that discussing this point with my Christian students usually leads nowhere, and this should not be surprising. The point of the exercise is to establish the place of reasoning and interpretation in deciding morality; students who are just trying to grasp this unfamiliar point with respect to the Bible are neither ready nor able to hear that all of Christianity has a human history. Instead, the discussion of why we accept one precept rather than the other can lead to a quite open-ended discussion covering the basis of morality. Students can be encouraged to bring up just about anything, and to feel free to question each other with respect to their ideas.

The beauty of this exercise, of course, is that it neither displaces the Bible as a source of morality nor enshrines it as one, thereby "satisfying" communities of meaning that hold to it, and those who are against it. The difference between these communities can be crossed by the very fallibilism of a realist epistemology;

both communities can learn that they make ethical judgments in a complex reasoning process that includes their historical and social location, their individual background, and those sources they choose to consult for ethical guidance. I felt it important to explain to the class that the necessity of interpretation and reasoning does not prove the Bible wrong, nor does it show that it is no aid in establishing moral precepts. This both clarifies the point for students and also provides some reassurance for students who may feel the ground is being pulled from under them. It is also helpful to have, maintain, and show a respectful attitude towards both faith and the Bible; arrogance or a heavy dose of Socratic irony will alienate your believers and shut down the exercise before it can begin. It is also disrespectful and undemocratic—this is where many of our students are, and we need to meet them there and take them further.

This raises two other points that I want to mention briefly before I go on and deepen this inquiry by discussing issues of oppression in relation to a fundamentalist Christian faith. The first concerns which community of meaning I align myself with in the classroom during this exercise. I have found that telling students explicitly my own story—that I used to be deeply Christian, but am now at most agnostic—is unhelpful. Frankly, I think it alarms or scares them; perhaps it confirms their fears that once questioning begins, it will never stop until all faith and belief is rejected. Hence, I have found it best to reiterate what I just said above: respect for the Bible as a document, and a firm commitment to moral reasoning and experience as the foundation of our knowledge.

Second, this exercise will not succeed with all believers. For some, the point about interpretation may be either too subtle or too philosophical, but for most who resist, the problem comes from what is perceived as an attack on the sacred. Some people will not want to give up access to what they believe to be divine authority and a connection to ultimate reality. The religious narrative and community is the profound home in which they live, and the idea of examining its construction is simply too alien to grasp, or too threatening to engage. In the end, I don't believe there is any way to actually push students beyond where they are ready to go without a kind of violence that would ultimately trigger defenses that are even harder to get past.

Nonetheless I must confess that I was surprised both by the willingness of students to discuss this question (as if it had always been on their mind, but they didn't know how to ask it) and also the respect and engagement they showed each other. Two processes facilitated this: (1) they discussed the questions raised in their own communities of meaning before classroom discussion. This allowed everybody to speak in a non-threatening forum, and to not feel alone with their opinions. (2) I sorted the students so that they were sitting together in their groups (monotheists here, atheists here, etc.). This gave the class just a hint of a "game-like" atmosphere; nobody was *attacking* anybody, everybody was just engaged in getting different opinions out and discussing them. Laughter, here as everywhere, was the most helpful tool.

Liberatory Practice in the Classroom

This method can be directly adapted to show how biblical authority can give us indirect knowledge about questions of oppression and identity. This requires some care, since I do not want students to walk away from class thinking that the bible endorses slavery, racism, homohatred and sexist gender roles. Nor ultimately do I want them to think that the Bible is neutral on such matters. Since they often consult the Bible for moral guidance, my goal is first, to show them that such consultation is an interpretive project, and second, that there is a biblical basis for liberatory thinking. Doing this requires students to see that we can treat the Bible "holistically" as a document that stresses loving kindness and care for the downtrodden. The contrast is between simplistic readings that look only at individual passages, and a complex interpretive project that tries to distill broad guidelines for reasoning about ethical and political issues.[6]

I begin this project with a summary of what we have learned, stressing the ideas of interpretation and reasoning about scriptural authority. I point out that in reading the Bible, only if we interpose our own beliefs and ideals and think critically can we decide what lessons are to be learned from a seemingly sacred text. Then, I begin with a quite strong example: biblical teachings about slavery.

History is important here. Students are often aware that both advocates of slavery and abolitionists used biblical authority to bolster their arguments. I discuss this fact with the students, and I also stress the differences between the ancient slavery of the Old and New Testaments and U.S. slavery. Then I pose the question: what does the Bible teach us about slavery? The focus here is twofold: (1) to stress again the importance of interpretation and reasoning and (2) to suggest that even though we must engage interpretation, this does not suggest that biblical authority does *not* offer a genuine ethical guideline, which of course is a condemnation of slavery.

Slavery, of course, is a useful example. There are regulations in the Old Testament covering the trade of slaves, and Jesus and Paul both speak of slavery without condemning it. Clearly, this can be taken as an implicit acceptance of slavery. However, this is only an *implicit* claim, and what this means must be stressed: just because there is an implicit acceptance, is this the same as an explicit moral command?

Having stressed these points, we begin with four passages: *Exodus* 21:1–11, I *Corinthians* 7:17–24, *Galations* 3:28, and *Matthew* 22:34–40.[7] The first two contain implicit acceptance of slavery as fact, the third claims that all are alike who have faith in Christ (including slave and master), and the final is the "Golden Rule." Since some of my atheist students are undoubtedly hostile to this exercise, and resistant to spending so much time on the Bible, I stress the importance of this as an exercise in textual interpretation and in the creation of arguments. By limiting ourselves only to these four passages, we can begin to see just how complex interpretation can be. Even these seemingly straightforward sentences involve a lot of reasoning.

My students' responses to this exercise are fascinating. Universally, they discuss the three passages that explicitly mention slavery and try to find a way to show that they do not *really* condone slavery. Particularly, they focus on the shift (so typical of Paul's Epistles) from the literal to the metaphoric use of the word "slave" in the *Corinthians* passage. I encourage this discussion, of course, because it raises the question of how we decide metaphoric from literal uses and what this shows about the actual activity of reading and interpreting.

In contrast they almost totally ignore the Great Commandment of *Matthew* 22. By its own structure, this command should supersede all other and be one of, if not the central, criteria of morality from the Christian perspective: here is Jesus claiming that all commandments derive from the one to love. This would seem to show that deciding the question of slavery really should turn on the question of love. If ever there was an explicit divine command, this could be it. But because the word "slavery" does not appear in the passage, they do not regard it as concerning slavery.

Naturally, this set up an excellent discussion of how we can use general moral principles to decide specific moral questions, and how this was itself a process of reasoning and interpretation. At the end of the discussion, I took control of the class and argued to them that the Bible, while explicitly endorsing slavery, contained an ultimate message that was antislavery, even if it required some work to *see* this. I did this to model the moral reasoning process for them.

Slavery works well as a first example, because students universally believe it to be wrong. In the end, they all find ways to get around these passages and make them square with their prior beliefs. It also works because students will often respond to the biblical passages endorsing slavery by historicizing them: that was how they lived *then*, we don't live that way *now*. This sets the stage for a far more controversial question: biblical authority regarding homosexuality.

Here I used a variety of passages to try and display the historicality of biblical sexual mores. I used the standard passages from *Romans* and *Leviticus*, and then also added a line from *Deuteronomy* advocating death for heterosexual adulterers, a passage from *I Corinthians* requiring women to cover their heads in worship, and another passage from *Romans*, Paul's own statement of the love commandment.[8] My method was the same, discussion in groups followed by class discussion. However, here was a place where I used my power as a teacher. I did stress the importance of expressing their opinions clearly and with argument, although I also pointed out that (1) there were gay and lesbian students in the class, and (2) the issue was not strictly the rightness or wrongness of homosexuality per se, but rather the biblical teaching concerning it. I provided some historical context for the claims about homosexuality in the Bible[9] and had ready at hand a discussion about usury, or the charging of interest. This is helpful, since Christians once considered it a terrible sin, and the Bible is again clear that it is wrong to charge interest, except to people who were not members of the Israelite tribe (*Exodus* 22:25, *Deuteronomy* 23:19–20, *Nehemiah* 5:10). Again, as with slavery, I attempt to show that the Bible promotes tolerance and acceptance of homosexuality, although this is obviously a harder case to make to my students.

Predictably, this aspect of the class did not go as well. Students took longer to warm up to a full discussion of this question, perhaps because my own pro-gay stance scared them (my sexual orientation is fairly well known on campus) and perhaps because they were not accustomed to a systematic investigation of this "hot-button" issue. Many of them also failed to see the historical point about homosexuality at all. But the resistance provides several openings for helpful instruction. The point I and the non-homophobic students stressed was simply one of logical consistency: if biblical endorsement of slavery could be historicized, and if biblical commands that women cover their heads was a specific historical and cultural situation, why could not biblical condemnation of homosexuality be historicized and seen as a specific historical issue?

This was not an attempt to get them to relinquish homophobia in blinding light of reason; rather it was again a point about fallibility and interpretation. If historical situations structured both the writing and the reading of the Bible with respect to some issues (slavery, or gendered worship practices), they must surely be involved in other issues. If not, then we need a quite convincing account of why *one* issue escapes history and hermeneutics altogether. Here as everywhere making knowledge claims about ethics was not a matter of simply making assertions or reading rules off from the big, biblical book of ethics, but a matter of reasoning and justifying knowledge claims to the best of our ability. Some individuals in my class had experiences that led them to question directly many of the assumptions other students held, and the open dialogue allowed all of these voices to come forward. My more sophisticated students, both believers and non-believes noted that the famous Romans passage seemed less a condemnation of sexual sin and rather a condemnation of idolatry and worldliness.

As a capstone to this unit, I require a dialogue writing assignment from my students. Students were allowed to pick one of three pairs of theses:

1. a. The Bible claims that slavery is morally wrong.
   b. The Bible claims that slavery is morally permissible.
2. a. The Bible claims that homosexuality is morally wrong.
   b. The Bible claims that homosexuality is morally permissible.
3. a. The Bible's moral claims are obvious and require no interpretation or historical context.
   b. The Bible's moral claims require both historical context and interpretation.

Students constructed a dialogue essay in which they presented full, complete arguments for each viewpoint, with substantial "back-and-forth" among the dialogue participants. The point here is again to force the dialectical message: all claims to knowledge require justification, and we must work through all sides to a position in order to understand it. While I am queasy about requiring my students to argue pro-slavery and homophobic positions, the alternative was to allow them to choose their own position on these matters, and I am certain some of them would have written some quite homophobic essays. In this way, students with homophobic attitudes were at least required actually to think through the anti-homophobic position, and students free of homophobia were given an opportunity to sharpen their own anti-homophobic views in developing an argument. My Muslim students were allowed to write papers on similar issues in their faith and based upon the Koran. As it turned out, the essays I received were outstanding—both fun to read and quite thoughtful, and they provided some opportunity for me to engage in an individual dialogue with my students.

Conclusions

These exercises succeed, in my view, because they use the differences among the communities of meaning in the classroom as a starting point for creating knowledge, and for modeling genuine philosophical and ethical thinking. The believers feel that some of the most cherished aspects of their identity have been engaged, although critically, while the non-believers can feel that some of their concerns about biblical authority have been raised, in a fashion that demonstrates fairness to believers. Believers, at least thoughtful ones, did learn something about the complexities of biblical interpretation, and have seen the Bible used as text arguing in favor of liberation and progressive causes. Non-believers saw that their concerns about religious authority were legitimate, but that they were not given the space to simply construct their own, non-religious authority. I believe this provides the best, practical solution to the dilemma I opened with: we do not have to question religious beliefs as such, nor do we have to simply allow students their religious beliefs. Instead, in a spirit of fallibilist hermeneutics, we can proceed from a realist standpoint that gives us the opportunity both to question and to work from deeply held religious beliefs, to respect them and to scrutinize them. Teachers can maintain both their progressive ideals and critical tools, and also talk religion with their students.

Indeed, these exercises did not end the discussion of religion in my class. One result of beginning the class with these discussions was continued engagement with the religious viewpoint and with divine command theory throughout the rest of the class. This was helpful in unanticipated ways: Kant's categorical imperative could now be compared to a sort of divine command without divinity; the religious roots of Beauvoir's secular, existential ethics could be seen and discussed. At times, I regretted this, because the Bible would come up in strange ways, but I always had to remind myself that this was a sign of some success—students felt engaged in that unique place from which they create their values and their knowledge.

We do not need to be afraid of religious identities and religious claims in the progressive classroom; rather we need to engage them as one among other sources of knowledge and value. Rather than run and hide from these questions (my old strategy) I now welcome this as a place to engage students who feel otherwise disconnected. In teaching environments like mine, religious belief is one of the best tools we have—and we must use it!

Notes

1. This is post-positivist realist understanding of identity, as detailed in the essays contained in Paula M. L. Moya and Michael Hames-García, *Reclaiming Identity* (Berkeley: University of California Press, 2000).
2. To me, one of the best articulations and defenses of this epistemic program is Sandra Harding, "Rethinking Standpoint Epistemology: What Is Strong Objectivity?" in *Feminist Epistemologies*, ed. Linda Martín Alcoff and Elizabeth Potter (New York: Routledge, 1993), 49–82.
3. The Euthyphro problem, found in the Platonic dialogue of the same name, questions whether a divine command can function as a rational ethical precept, given that whatever divinity commands would by its very nature then be good, no matter how irrational or repellent. The central question can be paraphrased simply as: is something good because divinity commands it, or does divinity command it because it is good?
4. I was inspired to approach this epistemologically rather than metaphysically by Anthony Ellis's discussion of religious pedagogy. Anthony Ellis, "Morality and Scripture." *Teaching Philosophy*, 19.3 (1996): 233–246.

5. I am thinking here of Heidegger's familiar discussion of the Hermeneutic circle in Martin Heidegger, *Being and Time*, trans. John Macquarrie and Edward Robinson (San Francisco, CA: Harper and Row, 1962), 188–195.
6. In this way, I am combining the more traditional approach of somebody like Ellis with something that functions more like liberation theology, which stresses the construction of theology collectively and from the grassroots.
7. Here are the passages (NIV):

> If a man beats his male or female slave with a rod and the slave dies as a direct result, he must be punished, but he is not to be punished if the slave gets up after a day or two, since the slave is his property. *Exodus* 21:20–21

> Nevertheless, each one should retain the place in life that the Lord assigned to him and to which God has called him. This is the rule I lay down in all the churches.... Each one should remain in the situation which he was in when God called him. Were you a slave when you were called? Don't let it trouble you—although if you can gain your freedom, do so. For he who was a slave when he was called by the Lord is the Lord's freeman; similarly, he who was a free man when he was called is Christ's slave. You were bought at a price; do not become slaves of men. Brothers, each man, as responsible to God, should remain in the situation God called him to. *I Corinthians* 7:17–24

> There is neither Jew nor Greek, slave nor free, male nor female, for you are all one in Christ Jesus. *Galations* 3:28

> One of them, an expert in the Law, tested him with this question: "Teacher, which is the greatest commandment in the Law?" Jesus replied: "Love the lord your god with all your heart and with all your soul and with all you mind. [DT 6:5] This the first and greatest commandment. And the second is like it: Love your neighbor as yourself" [Lev 19:18] All the Law and the Prophets hang on these two commandments. Matthew 22:35–40

8. Here are the other passages I used (NIV):

> If a man lies with a male as with a woman, both them have committed an abomination; they shall be put to death. *Leviticus* 20:13

> If a man is found lying with the wife of another man, both of them shall die, the man who lay with the woman, and the woman. *Deuteronomy* 22:22 [It continues: So you shall purge the evil from Israel.]

> Therefore God gave them over in the sinful desires of their hearts to sexual impurity for the degrading of their bodies with one another. They exchanged the truth of God for a lie, and worshipped created things rather than the Creator.... Because of this, God gave them over to shameful lusts. Even their women exchanged natural relations for unnatural ones. In the same way, men also abandoned natural relations with women and were inflamed with lust for one another. *Romans* 1: 24–27

> The commandments...are summed up in this one rule: "love your neighbor as yourself." Love does no harm to its neighbor. Therefore love is the fulfillment of the law. *Romans* 13:9–10

> And every woman who prays or prophesies with her head uncovered dishonors her head—it is just as though her head were shaved. If a woman does not cover her head, she should have her hair cut off....A man ought not cover his head, since he is the image and glory of God; but the woman is the glory of the man. *1 Corinthians* 11:5–7

9. Some good sources are Louis Crompton, *Homosexuality and Civilization* (Cambridge, MA: Harvard University Press, 2003) and John Boswell, *Christianity, Social Tolerance, and Homosexuality* (Chicago: University of Chicago Press, 1980).

REFERENCES

Boswell, John. *Christianity, Social Tolerance, and Homosexuality.* Chicago: University of Chicago Press, 1980.

Crompton, Louis. *Homosexuality and Civilization.* Cambridge, MA: Harvard University Press, 2003.

Ellis, Anthony "Morality and Scripture." *Teaching Philosophy*, 19.3 (1996): 233–246.

Harding, Sandra "Rethinking Standpoint Epistemology: What Is Strong Objectivity?" in *Feminist Epistemologies*, edited by Linda Martín Alcoff and Elizabeth Potter. New York: Routledge, 1993.

Heidegger, Martin *Being and Time.* Translated by John Macquarrie and Edward Robinson. San Francisco, CA: Harper & Row, 1962.

Macdonald, Amie "Feminist Pedagogy and the Appeal to Epistemic Privilege" in *Twenty-First Century Feminist Classrooms*, edited by AmieMacdonald and Susan Sánchez-Casal. New York: Palgrave Macmillan, 2002.

Macdonald, Amie and Susan Sánchez-Casal. *Twenty-First Century Feminist Classrooms.* New York: Palgrave Macmillan, 2002.

Moya, Paula and Michael Hames-García *Reclaiming Identity.* Berkeley: University of California Press, 2000.

10

POSTETHNIC AMERICA? A MULTICULTURAL TRAINING CAMP FOR AMERICANISTS AND FUTURE EFL TEACHERS

Barbara Buchenau, Carola Hecke, Paula M. L. Moya, and J. Nicole Shelton

North American ethnic diversity—with its conflictive roots in the histories of migration, settlement, slavery and conquest—has always been what David Hollinger calls "a major preoccupation in American life" (101). According to Hollinger, the late twentieth century saw a considerable shift in the way this topic is discussed in the public sphere. He argues that while diversity generated little in the way of enthusiasm throughout the history of the United States, there is today a general consensus that it is a crucial and promising aspect of U.S. society (Chapter 4). In recent decades, he suggests, educators, employers, journalists, and politicians have come to embrace diversity as a "national value" (142). Hollinger's stance toward this value, however, is ambivalent. While cherishing dissent as one of the core values of the multiculturalist stance, he argues for the necessity of containing what he sees as the excessively ethnocentric forces in the multicultural debate. Carefully evaluating the potentials and the pitfalls of what he calls the multicultural "doctrine", Hollinger argues in favor of a multiplicity of malleable, epistemically unimportant identities. This multiplicity, he hopes, will help us to move away from unproductive dissent toward a pluralistic consensus—toward, that is, a "postethnic America". Hollinger's critique of multiculturalist excesses, echoed by scholars in the United States and in Germany, marks a decisive point in the public debate about the challenges of diversity for public education. Although the end of the twentieth century saw the demise of many affirmative action programs, it also witnessed a revitalized debate about the continuing, if altered, social significance and epistemic salience of politicized group membership in U.S. American postmodern society.

This contended vision of a "postethnic America" was the topic of two student conferences held at Göttingen University in Germany in the summer of 2007. These conferences, funded by the committee for the allocation of tuition fees of Göttingen's school of humanities and substantially supported by the English department, and especially the Teaching Methodology program, brought together future English as a Foreign Language (EFL) teachers, master's students of North American literatures, and German as well as U.S. educators—all actors in the cultural and political arena. The conferences were the culmination of a year-long teaching and research collaboration involving scholars, methodologies, and theories from both Germany and the United States. The participants' common goal was to jointly address the U.S. American debate about ethnic diversity, its repercussions in recent literature and its possible implications for an increasingly diverse German society. By establishing a special focus on this most prominent and contested vision in the national multicultural debate—the ideal of a "postethnic America"—the student conferences sought to delineate appropriate strategies for teaching about U.S. American diversity in EFL and literary studies classrooms in German high schools and universities.

The hosts of the two Göttingen student conferences, Barbara Buchenau (American Studies, Göttingen) and Carola Hecke (EFL Teaching Methodology, Göttingen), share an interest in recent U.S. American fiction and its contribution to the ongoing debate about the future of an increasingly diverse society. But in putting together this effort, they also wanted to respond to Peter Freese's call for a rapprochement between two fields of inquiry that share an interest in cultural diversity: American Studies and EFL Teaching Methodology ("American Studies," "Beitrag").[1] In Germany, the two disciplinary fields, in spite of overlapping concerns for U.S. American literature and culture, have increasingly lost touch with each other over the last two decades. As Freese sees it, the gap between the two is due largely to two trends in American Studies which effectively decenter both the field and its object of study.[2] On the one hand, postmodern literary theories have discredited the search for reliable meaning thus complicating the use of literature in the EFL classroom; on the other, transnational interests in borderlands and minority perspectives have effectively undone former master narratives about "America". Not surprisingly, this unsettling of stable borders and secure objects of study has had complex and difficult implications for an EFL teaching traditionally geared toward discussing definable characteristics of the United States; (Freese, "American studies" 220; "Beitrag," 168–171). By and large, Freese doubts that recent trends in American Studies could be useful for EFL classes in German high schools. And, although he briefly mentions that the multicultural debate in the United States might matter to a culturally, racially, religiously diverse German student body, he does not explore the implications of this suggestion ("American National Identity," third but last page).

Clearly, Freese's view of the two fields is concerned with the differences between disciplines; more importantly, it is driven by a critique of recent developments in American Studies. But the organizers of the Göttingen conferences hold that the gap between literary and cultural studies on the one hand, and foreign language teaching on the other, might be beneficially addressed if multiculturalist struggles and debates are taken as resources rather than obstacles in educational settings which highlight U.S. American national characteristics. In contrast to Freese, Buchenau and Hecke held that multiculturalist struggles are of pivotal importance for increasingly diverse EFL classes in Germany, both because German schooling does not effectively provide equal opportunities to minority students (Muñoz, Karakaşoğlu-Aydin, Fornefeld), and because the rise of racist hate crimes in recent years places additional obligations on educators and all actors in the public sphere to address, rather than to dismiss, diversity and its challenges.

In the context of these institutional, educational, and political challenges, Buchenau and Hecke sought to develop a viable venue that might help "close the gap" between teaching U.S. American cultures and texts in higher and in secondary education. They devised a teaching experiment involving a year-long joint venture that brought together perspectives from the fields of U.S. American literature and cultural theory (Barbara Buchenau and Paula Moya), English teaching methodology (Carola Hecke), and social psychology (J. Nicole Shelton). The venture placed multicultural theories and practices center stage and involved two academics from the United States—Nicole Shelton (Psychology, Princeton) and Paula M. L. Moya (English, Stanford), with expertise in intergroup contact and identity respectively—as invited speakers and mentors for the student conferences. In addition to investigating the usefulness of the ideal of a postethnic America, the project was undertaken with the goal of encouraging graduate students of the American Studies and English Teaching Methodology programs at Göttingen University to identify and discuss multicultural literary and educational reappraisals of diversity while developing professional expertise in a demanding academic setting.

This paper presents the presuppositions, questions, and findings of this particular effort in German and U.S. American higher education to study and teach about the value of diversity, the ongoing conflicts that ensure its vitality, and its implications for an increasingly globalized world. Grounding our conclusions in the experiences gathered during the cooperative teaching experiment and the two student conferences, the four co-authors of this paper argue for identity-based, multiculturalist emphases of difference—in its various shapes of culture, religion, race, class, gender, and ability—rather than for models that envision a postethnic "America." Especially in the context of the German EFL classroom, we see two incentives to engage in a close reading of the American multicultural debates in our work with cultures, languages, and literatures:[3] first, research dedicated to social justice and the recognition of minority identities has come up with pro-diversity and pro-identity classroom strategies that can be meaningfully incorporated into German classroom designs which focus on intercultural learning targets (esp. Moya, *Learning from Experience*, 136–174, and "Identity;" Steele, "Race," 75–78; Aronson/Steele 450–453; Sánchez-Casal/MacDonald; Graham/Hudley).[4] Second, central goals of EFL teaching such as intercultural competence, change of perspective, and empathy are themselves closely related to issues of diversity and identity (see Antor, esp. 119–120). Over the course of this paper, we argue that discussing the complex debates concerning diversity and identity too scantily and possibly dismissively in the EFL classroom can inadvertently obstruct one of the prime goals of EFL teaching—the difficult and sometimes painful negotiation of our own perspectives with those of others. It is this negotiation, we contend, that can help us to recognize our own unwilling and unwitting contributions to racial and ethnic discrimination in both Germany and the United States.

INSTITUTIONAL CONTEXTS OF TEACHING DIVERSITY

Obstacles to closing the gap between German EFL Teaching Methodologies and American Studies derive from their distinct approaches to cultural diversity and from separate institutional histories. Cultural diversity has always been a topic in those American Studies courses and EFL classrooms in Germany dealing with European settlements in North America, slavery and segregation, and large-scale immigration from Europe and later from Latin America and Asia (see Freese, "American Studies"). Even so, the *Landeskunde/American Culture and Institutions* curriculum within EFL teaching has tended to treat these issues with little critical rigor, since it traditionally has aimed at minimizing potential conflict with the target culture (Volkmann, "Aspekte"). Only recently—in the context of the paradigm shift from communicative to intercultural communicative competence,[5] and with a heightened awareness of the importance of cultural diversity in German society—have EFL classrooms placed a greater emphasis on cultural difference in the United States.[6] Current EFL methodologies approach difference both as a conflictive preoccupation *and* as a U.S. American societal good. In the context of the intercultural paradigm, EFL training has come to be seen as one of the primary training grounds for the ability to move in a globalized, culturally complex world—one in which individuals are no longer representatives of national cultures, but are instead aligned with a multiplicity of sometimes conflictive cultural identities (see Antor; Volkmann, "The Global Village"; Roche, 47–52).

To understand better why cultural difference is now much more likely to be seen as a resource in German EFL classrooms, it helps to understand its institutional development. Over the last fifty years, the field of foreign language teaching in Germany has undergone a series of paradigm shifts. Before World War II, EFL teaching sought to provide primarily upper-class German students with the kind of cultural capital that had come to be regarded as a crucial tool for the reproduction of social hierarchies (Bourdieu, "Les trios états," and *Soziologie*); English was taught for the purpose of moving the students toward a better appraisal of "World Literature." But because this goal was assumed to be appropriate for the upper classes only (Bredella, 93), EFL teaching reoriented itself in what was assumed to be a more pragmatic direction when it was introduced into secondary schools that prepared students for vocational training rather than for the academia. This reorientation involved a move toward a "tourist" mode of thorough linguistic immersion (Bredella, 94, translation ours). The tourist ideal shaped EFL teaching until the late 1970s, when it was attacked for making false promises (only a minority of the students would ever travel to English-speaking countries), and for failing to prepare students adequately in reading comprehension, composition, and interpretative skills. The tourist mode was also faulted by its critics for having too marked focus on a kind of area studies knowledge that does not allow for critical assessments (Bredella, 94; Delanoy/Volkmann, 12).

In the 1980s, foreign language teaching followed the cultural turn when it moved away from area studies and redefined its ulterior goal. Present day EFL classrooms consider the formation of intercultural competences and intercultural communication skills to be a central part of language acquisition (see Roche). Ideally, accomplished EFL learners are prepared to be sojourners and migrants rather than tourists (Byram/Esarte-Sarries; Byram).[7] In this context, the term *intercultural communicative competence* denotes the awareness that meaning is made, and "transcultural harmony" established, through the interaction and negotiation of ideally self-reflexive people with distinct cultural and linguistic backgrounds (Roche, 4, translation ours). Accordingly, *intercultural communication skills* should ensure that various cultural practices are able to coexist in a productive, self-reflexive manner (see Delanoy/Volkmann, 12–13; Roche, 154–155).

Meanwhile, American Studies' participation in the cultural and linguistic turns in the humanities has had strong, but intensely difficult, implications for EFL teaching in Germany. Recent trends in literary and cultural studies left none of the literary canons intact, seriously unsettled the well-wrought *close reading* methods of literary analysis, and fostered an increasing usage of theoretical terminology; these trends did not translate easily into clear-cut recommendations of what German EFL students should be learning in secondary schools.[8] In addition, although the new EFL intercultural learning paradigm is often linked to "ethical and moral perspectives [that seek to] further a democratic agenda," it tends to be "less overtly political and less critical of dominant ideologies" than the cultural studies approaches favored by American Studies (Delanoy/Volkmann, 13). Another shift in American Studies in Germany that proved to be even more controversial in terms of its applicability in the EFL classroom was the turn toward minority literatures and cultures. Peter Freese is possibly the most pronounced critic of the applicability of politicized multiculturalist debates to the German EFL classroom. But Freese is by no means the only critic; his major argument—that a scholarly engagement in minority struggles potentially sidelines analyses of larger, globally influential, and educationally relevant trends in U.S. American culture and society ("Beitrag," 170)—is supported by other voices in the field of American Studies—most notably Winfried Fluck.

American Studies in Germany has its own unique history that shaped its development as a field. The interrelated disciplines assembled under the name of American Studies did not emerge in Germany until after World War II. Up to this point, U.S. American literature had been taught by philologists with strong investments in British literature and culture; not surprisingly, their Anglo-centric approach fostered readings that centered on the supposedly provincial and derivative aspects of literature written in the United States (Fluck, "Kultur," 698–699). The German Association of American Studies, founded in 1953, emerged in the context of U.S. American and German policies of re-education and re-orientation; the founders imagined that a scholarly engagement with the United States could provide avenues of envisioning a life beyond fascism (Grabbe, 163). Suggesting a strong nexus to institutionalized re-education, the foundational documents of the association emphasized the need to inform as many Germans as possible about the history, literature, and politics of the United States. The first generation of American Studies scholars in Germany included philologists interested in the democratizing power of American culture and scholars who had fled fascist Germany and who now brought experiences gathered during their American exile to bear on a Germany under reconstruction. However, a significant number of postwar American Studies scholars had to come to terms with the fact that they had held educational positions in Nazi Germany (Grabbe, 165–167). Thus, postwar American Studies in Germany had at its origin a set of complex and contradictory ideological baggage.

Recently, the shift toward post- and transnational paradigms in American Studies in Germany and elsewhere has fostered a redefinition of non-American American Studies research as more explicitly comparativist, relational, or intercultural (e.g., Grabbe, 183). This trend has had the effect of vitalizing German research in minority studies, since fields such as Black and Chicano/a studies are recognized as being of pivotal importance for both transnational and national imaginaries. At the same time, numerous U.S. American and European calls for the reappraisal of the inter- and transcultural dimensions of an (American) age of globalization have been wedded to hopes that American Studies might move beyond multiculturalism (e.g. Grabbe, 183–184, Freese, "Beitrag," 169). Winfried Fluck, for one, has argued that the anti-hegemonic, political, and multiculturalist bent in American Studies and the humanities at large might not amount to much more than another contribution to "an age of expressive individualism." Such an age, writes Fluck, thrives on "radical dehierarchization to eliminate cultural restrictions on self-empowerment," even as it needs "the cultural construction of difference to escape from the consequences of radical equality" ("The Humanities," 221). Culture, Fluck suggests, has been largely redefined as a major arena for "self-realization, self-assertion, and self-fashioning" (220). In Fluck's estimation, this move has diminished the overall significance of the field of American Studies: "By turning intellectual work into imaginary role-taking, the attractiveness of literary and cultural studies for the individual has increased, while their importance and social relevance have decreased" (221).

Fluck's critique of the individualistic trend in recent research in the humanities has little in common with Freese's potentially disparaging review of multiculturalist debates in the United States. Read in conjunction, however, the positions of both Peter Freese and Winfried Fluck make clear that multiculturalist agendas encounter significant critique in both EFL and American Studies contexts in Germany. While Freese is primarily concerned that multiculturalism might spell fragmentation, Fluck worries about the usefulness of a critical inquiry that foregrounds individual oppression. Neither scholar's concern, we suggest, adequately addresses the possibility that engagements with the cultural and literary dimension of diversity acquire enormous social significance once they self-reflexively and self-critically account for the institutional frameworks and methodological contexts in which they operate. Nor do they consider the possibility that difficult issues related to diversity—such as colonization, migration, slavery, and segregation—should be considered significant for *how* teachers in German secondary and higher education teach about cultural, racial, and ethnic differences in the United States.

The separate but related histories of EFL Teaching Methodologies and American Studies in Germany suggest that the development of fields of inquiry, as much as the design of curricula and textbooks, follow certain political logics. At the same time, scholarly research and teaching in both fields are subject to pragmatic choices—choices made largely in view of the need for financial support and the ability to forge workable coalitions. Teaching and research take place in specific institutional and interpersonal contexts, and are always geared toward multiple audiences. Consequently, even those research and pedagogical methods, which endeavor to change these contexts radically, are always at least partially complicit with the very structures they seek to overcome. This confluence of sociopolitical critique and complicity, and of idealism and pragmatism, places educators at a difficult and often lonesome crossroads. Individual educators rarely have time or resources to develop pedagogies that can simultaneously educate for citizenship effectively while engaging with issues of social justice and conducting a radical historical critique.[9] It was for this reason that Barbara Buchenau, the initiator of this project, joined by Carola Hecke, approached her colleagues Paula Moya, and Nicole Shelton with the idea of putting together a project that worked across disciplinary and national borders to investigate answers to important questions about how to understand, as well as how to teach about the workings and epistemic significance of racial and ethnic identities and the significance and dangers of ethnic and racial stereotyping.

POSTETHNICITY?—A COOPERATIVE PROJECT OF THE AMERICAN STUDIES PROGRAM AND THE ENGLISH TEACHING METHODOLOGY PROGRAM

The idea for the cooperative project began as a pragmatic one that has implications for institutional politics, since it calls for better educational collaborations in the programs involved: while Carola Hecke was reviewing material—novels, plays, films—for her class on teaching U.S. American multiculturalism in German high schools (*Gymnasien*), Barbara Buchenau was designing a course that investigated multicultural theories and their repercussions in recent literature in a manner that would encourage students to see connections between the multicultural debate in the United States and German discussions and educational politics concerning the multicultural or intercultural future of an increasingly diverse society.[10] From this common interest in the various shapes of ethnic and cultural diversity in recent literature Buchenau and Hecke developed the plan for two separate courses taught during the same semester culminating in two joint student conferences where students in both courses could present their respective findings to a larger, informed and engaged audience, and where we could learn from each other as we went along.

The Joint Teaching Experiment

In response to the general trends in their respective fields, the two courses investigated U.S. American literature and multiculturalism, together with their implications for German public education, from related but distinct perspectives. The "Teaching American *Landeskunde* [American Culture and Institutions]" EFL course prepared students for their future work as teachers of English in German secondary, college-preparatory education.[11] Its approach to language, literature and culture sought to tackle the teaching of U.S. American multiculturalism in the EFL classroom in ways that could encourage students in high school to move beyond their own perspectives, to see the world through somebody else's eyes, and thus to challenge productively their own assumptions. The idea was to foster cultural encounters in the EFL classrooms that would help students become active members in a culturally and ethnically diverse society.

Following the intercultural learning paradigm, Carola Hecke's training toward future EFL teaching sought to do two things simultaneously: it delineated teaching strategies which aimed at an understanding of certain aspects of "the American mentality" (ethnic diversity and multiculturalism), and which also targeted the social skills of the EFL students. Intercultural competence theories generally assume that studying ethnic diversity and multiculturalism in the United States encourages students to interact more successfully with people of different cultural backgrounds in Germany (see Byram, 33, 44; Doyé, 13; F. Lenz). In this context, literature is assumed to be particularly well suited for fostering intercultural understanding on the cognitive, affective, and pragmatic levels.[12] Accordingly, Carola Hecke's Teaching Methodology students were asked to design high school teaching units that used recent poetry, fiction, drama, and film for the goals of intercultural learning.

Barbara Buchenau's American Studies course, "Postethnic America? Progressive Identity Politics in Recent American Literature," critically reviewed U.S. American debates about diversity through both literature and theory, in addition inviting students to consider the relevance of these texts and theories for their own lives—their experiences with diversity and discrimination, and their future work in German multicultural classrooms or the public sphere. Students considered the significance of "voluntary affiliations" to social, racial, ethnic or cultural communities and the reported demise of identity struggles (Hollinger 105–129, quote 124). They took as a starting point three varied perspectives on multiculturalism and their distinct conceptualizations of diversity: David Hollinger's insistence on a "revocable consent" to group membership (118); Kwame Anthony Appiah's suggestion that we deemphasize culture and instead highlight personal identity in order to promote a "rooted cosmopolitanism" where our special obligations provide ethical guidance (213); and Paula Moya's argument for a non-essentialist, culturally aware belonging that understands social and cultural identity as "politically and epistemically important" to all endeavors which seek to understand some of the most "fundamental aspects of U.S. society" (*Learning from Experience*, 13). The students combined these conceptualizations of diversity with an introduction to psychological research on the pervasive impact of prejudice on social interaction (see the various publications by Shelton et al.; Richeson et al.; Steele; Steele/Aronson and Norton et al.), before they turned to critical readings of recent novels about discrimination and their interventions into multicultural struggles.

In light of Moya's careful delineation of the elements of a non-divisive and non-domineering multicultural education (*Learning from Experience*, Chapter 4; "Identity"), Buchenau and her students came to regard literary texts as necessary ingredients of successful training in the field. In both "Learning How to Learn From Others" and "What's Identity Got to Do with It?" Moya argues that studying the concepts, as well as the contents, of cultures and identities should be a fundamental part of all successful multicultural efforts. Cultures and their correlative identities, Moya contends, are fields of moral inquiry, and, as such, can be seen as resources providing us with potential alternative ways of living in and relating to the world (*Learning from Experience* 158–170; "Identity"). And because cultures and identities have value not in and of themselves, but rather in relation to specific (and constantly evolving) socioeconomic arrangements and geographic contingencies, teachers and students alike need to be able to recognize the malleability of cultures and identities, even as they identify the precise relationship of specific cultures and identities to structures of inequality in the larger global society ("Identity"). Literature, Moya reminds us, is a historically and culturally embedded representational form "that makes active use of the imagination." As such, it can be a way of "reflecting on and grappling with a society's contradictions" ("The Dialogic Potential"). Realizing that literature can be an important but inherently difficult player in public debates about the meaning and effects of cultural and racial diversity, Buchenau and her students asked a series of questions regarding the conscientious use of literature in classrooms aiming at multicultural or intercultural learning: Which texts lend themselves to enlightening changes of perspectives? How can cultural critics meaningfully incorporate the divergent perspectives of a heterogeneous readership into the already heterogeneous process of textual interpretation? How do we approach texts that essentialize cultures or disparage appropriate social identities? And finally, can we identify texts that might produce intercultural misunderstanding on one or more of their textual levels of narrative transmission?

In order to create a common ground for the conferences, Buchenau and Hecke agreed on a few texts that would be read in both courses; they also identified an area in which their courses would overlap in theoretical concerns. The choice of "shared" literary texts was guided by two questions: The first addressed the question of what future teachers need to know in order to prepare properly their students for the *Zentralabitur,* which is the state-run final exam that is taken after eight years of secondary education.[13] Here the choice was T. C. Boyle's 1995 novel *The Tortilla Curtain,* since Lower-Saxony and a few other federal states at the time had identified this novel as the key text for the negotiation of ethnic diversity in the final exam of the college-preparation level of secondary schools.[14] Second, they asked which novel might best negotiate diversity in a meaningful manner and is a challenging but feasible read for advanced high school students? Here the choice was Toni Morrison's 1987 novel *Beloved*. For their common theoretical concerns they asked: Which issues in the debate of multiculturalism are particularly conflictive? How do they matter in the German EFL classroom and for the *German multicultural project* at large? Two contested issues in particular—stereotype and identity—have lost none of their salience despite extended efforts to discourage stereotyping and to recognize and respect formerly embattled identities. As research in the humanities and the social sciences indicates, these issues can only be tackled meaningfully if the fields of primary, secondary, and higher education are involved, since schools and universities play a major role in the formation, transformation, and redefinition of stereotypes as well as identities (see Graham/Hudley, Steele, "Race," and "Stereotyping;" Moya, *Learning from Experience*, and "Identity").

Closing the Gap: Preliminary Observations

In the course of conducting the teaching experiment, several salient distinctions between the fields of American Studies and English as a Foreign Language Teaching Methodology in Germany became evident. Buchenau's and Hecke's teaching cooperation was based on a shared interest in concepts such as difference and diversity, but, as noted previously, *difference* and *diversity* take similar and yet distinctive shapes in the two fields. In American Studies, "diversity" is the key term for a recognition of difference that foregrounds the individual rather than the community, whereas "multiculturalism" is more often used for publicly and institutionally-supported community-oriented critiques, practices, and designs that embrace difference as a communal asset on the path toward social and cultural heterogeneity and equality (see G. Lenz, "Multicultural Critique," and "Dialogics"; Sanchez). In EFL Teaching Methodology, by contrast, diversity is not a key term. Here, the recognition of difference (of groups rather than individuals) is framed under the key terms "foreignness" and "alterity"; both terms are profitably conceptualized as thoroughly relational and culture-based. In EFL teaching, the term "interculturality" is much better established than the term "multiculturalism"; interculturality is the central concept used to envision social settings that foster community building and support the accommodation, rather than the incorporation, of difference (Bredella et al., esp. xxx–xxxvii).

A second distinction between the two fields in German universities involves the relatively apolitical profile of intercultural learning as compared to an American Studies "cultural studies" approach. The idea of intercultural learning was formed in the context of governmental policies that sought to respond to immigration and, less directly, to globalization. As such, intercultural learning has always been related to—and ambivalent about—concepts such as nation, national culture, homogenous societies, and non-hybrid populations (Bukow, 91–93). Its goals point in the same general direction as, but are not identical with, those of German theories of multiculturalism. Generalizing a bit, one can say that intercultural learning focuses on enabling *interrelations* and *interactions* of potentially conflictive groups and peoples, whereas multiculturalism tends to concentrate on improving *institutional structures* on the one hand, and strengthening the *individual* in cases of racial discrimination on the other.[15] Moreover, the relationship between proponents of the two different approaches can be uneasy and even combative. The governmentally funded essay collection *Interkulturell Denken und Handeln* (Thinking and Acting Interculturally), for instance, conceptualizes multiculturalism in terms of a political struggle for minority rights that has lost its potential for social change (Nicklas et al.). It portrays U.S. American multicultural policies as being caught up in a rhetoric of recognition that provokes rather than alleviates "discrimination and nationalism" (Demorgon/Kordes, 30, translation ours). By contrast, critics of intercultural learning such as Georg Auernheimer and Wolf-Dietrich Bukow suggest that government-sponsored intercultural agendas often fail to come to terms with German "structures of inequality," their repercussion in the national imagination, and the concomitant history of "collective experiences" (Auernheimer, 14, translation ours; compare Bukow, 92–99). In addition, Cristina Allemann-Ghionda has pointed out that the public debates about intercultural interactions are extensively shaped by "a historical stigma" which controls and ultimately curtails public engagements with "difference" (33, translation ours). In Germany, concepts such as race and difference call to mind the murderous racist logic of the Nazi regime. For this reason, Germans have developed a "forced habitus," grounded in an unease with public recognitions of (especially visual) difference, that is noticeable in all political and educational practices that seek to design anti-racist, productive ways of recognizing difference (Allemann-Ghionda, 35, translation ours). This habitus has legitimate historical roots and it needs to be addressed in its own right. It is one reason why educational policies and strategies developed in U.S. American sociopolitical contexts cannot be easily translated into recommendations for classrooms in Germany.

Buchenau's and Hecke's endeavor to close the gap between their respective fields and to become better scholars, teachers, and (inter)cultural workers in German classrooms brought their field-specific and disciplinary differences and similarities to the fore. The cooperation allowed them to engage in ongoing educational interactions and critical analyses that highlighted the limitations and strengths of their respective frameworks, while their focus on recent literature and its real-world cultural implications promoted an understanding of how disciplinary frameworks participate in the shaping of all scholarly conclusions about what might be the best ways of being in the societies the scholars themselves live and work in.

The Student Conferences

Drawing on what they had learned in Buchenau's and Hecke's courses, and under the guidance of their mentors, Nicole Shelton and Paula Moya, the Americanist and the English Teaching Methodology students negotiated the overlapping terrain of intercultural and multicultural designs, reconsidered the role of identity politics in societal struggles for social justice (Alcoff et al.), and investigated the epistemic significance of identity for teaching literature in the German classroom. The conferences created a generative environment in which students were able to develop presentation and discussion skills, apply what they learned in new contexts, shape their learning environments and the topics to be studied, and connect their course work to their own independent studies and experiences outside academia. In view of the use of literature in the EFL classroom, we paid special attention to the function of stereotypes and prejudice in literature and to literary interventions into multicultural theory and its politics, investigating how postmodern literature redesigns identity as slippery, yet persistently present, and interrogating the powerful connections between literature and a society's contested goals. Having outside mentors from the United States—both of whose work had been closely studied by especially the Americanists among the students—contributed nicely to the students' experience of professionalization.

For our first workshop, "Postethnicity? North American Theories and Literary Practices in and outside of German Multicultural Classrooms," Nicole Shelton provided psychological expertise on how multicultural theories can be used to understand daily inter-racial interactions between students and teachers as well as people more generally. Citing recent research, Shelton's talk "Divergent Attributions, Divergent Experiences: Whites and Ethnic Minorities in Interracial Interactions" showed that "color-blind" theories foreground an approach that avoids noticing visual differences and downplays cultural differences without actually doing away with the social impact of either visual or cultural differences. By contrast, another set of theories acknowledges the profound impact visual and cultural differences have on social interactions, seeks to assess this impact and generally assumes that cultural differences can be made to matter in a positive way—these are "multicultural" approaches in the stricter sense (see Markus et al.; Plaut; Richeson/Nussbaum). This is an important distinction because while the multicultural approach provides more resources with which to combat prejudice and stigma, both approaches have potential disadvantages. For example, while colorblind approaches stifle effective intercultural communication and allow prejudicial behavior to continue unmarked (Markus et al.), poorly implemented multicultural approaches can exacerbate existing racial and cultural tensions (Moya, "Identity," 148–158). Providing empirical evidence for the epistemic significance of identities and the complexities and pitfalls of interracial interactions (Shelton; Shelton et al.; Richeson et al.), Shelton pointed to psychological research showing that minorities tend to embrace a multicultural, differential logic, while people who belong to the mainstream often support colorblind, universalizing concepts of social interaction (see Plaut; Markus et al.). It was this particular focus on the many divergences in approaches to diversity and intercultural/-racial misunderstanding that set the tone for the students' presentations on multicultural theories, intercultural methodologies and the sociocultural work of novels such as Toni Morrison's *Beloved* (1987), Edward P. Jones' *The Known World* (2003), and Philip Roth's *The Human Stain* (2000).

Within the mentoring framework set by Nicole Shelton, the conference participants were able to distinguish literary and educational theories and practices that downplay differences from those that emphasize differences (of culture, race, ethnicity, class, gender) in order to turn them into communal assets. Students were thus empowered to begin to evaluate the theories to which they had been exposed in their classes within the contexts of German society and the German educational system. The German suspicion against all emphases of difference—visual, cultural or religious—that is the legacy (and a rejection) of the country's Nazi past, has meant that very few classroom strategies seek to take advantage of the increasing diversity of the student body in German schools and universities. More often than not, the heterogeneity of the student body in German high schools and universities has been regarded as an obstacle to successful teaching rather than as a resource (see Hu). Lothar Bredella's insights into the link between intercultural learning and the multicultural debate notwithstanding, concepts are widely missing that take the diversity of German classrooms seriously into account.[16] Taking this history into consideration, Nicole Shelton invited the student presenters to make sense of interracial interactions in the novels they read in ways that allow us to see and to ask for the many instances in which theories and experiences do not overlap and merge, but rather diverge (see "Divergent Attributions"). As the students and their teachers moved through the literary texts and the theoretical debates about identity and stereotyping, everyone became more self-conscious about their own behavior regarding intercultural interaction. One of the teachers overheard one student ask another, in all seriousness, "Are we stereotyping here?"

On the grounds of the explorations undertaken at the first conference with Shelton, Buchenau and Hecke and their students were able to come up with two preliminary findings about recent designs for EFL teaching as well as for the promotion of diverse societies: (1) the ideal of a "postethnic America", despite its many strengths, needs to be revised, especially as it is perpetuated in German EFL classrooms. This is so because a respectful recognition of differences, visual as well as cultural, can have positive effects on social interaction and on the learning processes in general, whereas efforts to ignore differences and the structural inequalities accompanying them often intensify ignorance and social inequalities; (2) German concepts of *Fremdverstehen/foreign intercultural understanding*, we realized, do seek to problematize and denaturalize their students' identities through intercultural contact (Bredella, Krumm).[17] But although theories of intercultural learning recognize and cherish cultural differences, they tend to ignore the social salience of visual differences, leaving teachers and students with little guidance on how to deal with visual stereotyping and often severely hampered forms of social interaction across the socially and culturally produced lines of visual differences. Moreover, while intercultural learning correctly acknowledges the existence and the salience of cultural stereotypes (see Burwitz-Melzer, Husemann), additional research is needed if we want to move toward a better understanding of how those stereotypes should be addressed, and how teachers can make sure that stereotyping is not the primary way of dealing with cultural differences in the classroom or in the preparation of educational materials.

During our second conference, we drew on these findings and accordingly entitled the venue "Postethnicity?—Identity Politics Reconsidered: North American Theories and Literary Practices in and outside of German Multicultural Classrooms." This time, Paula M. L. Moya brought her theorizing on identity and multicultural literature to bear on the research presented in two interrelated fields. Carola Hecke's students had designed various teaching units for literary texts—including T. C. Boyle's *The Tortilla Curtain*—that had been chosen for their potential to train intercultural competencies, whereas Barbara Buchenau's students had used multicultural theories along with literary studies methodology to probe the capacity of *The Tortilla Curtain* and Gloria Anzaldua's *Borderlands/ La Frontera* to address critically the political and epistemic salience of identity.

In her presentation, Moya discussed some of the possibilities and pitfalls of teaching multicultural literature. She began by arguing that good multicultural novels are rich in the potential for anti-bias pedagogy for at least the following reasons: (1) reading is a practice involving a person's intellectual and emotional engagement with a text; (2) reading expands a reader's horizon of possibility for experiential encounters; (3) novels might be describe as assemblies in which many, often contending voices speak. As such, they are heteroglossic textual mediations of complex social relations; (4) novels can work as a form of moral exploration. According to Moya, the novel form's constitutive heteroglossia allows for the possibility that a reader will engage dialogically at a deep emotional level with the difficult questions around race, culture, and inequality raised by good multicultural novels. Such a dialogic interaction can, she suggested, prompt a reader to question and then revise some of her assumptions about structures of racial and economic inequality and how they are sustained. Good multicultural literature, Moya argued, has the potential to powerfully implicate its readers and make them examine their own relationship to economic and social structures that reinforce racial and cultural hierarchies. She pointed out that while questioning oneself and the privileges attending to one's social location does not lead *ipso facto* to epistemic or empathic growth, the former is at least a precondition for the latter (Moya, "The Dialogic Potential").

Moya complemented her argument in favor of using literature in multicultural educational projects with a strong warning about the potential pitfalls of teaching literature (such as Boyle's *Tortilla Curtain*) that depicts—but does not adequately represent the interests of—people marginal to the structures of power in the societies they live in. She noted that a particularly serious challenge facing teachers who use literature to teach intercultural understanding is the possibility that a reader will believe that she consequently "understands" the people and culture she has read about in one—perhaps not very good—novel. Reading about cultural "others" in novels can lead some readers to make misinformed judgments about people different from themselves, particularly if the novelistic representations are distorted or rely on stereotypes. Consequently, Moya argued, responsible teachers of literature have to pay close attention not only to the kinds and quality of representations provided in the novels they choose, but also to the pedagogical methods they employ when teaching those novels ("The Dialogic Potential"). Moya's warning fleshed out a conviction about the dangers of using literature in multicultural educational initiatives that had been slowly forming in many participants of the venture throughout the term.

Literature, both in an American Studies and in an English Teaching Methodology context, is often approached as something that the philosopher Kwame Anthony Appiah calls a "good"—something that, because it is not one-dimensional or driven by ideology, might help us to envision the resolution of conflicts (*The Ethics*, 120). Yet in an essay published in *The New York Review of Books*, Appiah questions whether cultural awareness (trained, for instance, by reading multicultural literature) can solve social conflict and establish or strengthen recognition and respect.[18] His intervention seriously questions the optimistic stance of multicultural readers and guidebooks to intercultural competencies, since his blunt assessment that "[p]roximity, spiritual or otherwise, is as conducive to antagonism as it is to amity" (*The Ethics*, 256) reminds us that knowledge of another culture does not immediately, or even necessarily, lead to empathy and mutual understanding. In *The Ethics of Identity* Appiah further questions whether a society, like the one in the United States, that foregrounds individual freedom and that places the well being of the individual above all should define culture "as a resource, or good" in the first place. According to Appiah, culture in the United States is a rather "thin gruel" which cannot easily sustain the need for belonging (115). America's cultural, linguistic, and religious diversity is not great, in Appiah's estimation, since "most Americans share" not only English as a daily language, but also the possession or consumption of iconic consumer goods and the conceptualization of religion as an "essentially private" faith and practice (116). Accordingly, Appiah wonders "whether there isn't a connection between the thinning of the cultural content of [minority] identities and the rising stridency of their claims," a connection he sees as rooted in the individualist and liberalist base of many multicultural theories (117). He notes that culture cannot be regarded as only a "good" both because cultural membership mandates behavioral restrictions in addition to providing a sense of belonging, and because historical moves toward diversity have not always been conducive to greater social justice (120).

Appiah's skeptical judgment about the benefits of multicultural educational efforts is provocative on numerous counts. For instance, the authors of this paper do not agree that culture in the United States is either as "thin" or as homogeneous as Appiah suggests, nor do we agree that multiculturalist theorizing is always driven by individualism and liberalism. Furthermore, he underestimates the epistemic significance of minority identities, and fails to see how the perpetuation of the status quo by members of the mainstream is also driven by identity considerations. Nonetheless Appiah's doubtfulness regarding culture as a social good is a salutary reminder about the dangers of a too-quick embrace of the potential of literature—which is, after all, a cultural product—to provide guidance in the resolution of social conflicts. His warnings, when complemented by Moya's discussion of the pitfalls of teaching literature for the purpose of cultural understanding, provoked us all to become more specific about the role that we see literature playing in both the hardening and the resolution of social conflicts. It also motivated us to employ literary analysis as a way of illuminating the contradictory meanings, and social work, of literature.

Just two analyses carried out during our venture might illustrate this point. For example, Maria Hesse, a graduate student presenter at the first conference, drew the participants' attention to the "mesmerized gaze" of the narrator Nathan Zuckerman in Philip Roth's *The Human Stain* (2000), arguing that this gaze—wherever it become's the reader's primary access to the events of the story—cuts both ways, as it stereotypes both the narrator and the object of his fascination. Gazing through Zuckerman's eyes, we see the protagonist Coleman Silk, an undercover African American gone Jewish, transform into a sexually very active senior citizen living the carefree, unconventional life that the narrator could and would never dare to attempt. In the light of the racialized debate in the novel, Hesse showed that Zuckerman's focalization oddly inflects the tone of larger parts of the narrative, evoking contexts reminiscent of modernist primitivism and its self-referential, Othering infatuation with African and African American cultures (see Lemke). As we would like to suggest, analytic moves such as Hesse's can help us to clarify how specific novels contribute to ethnic and racial stereotyping. In *The Human Stain* we are thus made aware that the essentializations and dismissals are primarily written into the various levels of narrative transmission. This analysis signals that the way we see and the way we talk about our experiences are at the heart of intercultural *mis*understanding. Similarly, Rebecca Scorah, a graduate student presenter at the second conference, gave a paper on T. C. Boyle's *The Tortilla Curtain* that pointed to its potential to disrupt productive intercultural communication. Using the techniques of literary criticism, Scorah began by identifying the novel as a melodrama. She then showed that, as a melodrama, the novel allows only a highly limited, ideologically loaded representational space to its Mexican and Mexican American characters. Stereotyped as they are, the text offers its readers no possibility of assuming an immigrant perspective or of empathizing with these ridiculously ill-fated characters. On the other hand, the white racist in the plot is also too easily ridiculed to allow any sustainable insight into the pervasiveness of racist thought and its many disguises. The outcome is a thorough distortion of non-Anglo Saxon cultures in California and of the larger conflict over illegal immigration in the United States. The assignment of this particular novel in German EFL classes in the curricular context of "ethnic diversity and immigration" is thus ill-advised. Students who read the novel with the idea that they are learning something real and important about the way immigration and cross-cultural interaction occurs in the United States might end up drawing dangerous conclusions for their comparative assessments regarding the German debate on immigration and diversity.

Other presentations from both conferences similarly suggested that a study of stock characters and narrative trajectories—as in the melodrama with its clash of one-dimensional good and bad characters or in the sentimental novel with its "protagonist in distress" pattern—can help us to clarify why texts do not easily or always lend themselves for multicultural ends and what is the political work done by certain characterizations, plot patterns or narrative structures. Our explorations prompted us to trust literature less immediately, and this stronger critical stance pushed us in our discussions to analyze the specific ways in which some literary texts do not lend themselves to the heteroglossic ideal of literature. Our experiences in both the courses and the conferences thus contributed to our ability to review the ideals of postethnicity and interculturality, consider their repercussions in the logic of our respective institutional contexts, develop an awareness of our own biases, and critically assess the complex and often dubious role of literature in the conflictive discourse on diversity and social justice. In the presentations and discussions, we discussed the way textual worlds conceptualize diversity, (trans-)form identities, and disperse or perpetuate prejudices and stereotypes. Accordingly, we concluded our joint venture on a cautionary note—only some literary texts and only some ways of reading trigger the kind of moral exploration that will help us to be skillful contributors to our multicultural societies. Texts and the way we approach them, need to be assessed very carefully for the adequacy and accuracy of their representations, their inclusion of multiple perspectives, and their likelihood of opening up, rather than closing down, productive intercultural and interracial communication and interaction.

SUGGESTIONS FOR TEACHING DIVERSITY

Our Göttingen training camp on multiculturalism and intercultural learning suggests that a more rigorous approach to the selection of and approach to teaching literature is needed if we want to close the gap between EFL teaching and American Studies, and improve our knowledge of multiculturalism in the United States and its teachability in the German EFL classroom. To begin with, the practitioners in each field need to learn the best practices of the other: teaching methodologists and American Studies critics should continue and possible intensify their joint improvements of the techniques of literary criticism that allow them to tease out the various levels of ideological transmission in a work of literature. Scholars in both fields would do well to explore the possibilities and limits of textual irony, to analyze how symbolism works at an often subconscious level to persuade the reader, and to understand how generic conventions (such as melodrama) and narrative strategies (such as focalization and point of view) make possible and even shape literary meaning. Learning more about how literature actually works can help students in both fields develop pedagogical methods that will enable, rather than shut down, productive intercultural learning in the classroom.

In addition, even though explicit engagements with diversity in the German classroom are rather rare, three educational concepts derived from EFL theories of intercultural learning and employed during our venture—*student-orientation*, *openness,* and *action-orientation*—can help Americanists and Teaching Methodologists alike to conceptualize diversity as a resource in the classroom (see Buchenau/Hecke). *Student-orientation* encourages educators to pay special attention to the interests, needs, and skills of the learners in their design of the coursework (Haß, 308) by designing classes around topics that have repercussions in the students' own life. If, for instance, coursework is structured around a topic such as *Growing up in a Multicultural Society* students might find intriguing parallels to their own multicultural lives. By linking a U.S. American topic to the German context, the issues to be studied can be conceptualized as being less remote and more meaningful to the students' own experiences. As a result, students can develop a strong personal interest in the topic (Seletzky). Discussions throughout the coursework can be brought to bear on real-life experiences. Apart from this social-affective result, the potential parallels between the object of study and one's own sociocultural contexts facilitate the cognitive processing of the subject matter. Similar goals are at the core of *openness* as a teaching methodology. With the aim of achieving openness, students bring in their own ideas, shape their learning environment by deciding on topics and material, and take responsibility for their own learning progress. In these setups, authentic material from the target country is to be integrated into the coursework (Haß, 211). In units about diversity in the United States, students can share their own experiences regarding diversity with the class, thus establishing a relation between the topic, the material to be studied and the students' own lives. Once the connection is made explicit, students can approach their material with greater care, reflecting on its impact on their personal attitudes. This move toward self-reflexivity and critical awareness is not automatic. It needs to be encouraged by tasks that initiate a critical reflection of the material under review (Nünning, "Fremdverstehen"). Much like *openness, action-oriented* language and literature teaching seeks to promote the students' interactions in the foreign language. Here, however, Pestalozzi's call for the integration of head, hand, and heart in education is central to the teaching method (see Weskamp, 75–76). Creative tasks such as acting out passages from a literary text, rewriting it from the perspective of a character, or illustrating the character's feelings visually are action-oriented approaches to literature (see Surkamp). These tasks encourage students not only to reflect carefully on the explicit and implicit meaning of a text. They also encourage the students to refer back to their own life experience in order to close information gaps—a proceeding that fosters a personal connection between the literary text and its respective reader.

Openness, student-orientation, and action-orientation can facilitate comparativist approaches to diversity in Germany and the United States. At the same time, these teaching strategies have their drawbacks: too much emphasis on the students' personal experience might prove distracting to the critical analyses of literary texts, their U.S. American sociocultural contexts (see Surkamp, 101), and their performative contextualizations in German classrooms. Accordingly, methodologies have to be developed that encourage students to critically reflect on the kinds of correlations they draw to diversity in Germany. Also, not all classrooms lend themselves to these strategies: similarities between the students' lives and the situation in a foreign country are not always present. And restrictive timeframes curtail the room for comparisons and creative adaptations. In higher education, it is often the professors who are wary to incorporate creative approaches, since these methods are rarely taken to have sufficient academic rigor, even though it is readily acknowledged that textual meaning is created in the process of reading (see Iser). Creative tasks—such as acting and drawing—can initiate and intensify these processes as they stimulate the readers' imagination and help activate their previous knowledge. The products of the creative tasks might then serve as an appropriate starting point for critical reflections on the kinds of intercultural meaning created when students raised in multicultural Germany read literature engaging with diversity in the United States.

Finally, American Studies scholars and EFL Teaching Methodologists alike need to be very careful in their assessments of the intercultural potential of literature in the classroom. Especially when choosing a novel that depicts a politically volatile situation (such as immigration) for the purpose of teaching intercultural understanding, we need to ask whether the novel represents the situation in a way that is both adequate and accurate. We can do this first of all by considering the diversity and quality of the representations provided in the novel: Does the novel contain different voices and present different views? Does it include the voices and perspectives of those people at the center of the debate? Are those voices and perspectives realistic, or are they imbued with negative stereotypes? When the representations of a minority group are overwhelmingly negative, does the novel point (e.g., through the use of irony or satire) to other ways of representing the issue? Finally, who gains and who loses according to the particular set of representations in a given novel?

Of course, we know that insofar as both EFL teachers and American Studies scholars teach within a complex set of institutional contexts, high school and university teachers do not always have control over which texts they are required to assign to their students. Consequently, there will be instances when teachers are presented with the task of teaching a novel or other work of literature that does not do an adequate job of representing the breadth and depth of a complicated and politically charged issue like immigration. In such cases, we need to help our students to develop pedagogical strategies that can enable them to critically evaluate and teach the literature their own students are reading. In addition to promoting student-orientation, openness, and action-orientation, we can teach our students to bring into the pedagogical situation supplemental material that will situate the particular issue under consideration in relation to a wider economic, political, and ethical context. In this way, we can meet some of the challenges posed to us by Moya in her presentation at the second conference: we can make visible what has been invisible, work toward countering exclusively negative images with some positive ones, seek to represent all sides of the issue, and create mechanisms for allowing the previous silenced participants of the conversation to speak for themselves ("The Dialogic Potential").

Conclusion

The challenges to multiculturalism posed by leading scholars on both sides of the Atlantic are worth taking seriously, and have been instructive for us as scholars and teachers in the fields of American Studies and EFL Teaching Methodology. Peter Freese's criticisms of some of the theoretical excesses of American Studies' research reminds us that a search for reliable (not absolute!) meaning is an important part of any knowledge-producing endeavor, and that even as we strive to avoid reductionism and essentialism, those of us who teach in American Studies and EFL classrooms need to be able to identify distinctive characteristics of the United States that they can convey to their students. Similarly, Winfried Fluck's hesitations about certain strains of U.S. American multiculturalism are a salutary reminder that the approach to intercultural learning we promote should avoid both moral relativism and individual solipsism. Moreover, we can concede Kwame Anthony Appiah's point that knowledge of another culture does not automatically or necessarily lead to empathy and mutual understanding while still holding on to our contention that it *can and sometimes does*—and that without some measure of knowledge, intercultural understanding will never occur. Consequently, the task is not to avoid using literature for intercultural learning in the German classroom, but rather to choose carefully the texts we teach and to develop effective strategies for teaching them.

The multicultural approach that we—Buchenau, Hecke, Moya, and Shelton—put into practice in our Göttingen teaching and research collaboration is best described as critical, evaluative, and outward looking. Drawing on the insights and methodologies of scholars in both EFL Teaching Methodology and American Studies fields, we investigated—through empirical psychological research, literary critical analyses, and conversations with each other and our students—how racial and cultural identities actually work. We paid attention to how identities, together with their representations in the public sphere, inform social interactions, influence individuals' life chances, shape their bearers' perspectives, and sort different people into social roles. Thus, our approach to racial, ethnic, and cultural identities neither isolated the individual from society, nor did it privilege individualism as a concept. Rather, it considered the roles, behaviors, and well-being of entities we might call "individuals-in-community," or "individuals-in-context." Finally, we did not foreground minority identities for their own sake, but rather for the purpose of developing a more complete and accurate understanding of U.S. American literature and society.

In sum, our Göttingen collaboration led us to conclude that the ideal of a "postethnic America" as envisioned by David Hollinger is neither an accurate description of current United States society, nor a realistic vision of a desirable future. Through our work with both historical and contemporary literary and cultural representations of the United States, we came to understand that identities and cultures are not inherently good or bad. Identities and cultures—malleable and historically-situated as they are—can be, and often are, mobilized for productive as well as destructive purposes. Only by understanding and working with identities, we contend, can scholars and teachers figure out which identities in which contexts work to support the status quo, and which might be mobilized for radical, or even moderate, progressive social change. Only in this way can we become better scholars, more effective teachers, and more powerful actors in the several cultural and political arenas we inhabit.

Notes

Address correspondence to the initiator of the project, Barbara Buchenau, Department of English, Göttingen University, Käte-Hamburger-Weg 3, 37073 Göttingen, Germany (bbuchen@uni-goettingen.de).

1. Peter Freese is a German scholar of Teaching Methodology *and* American Studies.
2. Peter Freese, "American Studies and EFL-Teaching in Germany," *Amerikastudien/ American Studies*, 50.1/2 (2005): 220. Freese is quoting Leslie Fiedler, "Cross the Border—Close that Gap: Postmodernism," in *American Literature since 1900*, ed. Marcus Cunliffe (London: Barrie, 1975), 344–366.
3. Multiculturalism is an ever expanding and thus often diluted term. For the sake of clarity, we will concentrate on the straightforwardly progressive, multiculturalist conceptualizations which "aim at equality and...'recognition'" without being intrinsically separatist. We would furthermore like to limit the term to approaches to social interaction which ground their appreciation of difference in a general suspicion of cultural hegemonies and the universalist claims arising from them Juan Flores, "Reclaiming Left Baggage: Some Early Sources for Minority Studies," in *Identity Politics Reconsidered*, eds. Linda Martín Alcoff, Michael Hames-García, Satya P. Mohanty, and Paula M. L. Moya (New York: Palgrave Macmillan, 2006), 61.
4. Here, and in the following we use the term "minority" in a non-quantative sense as relating to the institutional or social limitation of power and choice. As Linda Martín Alcoff and Satya P. Mohanty point out, the term has a conceptual and a political dimension: "Conceptually, minority signifies the nonhegemonic, the nondominant, the position that has to be explained rather than assumed, or the identity that is not taken for granted but is on trial. Politically, minority signifies a struggle, a position that is under contestation or actually embattled, that does not enjoy equality of status, of power, or of respect" Linda Martín Alcoff and Satya P. Mohanty, "Reconsidering Identity Politics: An Introduction," in *Identity Politics Reconsidered*, ed. Linda Martín Alcoff, Michael Hames-García, Satya P. Mohanty and Paula M. L. Moya (New York: Palgrave Macmillan, 2006), 7–8.
5. Intercultural communicative competence is based on factual knowledge about the target culture(s), respect regarding cultural differences, and the ability and willingness to successfully interact with people belonging to other cultures than one's own.
6. For a consistent approach to multiculturalism and its consequences for intercultural learning see Kersten Reich, "Verstehen des Fremden in den Kulturen und situiertes Lernen: zu Grundsätzen einer interkulturellen Didaktik," in *Fremde Kulturen verstehen— fremde Kulturen lehren*, ed. Heinz Antor (Heidelberg: Winter, 2007), 71–90.
7. Lothar Bredella rightly emphasizes that this shift has strong implications for the German classroom: "If we adopt the migrant as our model for foreign language teaching, this step will take effect also for students, who do not 'migrate,' since they will encounter migrants in their own culture; and they will need to develop understanding and respect for them in order to co-exist with them in a multicultural society" Lothar Bredella, "Zielsetzungen interkulturellen Fremdsprachenunterrichts," in *Interkultureller Fremdsprachenunterricht*, ed. Lothar Bredella and Werner Delanoy (Tübingen: Narr, 1999) 94, translation ours.
8. Peter Freese, "American Studies and EFL-Teaching in Germany," *Amerikastudien/ American Studies*, 50.1/2 (2005), for suggestions that treat the cultural turn in literary studies as an asset for EFL classrooms see Werner Delanoy and Laurenz Volkmann, "Cultural Studies in the EFL Classroom," in *Cultural Studies in the EFL Classroom*, eds. Werner Delanoy and Laurenz Volkmann (Heidelberg: Winter, 2006: 11–21).
9. Manuela Guilherme, *Critical Citizens for an Intercultural World* (Clevedon, UK: Multilingual Matters, 2002) addresses this conundrum in the title, its theory and its proposals for teaching strategies, but her focus is clearly not on situations when critique is compromised.
10. In Germany, the general consensus is the self-fashioning as an intercultural society Hans Nicklas, Burkhard Müller, and Hagen Kordes, eds., *Interkulturell Denken und Handeln. Theoretische Grundlagen und gesellschaftliche Praxis*, Bundeszentrale für politische Bildung (Bonn: Campus, 2006). From an international perspective, however, our future has also been envisioned as that of a multicultural Germany. Deniz Göktürk and Anton Kaes of UC Berkeley directed the so-called *Multicultural Germany Project* in order "to foster cross-disciplinary research that addresses Germany's changing cultural identity in

the era of mass migration and globalization" (http://german.berkeley.edu/mg/index. php, Last accessed January 22, 2009). See also Deniz Göktürk, David Gramling, and Anton Kaes, *Germany in Transit: Nation and Migration, 1955–2005* (Berkeley: University of California Press, 2007).

11. The appropriate German term *Lehramt an Gymnasien* does not have an English-language correlative, since the German educational system implements tracking in the shape of different forms of schooling and schools. The school form of the *Gymnasium* provides academically oriented secondary education; passing the final exam (*Abitur*) after nine, respectively eight years of secondary education enables the graduates to attend all institutions of higher education. In comparison to the average in the OECD-nations, Germany's *Gymnasien* and comparable school forms provide access to further academic training at universities at a rather low rate: whereas the OECD-average of students eligible to move on into higher education is as high as 49%, only 37% of the high school students in Germany are enrolled in school programs that will allow them to later attend a university Vernor Muñoz, "Implementation of General Assembly Resolution 60/251 of March 15, 2006, entitled 'Human Rights Council,' Report of the Special Rapporteur on the Right to Education, Vernor Muñoz, Addendum; 'Mission to Germany,' Febuary 13–21, 2006," ed. United Nations General Assembly (March 9, 2007) paragraph 23 <http://daccessdds. un.org/doc/UNDOC/GEN/G07/117/59/PDF/G0711759.pdf?OpenElement>. Last accessed January 22, 2009).

12. See Ansgar Nünning, "'Intermisunderstanding.' Prolegomena zu einer literaturdidaktischen Theorie des Fremdverstehens: Erzählerische Vermittlung, Perspektivenwechsel und Perspektivenübernahme," in *Wie ist Fremdverstehen lehr- und lernbar?* eds. Lothar Bredella, Franz-Joseph Meißner, Ansgar Nünning, and Dietmar Rößler (Tübingen: Narr, 2002). In an argument unrelated to the EFL classroom, Winfried Fluck, following Wolfgang Iser's insights into the self-reflexive processes triggered by the act of reading, argues that a literary text always represents two things simultaneously: the world of the text and the imaginative additions made by its reader, thus enabling us readers to live in two worlds simultaneously, and to view and review our own emotional and cognitive framework (Winfried Fluck, "California Blue. Amerikanisierung als Selbstamerikanisierung," in *Amerika und Deutschland: Ambivalente Begegnungen*, eds. Frank Kelleter und Wolfgang Knöbl (Göttingen: Wallstein, 2006), 63.

13. See note 11. Schooling in Germany is directed by the respective states rather than federally. Whereas teachers had long been in charge of designing the tasks for the final exam, state agencies of education have recently taken responsibility for their design.

14. For the thematic focus in Lower Saxony see http://www.nibis.de/nli1/gohrgs/ zentralabitur/zentralabitur_2006/02englisch.pdf and http://www.nibis.de/nli1/ gohrgs/zentralabitur/zentralabitur_2007/02englisch.pdf. Last accessed January 22, 2009. Under the general topic "The American Experience," subdivided into (a) "America: Vision of a New World," (b) "Immigration: Opportunities and Problems," and (c) "The U.S.A. as a World Power" Boyle's *The Tortilla Curtain* was the one text that was mandatory for all advanced EFL classes in their final year in 2006. In 2007, Lower Saxony moved much closer to the concerns of this paper with its general topic "American Identities" subdivided into the areas of (a) "The Making of 'Americans'," (b) "American Landmarks and Icons," (c) "Mainstream American Values," and (d) "Ethnic and Social Diversity: Sorrows, Hopes, Carreers." Once again, Boyle's text was the state-wide mandatory reading for all advanced EFL courses in high schools. Similar requirements exist for Bremen 2009, and Berlin 2008.

15. The stance of a paper published by the German *Bundeszentrale für politische Bildung*, a governmental institution providing material for political education, leaves little doubt that the differences between multicultural and intercultural theories and programs have already acquired a political salience of quite noteworthy proportion. For Jacques Demorgon and Hagen Kordes, multiculturalism in its very broad and diverse array (thus not delimited to the progressive type that we have delineated previously) is almost directly connected to bloody racial, ethnic, and religious strife, whereas governmentally

implemented intercultural agendas promise to offer non-violent, socially just solutions "Multikultur, Transkultur, Leitkultur, Interkultur," in *Interkulturell Denken und Handeln. Theoretische Grundlagen und gesellschaftliche Praxis*, eds. Hans Nicklas, Burkhard Müller, and Hagen Kordes. Bundeszentrale für politische Bildung (Bonn: Campus, 2006), 30. This is a dichotomization that is not particularly conducive to a productive debate about diversity in comparative frameworks.

16. As Lothar Bredella points out, concepts of intercultural learning that affirm cultural differences and that distinguish the students' culture from the culture encountered in the classroom potentially highlight and strengthen cultural differences within an ethnically diverse classroom. This focus on alterity, however, might limit the space within which migrant students can manoeuvre (99). Kersten Reich has proposed a well-crafted and detailed approach to establishing a multicultural base through 'situated learning' in German classrooms, but there are no concepts yet that would link the study of U.S. multiculturalism in American Studies and EFL classes to a possibly diverse target audience in the classroom itself.

17. *Fremdverstehen* is a central teaching goal of German foreign language education. The noun *Fremdverstehen* consists of the words *foreign* (*fremd*) and *understanding* (*Verstehen*), thus yoking together demands for the recognition of difference and alterity with an emphasis on understanding and empathy. It is generally seen to involve a cognitive as well as an affective and a conative component (Hermann-Brennecke, 55). Gisela Hermann-Brennecke, "Die affektive Seite des Fremdsprachenlernens," in *Englisch lernen und lehren*, ed. Johannes-P. Timm (Berlin: Cornelsen, 1998) 53–59.

18. "There is no conflict of visions between black and white cultures that is the source of racial discord. No amount of knowledge of the architectural achievements of Nubia or Kush guarantees respect for African Americans. No African American is entitled to greater concern because he is descended from a people who created jazz or produced Toni Morrison. Culture is not the problem, and it is not the solution." Kwame Anthony Appiah, "The Multiculturalist Misunderstanding," *The New York Review of Books*, 44.15 (October 9, 1997), 31.

References

Alcoff, Linda Martín, and Satya P. Mohanty. "Reconsidering Identity Politics: An Introduction," in *Identity Politics Reconsidered*, ed. Linda Martín Alcoff, Michael Hames-García, Satya P. Mohanty, and Paula M. L. Moya. New York: Palgrave Macmillan, 2006, 1–9.

Alcoff, Linda Martín, Michael Hames-García, Satya P. Mohanty, and Paula M. L. Moya, eds. *Identity Politics Reconsidered*. New York: Palgrave Macmillan, 2006.

Allemann-Ghionda, Cristina. "Warum war es nötig, eine 'interkulturelle' Bildung zu erfinden (II) und welche Bedingungen müssen erfüllt werden, um sie in Unterricht umzusetzen?" In *Fremde Kulturen verstehen—fremde Kulturen lehren*, ed. Heinz Antor. Heidelberg: Winter, 2007, 29–58.

Antor, Heinz. "Inter-, multi- und transkulturelle Kompetenz: Bildungsfaktor im Zeitalter der Globalisierung," in *Fremde Kulturen verstehen—fremde Kulturen lehren*. Heidelberg: Winter, 2007, 111–126.

Anzaldúa, Gloria. *Borderlands: The New Mestiza = La Frontera*. San Francisco, CA: Aunt Lute, 1987.

Appiah, Kwame Anthony. "The Multiculturalist Misunderstanding." *The New York Review of Books*, 44.15 (October 9, 1997): 30–36.

———. *The Ethics of Identity*. Princeton: Princeton University Press, 2005.

Aronson, Joshua, and Claude M. Steele. "Stereotypes and the Fragility of Academic Competence, Motivation, and Self-Concept," in *Handbook of Competence and Motivation*, eds. Andrew J. Elliot and Carol S. Dweck. New York: Guilford Press, 2005, 436–456.

Auernheimer, Georg. "Interkulturelle Kompetenz revidiert," in *Fremde Kulturen verstehen— fremde Kulturen lehren*, ed. Heinz Antor. Heidelberg: Winter, 2006, 11–28.

Bourdieu, Pierre. "Les trois états du capital culturel." *Actes de la Recherche en Sciences Sociales*, 30 (November 1979): 3–6.
———. *Soziologie der symbolischen Formen*. 1974. Frankfurt am Main: Suhrkamp, 1991.
Boyle, T. C. *The Tortilla Curtain*. London: Penguin, 1995.
Bredella, Lothar. "Zielsetzungen interkulturellen Fremdsprachenunterrichts," in *Interkultureller Fremdsprachenunterricht*, eds. Lothar Bredella and Werner Delanoy. Tübingen: Narr, 1999. 85–120.
Bredella, Lothar, Franz-Joseph Meißner, Ansgar Nünning, and Dietmar Rößler. "Grundzüge einer Theorie und Didaktik des Fremdverstehens beim Lehren und Lernen fremder Sprachen," in *Wie ist Fremdverstehen lehr- und lernbar?* eds. Lothar Bredella, Franz-Joseph Meißner, Ansgar Nünning, and Dietmar Rößler. Tübingen: Narr, 2002, ix–lii.
Buchenau, Barbara, and Carola Hecke. "Die Literatur zur eigenen Sache machen: Offener, fächerverbindender Unterricht in der universitären Fremdsprachenlehrerausbildung." *Literatur in Wissenschaft und Unterricht*, forthcoming.
Bukow, Wolf-Dietrich. "Vom interkulturellen Lernen zum lebenspraktischen Umgang mit Differenzen," in *Fremde Kulturen verstehen—fremde Kulturen lehren*, ed. Heinz Antor. Heidelberg: Winter, 2007, 91–110.
Burwitz-Melzer, Eva. *Allmähliche Annäherungen: fiktionale Texte im interkulturellen Fremdsprachenunterricht in der Sekundarstufe I*. Tübingen: Narr, 2003.
Byram, Michael, and Veronica Esarte-Sarries. *Investigating Cultural Studies in Foreign Language Teaching*. Clevedon, UK: Multilingual Matters, 1991.
———. *Teaching and Assessing Intercultural Communicative Competence*. Clevedon, UK: Multilingual Matters, 1997.
Delanoy, Werner, and Laurenz Volkmann. "Cultural Studies in the EFL Classroom," in *Cultural Studies in the EFL Classroom*, eds. Werner Delanoy and Laurenz Volkmann. Heidelberg: Winter, 2006, 11–21.
Demorgon, Jacques, and Hagen Kordes. "Multikultur, Transkultur, Leitkultur, Interkultur," in *Interkulturell Denken und Handeln. Theoretische Grundlagen und gesellschaftliche Praxis*, eds. Hans Nicklas, Burkhard Müller, and Hagen Kordes. Bundeszentrale für politische Bildung. Bonn: Campus, 2006, 27–36.
Doyé, Peter. *The Intercultural Dimension. Foreign Language Education in Primary School*. Berlin: Cornelsen, 1999.
Fiedler, Leslie. "Cross the Border—Close that Gap: Postmodernism," in *American Literature since 1900*, ed. Marcus Cunliffe. London: Barrie, 1975, 344–366.
Flores, Juan. "Reclaiming Left Baggage: Some Early Sources for Minority Studies," *Identity Politics Reconsidered*, eds. Linda Martín Alcoff, Michael Hames-García, Satya P. Mohanty, and Paula M. L. Moya. New York: Palgrave Macmillan, 2006, 53–68.
Fluck, Winfried. "The Humanities in the Age of Expressive Individualism and Cultural Radicalism," in *The Futures of American Studies*, eds. Donald E. Pease and Robyn Wiegman. Durham, NC: Duke University Press, 2002, 211–230.
———. "Kultur," in *Länderbericht U. S. A.*, eds. Peter Lösche and Hans Dietrich von Loeffelholz. Bonn: Bundeszentrale für politische Bildung, 2004, 698–787.
———. "California Blue. Amerikanisierung als Selbstamerikanisierung," in *Amerika und Deutschland: Ambivalente Begegnungen*, eds. Frank Kelleter und Wolfgang Knöbl. Göttingen: Wallstein, 2006, 54–72.
Fornefeld, Barbara. "Bildung von Menschen mit Behinderungen im interkulturellen Kontext," in *Fremde Kulturen verstehen—fremde Kulturen lehren*, ed. Heinz Antor. Heidelberg: Winter, 2007, 175–206.
Freese, Peter. "American Studies and EFL-Teaching in Germany." *Amerikastudien/American Studies*, 50.1/2 (2005): 183–229.
———. "Der Beitrag der *American Studies* zum fortgeschrittenen Englischunterricht in Deutschland," in *Neue Ansätze und Konzepte der Literatur- und Kulturdidaktik*, eds. Wolfgang Hallet and Ansgar Nünning. Trier: WVT, 2007, 167–182.

Freese, Peter. "American National Identity in a Globalized World as a Topic in the Advanced EFL-Classroom." *American Studies Journal*, 51 (Spring 2008) August 9, 2008 http://asjournal.zusas.uni-halle.de/103.html. Last accessed January 22, 2009.

Göktürk, Deniz, and Anton Kaes. *Multicultural Germany Project*. http://german.berkeley.edu:8002/mg/?gclid=CPnupIGOo5gCFQNItAody3-Wmg. Last accessed January 22, 2009.

Göktürk, Deniz, David Gramling, and Anton Kaes, eds. *Germany in Transit: Nation and Migration, 1955–2005*. Berkeley: University of California Press, 2007.

Grabbe, Hans-Jürgen. "50 Jahre Deutsche Gesellschaft für Amerikastudien." *Amerikastudien/American Studies*, 48.2 (2003): 159–184.

Graham, Sandra, and Cynthia Hudley. "Race and Ethnicity in the Study of Motivation and Competence," in *Handbook of Competence and Motivation*, eds. Andrew J. Elliot and Carol S. Dweck. New York: Guilford, 2005, 392–413.

Guilherme, Manuela. *Critical Citizens for an Intercultural World: Foreign Language Education as Cultural Politics*. Clevedon, UK: Multilingual Matters, 2002.

Hallet, Wolfgang, and Ansgar Nünning, eds. *Neue Ansätze und Konzepte der Literatur- und Kulturdidaktik*. Trier: WVT, 2007.

Haß, Frank. *Fachdidaktik Englisch. Tradition. Innovation. Praxis*. Stuttgart: Klett, 2007.

Hermann-Brennecke, Gisela. "Die affektive Seite des Fremdsprachenlernens," in *Englisch lernen und lehren*, ed. Johannes-P. Timm. Berlin: Cornelsen, 1998, 53–59.

Hesse, Maria. "Conceptions of the Normal and the Abnormal in *The Human Stain*." Talk given at *Postethnicity?—North American Theories and Literary Practices in and outside of German Multicultural Classrooms Conference*. Georg-August-Universität-Göttingen. June 24, 2007.

Hollinger, David. *Postethnic America: Beyond Multiculturalism*. New York: Basic, 2000.

Hu, Adelheid. "Identität und Fremdsprachenunterricht in Migrationsgesellschaften," in *Interkultureller Fremdsprachenunterricht*, eds. Lothar Bredella and Werner Delanoy. Tübingen: Narr, 1999, 209–239.

Husemann, Harald. "Stereotypes in Landeskunde—Shall We Join Them if We Cannot Beat Them?" In *Mediating a Foreign Culture: The United States and Germany; Studies in Intercultural Understanding*, ed. Lothar Bredella. Tübingen: Narr, 1991, 16–35.

Iser, Wolfgang. *Der Akt des Lesens: Theorie ästhetischer Wirkung*. München: Fink, 1976.

Jones, Edward P. *The Known World*. New York: Amistad, 2003.

Karakaşoğlu-Aydin, Yasemin. "Kinder aus Zuwandererfamilien im Bildungssystem," in *Bildung und Soziales in Zahlen. Statistisches Handbuch zu Daten und Trends im Bildungsbereich*, eds. Wolfgang Böttcher, Klaus Klemm, and Thomas Rauschenbach. Weinheim/München, 2001, 273–302.

Krumm, Hans-Jürgen. "Interkulturelles Lernen und interkulturelle Kommunikation," in *Handbuch Fremdsprachenunterricht*, eds. Karl-Richard Bausch, Herbert Christ, and Hans-Jürgen Krumm. Tübingen: Francke, 1995. 156–161.

Lemke, Sieglinde. *Primitivist Modernism. Black Culture and the Origins of Transatlantic Modernism*. Oxford: Oxford University Press, 1998.

Lenz, Friedrich. "The Concept of *Do's* [sic] *and Don'ts* in Intercultural Communication: Critical Considerations," in *Cultural Studies in the EFL Classroom*, eds. Werner Delanoy and Laurenz Volkmann. Heidelberg: Winter, 2006, 211–220.

Lenz, Günter H. "Multicultural Critique and the New American Studies," in *Multiculturalism and the Canon of American Culture*, ed. Hans Bak. Amsterdam: VU University Press, 1993, 27–56.

———. "Towards a Dialogics of International American Culture Studies: Transnationality, Border Discourses, and Public Culture(s)." *Amerikastudien / American Studies*, 44.1 (1999): 5–23.

Markus, Hazel, Claude M. Steele, and Dorothy M. Steele. "Color-Blindness as a Barrier to Inclusion: Assimilation and Immigrant Minorities." *Deadalus*, 129.4 (2001): 233–254.

Morrison, Toni. *Beloved*. London: Chatto-Windus, 1987.

Moya, Paula M. L. *Learning from Experience: Minority Identities, Multicultural Struggles*. Berkeley: University of California Press, 2002.

Moya, Paula M. L. "What's Identity Got to Do with It? Mobilizing Identities in the Multicultural Classroom," in *Identity Politics Reconsidered*, eds. Linda Martín Alcoff, Michael Hames-García, Satya P. Mohanty, and Paula M. L. Moya. New York: Palgrave Macmillan, 2006, 96–117.

———. "The Dialogic Potential of Multicultural Literature." Talk given at *Postethnicity?— Identity Politics Reconsidered: North American Theories and Literary Practices in and outside of German Multicultural Classrooms Conference*. Georg-August-Universität-Göttingen. July 17, 2007.

Muñoz, Vernor. "Implementation of General Assembly Resolution 60/251 of March 15, 2006, entitled 'Human Rights Council', Report of the Special Rapporteur on the Right to Education, Vernor Muñoz, Addendum; 'Mission to Germany', Febuary 13–21, 2006." Ed. United Nations General Assembly. March 9, 2007 <http://daccessdds.un.org/doc/UNDOC/GEN/G07/117/59/PDF/G0711759.pdf?OpenElement>. Last accessed January 22, 2009.

Nicklas, Hans, Burkhard Müller, and Hagen Kordes, eds. *Interkulturell Denken und Handeln. Theoretische Grundlagen und gesellschaftliche Praxis*. Bundeszentrale für politische Bildung. Bonn: Campus, 2006.

Norton, Michael I., Samuel R. Sommers, Evan P. Apfelbaum, Natassia Pura, and Dan Ariely. "Color Blindness and Interracial Interaction: Playing the Political Correctness Game." *Psychological Science*, 17.11 (2006): 949–953.

Nünning, Ansgar. "'Intermisunderstanding.' Prolegomena zu einer literaturdidaktischen Theorie des Fremdverstehens: Erzählerische Vermittlung, Perspektivenwechsel und Perspektivenübernahme," in *Wie ist Fremdverstehen lehr- und lernbar?* eds. Lothar Bredella, Franz-Joseph Meißner, Ansgar Nünning, and Dietmar Rößler. Tübingen: Narr, 2002, 84–132.

———. "Fremdverstehen und Bildung durch neue Weltansichten: Perspektivenvielfalt, Perspektivenwechsel und Perspektivenübernahme durch Literatur," in *Neue Ansätze und Konzepte der Literatur- und Kulturdidaktik*, eds. Wolfgang Hallet and Ansgar Nünning. Trier: WVT, 2007, 123–142.

Plaut, Victoria. "Cultural Models of Diversity: The Psychology of Difference and Inclusion," in *Engaging Cultural Differences: The Multicultural Challenge in Liberal Democracies*, eds. R. Shweder, M. Minow, and Hazel Rose Markus. New York: Russell Sage, 2002, 365–395.

Reich, Kersten. "Verstehen des Fremden in den Kulturen und situiertes Lernen: zu Grundsätzen einer interkulturellen Didaktik," in *Fremde Kulturen verstehen—fremde Kulturen lehren*, ed. Heinz Antor. Heidelberg: Winter, 2007, 71–90.

Richeson, Jennifer A., and J. Nicole Shelton. "When Prejudice Does Not Pay: Effects of Interracial Contact on Executive Function." *Psychological Science*, 14 (2003), 287–290.

Richeson, Jennifer A., and Richard J. Nussbaum. "The Impact of Multiculturalism versus Colorblindness on Racial Bias." *Journal of Experimental Social Psychology*, 40 (2004): 417–423.

Richeson, Jennifer A., and Sophie Trawalter. "Why Do Interracial Interactions Impair Executive Function? A Resource Depletion Account." *Journal of Personality and Social Psychology*, 88 (2005): 934–947.

Roche, Jörg. *Interkulturelle Sprachdidaktik*. Tübingen: Narr, 2001.

Roth, Philip. *The Human Stain*. London: Vintage, 2000.

Sanchez, George. "Creating the Multicultural Canon: Adventures in Post-nationalist American Studies," in *Post-nationalist American Studies*, ed. John Carlos Rowe. Berkeley: University of California Press, 2000, 40–58.

Sánchez-Casal, Susan, and Amie A. MacDonald, eds. *Twenty-first-Century Feminist Classrooms: Pedagogies of Identity and Difference*. New York: Palgrave Macmillan, 2002.

Scorah, Rebecca. "T. C. Boyle's *The Tortilla Curtain* as Melodrama—Consequences for the Classroom" Talk given at *Postethnicity?—Identity Politics Reconsidered: North American Theories and Literary Practices in and outside of German Multicultural Classrooms Conference*. Georg-August-Universität-Göttingen. July 17, 2007.

Seletzky, Martin. "Deutsch für multikulturelle Klassenzimmer in den U. S. A.: Ein Planungsraster für die Inszenierung interkultureller Lernprozesse in multikulturellen Lerngruppen," in *Interkultureller Fremdsprachenunterricht*, eds. Lothar Bredella and Werner Delanoy. Tübingen: Narr, 1999. 183–208.

Shelton, J. Nicole. "Divergent Attributions, Divergent Experiences: Whites and Ethnic Minorities in Interracial Interactions." Talk given at Postethnicity?—North American Theories and Literary Practices in and outside of German Multicultural Classrooms Conference. Georg-August-Universität-Göttingen. June 24, 2007.

Shelton, J. Nicole, and Jennifer A. Richeson. "Intergroup Contact and Pluralistic Ignorance." *Journal of Personality and Social Psychology,* 88.1 (2005): 91–107.

Shelton, J. Nicole, Jennifer A. Richeson, and Jessica Salvatore. "Expecting to Be the Target of Prejudice: Implications for Interethnic Interactions." *Personality and Social Psychology Bulletin,* 31 (2005): 1189–1202.

Shelton, J. Nicole, Jennifer A. Richeson, and J. D. Vorauer. "Threatened Identities and Interethnic Interactions." *European Review of Social Psychology,* in press.

Steele, Claude M. "Race and the Schooling of Black Americans." *The Atlantic Monthly,* 269.4 (1992): 68–78.

———. "Stereotyping and Its Threat Are Real." *American Psychologist,* 53 (1998): 680–681.

Steele, Claude M, and Aronson, Joshua. "Stereotype Threat and the Test Performance of Academically Successful African Americans," in *The Black-White Test Score Gap,* ed. C. Jencks and M. Phillips. Washington, DC: Brookings, 1998.

Surkamp, Carola. "Handlungs- und Produktionsorientierung im fremdsprachlichen Literaturunterricht." *Neue Ansätze und Konzepte der Literatur- und Kulturdidaktik.* Eds. Wolfgang Hallet and Ansgar Nünning, Ansgar. Trier: WVT, 2007, 89–106.

Volkmann, Laurenz. "Aspekte und Dimensionen interkultureller Kompetenz." *Interkulturelle Kompetenz. Konzepte und Praxis des Unterrichts.* Tübingen: Narr, 2002, 11–48.

———. "*The global village*: Von der interkulturellen zur multikulturellen Kompetenz," in *Fremde Kulturen verstehen—fremde Kulturen lehren,* ed. Heinz Antor. Heidelberg: Winter, 2007, 127–158.

Weskamp, Ralf. *Fachdidaktik: Grundlagen und Konzepte Anglistik und Amerikanistik.* Berlin: Cornelsen, 2001.

Zentralabitur-Themen Niedersachsen / topics for the state-run final high school exam in Lower Saxony: <http://www.nibis.de/nli1/gohrgs/zentralabitur/zentralabitur_2006/02englisch.pdf> Last accessed January 22, 2009. <http://www.nibis.de/nli1/gohrgs/zentralabitur/zentralabitur_2007/02englisch.pdf> Last accessed January 22, 2009.

11

THE USES OF ERROR: TOWARD A REALIST METHODOLOGY OF STUDENT EVALUATION

John J. Su

What does realist theory have to say about the evaluation of students' performance in the humanities classroom? My question is motivated by the concerns students have expressed to me with the grading systems in their courses, and my own sense that realist theory might provide guidance in addressing these concerns. Particularly in qualitative fields such as English literature, the field in which I work, there is a fairly common perception among students that grade determinations are subjective, owing more to how well their ideas correspond with their instructor's than to their ability to produce work that meets a coherent set of course objectives. And little wonder. In many courses, little or no written documentation is ever provided describing the course objectives, the criteria by which individual assignments are assessed, or the relative weight of assignments in the determination of final grade. Factors such as "class participation" may constitute up to a quarter of a student's final grade without ever being explicitly defined, much less presented in a manner that explains why such factors should be relevant in the determination of grades.

The growing interest in realist theories and realist pedagogy across the humanities and social sciences—sparked by the 2002 publication of Amie A. Macdonald and Susan Sánchez-Casal's edited volume, *Twenty-first-Century Feminist Classrooms: Pedagogies of Identity and Difference*—suggests that now might be an opportune moment to explore the possibility of a realist methodology of student evaluation. The project of "democratizing" the classroom, as articulated by Macdonald, Sánchez-Casal, and others, could only benefit from a system of evaluation that students consider to be clear, coherent, fair, and, to the extent possible, objective. The poststructuralist and postmodern theories that continue to circulate within many university classrooms have played a crucial role in helping students to understand the relationship between power and knowledge by providing the skills to question unstated assumptions and to reveal the instability of all truth claims; however, the rejection of the idea of objective knowledge on which such theories are based limits their potential usefulness for the formulation of university grading standards. The existence of a grading system presupposes that students acquire knowledge to varying degrees, and that instructors have the ability to discern with some degree of accuracy qualitative differences among students. Whatever doubts an instructor may harbor about the idea of grades and the notion of objective assessment, grades are an inescapable reality. And the project undertaken by realists to defend a more nuanced, post-positivist notion of objectivity despite the critique of positivism and its notion of a theory-independent objectivity may prove useful for the creation of better grading systems.

The relative dearth of scholarship on the topic of grading is due in no small part to the fact that it is a source of ongoing embarrassment for many college and university instructors. As Frances Zak and Christopher C. Weaver note, "grading puts us in the uncomfortable position of having to reconcile our authority over students with our desire to empower them" (xv). The ideals of democratizing the classroom seem at odds with the almost unfettered power granted to instructors in the area of grading. Unless a student files a formal protest, there is rarely any oversight on how individual assignments or final grades are determined. Thus, it is not surprising that a number of scholars categorically assert that grading is antithetical to teaching. Betty Garrison Schiffman, for example, insists that for those engaged in feminist teaching, "grades get in the way of learning" (58). Over the course of this essay, I will argue that such a perception assumes an artificial and unproductive distinction between learning and assessment. More immediately, however, I want to challenge the assumption that grading is inherently discriminatory. Historically, women and racial minorities were not excluded from positions of power primarily by means of uniform evaluation systems; rather, they were denied the opportunity to demonstrate their mastery of skills and knowledge in comparison to their white male peers. As Kathleen Yancey powerfully writes, "it is only because of *grades* that girls get some kind of fair shot in school...When we take grades away, we take away the one means currently available for girls and women to show what they do know" (quoted in Allison, Bryant, and Hourigan 7). Our students—all of our students—have the right to expect a fair evaluation of their ability to acquire skills and knowledge relative to their peers. As such, I will be very specific in this essay about my own grading practices, something I have rarely found in scholarship on student assessment. This naturally entails risk. Readers will no doubt discover my own errors and biases, but this is a necessary part of the process of creating a system that our students deserve.[1]

To achieve a more genuinely realist methodology of student evaluation will require significant rethinking of practices currently employed in many classrooms across the country. On the most basic level, this essay encourages a shift away from thinking of evaluation as "grading," a fairly narrow process characterized by identifying errors in students' work. Instead, evaluation in the present context will be seen as a more encompassing and dialogic process of helping students to use their errors in the pursuit of knowledge, a process that begins from the moment syllabi are distributed on the first day of class. The implications of this notion of evaluation can be encapsulated in six recommendations that will be explored more fully in the body of this essay:

1. to establish explicit criteria for evaluating even the most amorphous assignments;

2. to use the established criteria as a means of discussing the learning objectives for the course and how they can be mastered;
3. to identify the methodology of evaluation, and how the instructor's biases inform grading;
4. to offer guidance on how errors can be used to improve performance on subsequent assignments;
5. to develop assignments in clusters/series and to reward demonstrated progress in achieving learning objectives;
6. to rearticulate the criteria for evaluation at periodic intervals and in varying formats to assist different kinds of learners.

***

Perhaps the most significant objection to the project I am proposing would be the claim that a realist methodology of evaluation already exists in the natural sciences, and that it provides the only method for objective evaluation. Certainly, the sciences provide a very clear and coherent system of evaluation that is easy to justify. In most such courses (lab courses excepted), students are measured by the extent to which they demonstrate mastery over materials presented in lecture. Uniform written examinations provide the primary means of evaluation, so that all students are measured according to a common and at least putatively absolute scale. The exam results may be "curved" in order that the final grade distributions correspond to some historical or theoretical norm, but the rankings of students with respect to each other remain consistent. In other words, if one student scores 94 and another scores 87 on an exam, the first student is understood to have done better whether or not the final grade distribution determines that one or both received an A. This system of evaluation is considered objective in the sense that a single standard is applied to all students regardless of who they are, and that within this system the identity or opinions of the grader are irrelevant to the process of grading. The answer key is the final arbiter, and the question is simply: did the student get the correct answer or not?

The ease and apparent neutrality of this system lead to the widespread perception that all qualitative measures are subjective, and hence flawed. This conclusion, however, should be qualified on at least two grounds. First, the assumption that eliminating subjective judgment is possible and desirable is questionable. In many natural science courses, examinations offer the possibility of "partial credit" for student answers that demonstrate knowledge of the theories or principles necessary to answer a particular question even if the student did not get the final answer correct. But once the possibility of partial credit enters into consideration, the identity and perceptions of the grader do become relevant factors in the evaluation process. When I was a teaching assistant for university math courses, I made interpretive decisions on every examination I graded. The very fact that I had to make an interpretation opened up the possibility that another grader could have assessed the same student answer in a different way. Yet, neither students nor instructors typically perceive this possibility as representing a significant loss of objectivity. Judgment calls of this nature generally involve a fairly small percentage of cases and the adjustments to the final scores are often relatively insignificant; hence, the overall objectivity of the system seems unthreatened. Indeed, a mathematics examination that offers the opportunity of partial credit can be more objective than one that does not, despite the fact that it opens up greater opportunities for grader bias to affect the final result. The student who provides the correct formula for solving a particular mathematical problem but makes a computation error demonstrates a more sophisticated and accurate analysis than a student who simply skips the problem, and an objective evaluation would credit this difference. But to accept this point means to acknowledge that subjective judgments are not *necessarily* antithetical to objective evaluation. Subjective bias is not the problem per se so much as inconsistency in how such biases influence the evaluation of students.

The second and more substantial qualification to the idea that nonquantitative measures are flawed is that humanities courses could easily adopt such a system, but have not. Courses could be restructured to focus on the explicit content presented in the texts under study, and examinations could be designed to focus on questions whose answers can be verified by the texts themselves. The answer to the question "in what year was the Declaration of Independence signed in the United States" is no less verifiable than answers to questions of how to integrate an equation, to determine the surface area of a sphere, to measure the pH of a solution, or to calculate the amount of time it will take for a rocket moving at a given speed to travel from the Earth to Mars. This altered form for humanities courses would have the added benefit of reducing significantly the labor for instructors both in terms of course design and grading. Examinations could be designed using a multiple-choice format that could be electronically scanned in order to provide quick and accurate results.

That the majority of Humanities faculty have not in fact switched to such a system indicates not a lack of intellectual rigor or laziness; rather, there is a definite if often undertheorized sense that the sorts of skills taught in humanities courses are not amenable to the methodologies of evaluation that are so useful to colleagues in the natural sciences. Quantitative measures are ill-suited for assessing the ability of students to discuss complex concepts in coherent and systematic ways, to present arguable theses that are by their very nature subject to debate and therefore not absolutely verifiable, and to develop an awareness of the richness and diversity of human modes of thought. The fine discriminations characteristic of quantitative measures require the existence of an idealized template or "answer key" that would be unable to credit the various ways in which course objectives in many humanities classes can be met. Students can employ very different strategies, styles, and genres in fulfilling a critical essay assignment, for example, and assessing their performance according to an absolute scale with gradations as narrow as a single percentage point (as used in natural science classrooms) often cannot be accomplished with sufficient consistency to warrant such a system. The degree of imprecision here does not imply the absence of standards or that such standards are entirely relative, simply that measurements using such standards are often made in more qualitative terms.

The difficulty in importing evaluation systems from the natural sciences should not be entirely surprising: quantitative measures were designed within disciplines that traditionally assumed the possibility of creating absolute distinctions between facts and values. The widespread currency of the fact/value distinction, Hilary Putnam argues, has led to the perception that inquiries into questions involving values cannot be objective. In other words, different disciplines are not seen to provide different kinds of objectivities; instead, the natural sciences provide the only means of objective knowledge, and the humanities are trapped within the realm of subjective perceptions and feelings. Facts, according to this conception, are independently and empirically verifiable; values, in contrast, are typically inseparable from the cultures in which they exist, and their validity often cannot be verified by empirical means. Whether or not a rigid distinction between facts and values is defensible (and for Putnam it is not), it has had the effect of bracketing off from scientific consideration a whole host of questions, then, in disciplines from economics to philosophy (see *Renewing Philosophy*). And it is precisely these questions of value to which the humanities are devoted. To do justice to these questions requires instructors in humanities either to abandon the idea of an objective system of evaluation altogether or to rethink the positivist notion of objectivity that continues to circulate as the dominant understanding within the natural sciences and much of American society as well.

Realist theory, as articulated by so-called "post-positivist realists" including Satya Mohanty, Paula Moya, Michael Hames-García, and Linda Martín Alcoff, as well as their antecedents including Putnam, W. H. Quine, and Richard Boyd, provides a compelling inroads to the latter option. Rejecting the positivist idea of objectivity as unchanging, unmediated, and free of bias, these scholars characterize objectivity as historically situated and context-dependent.[2] On this understanding, the pursuit of a single transcendental, God's-eye view of the world is misleading and counterproductive. Individuals can describe more or less accurately the worlds they inhabit, but this capacity does not result from their ability to divorce themselves from their life experiences, beliefs, and identities. Quite the opposite, the ways in which individuals perceive themselves often help them to make sense of the world around them. Thus, individual perspectives and values do not necessarily inhibit objectivity but are in many cases its necessary precondition. As Moya argues, "identity categories provide modes of articulating and examining significant correlations between lived experience and social location" (*Reclaiming Identity*, 4).

The idea that an individual's subjective values and beliefs can facilitate rather than inhibit the acquisition of objective knowledge depends on a notion of error that will be crucial to this essay's understanding of student evaluation. The abandonment of the idea of a single objective standpoint by post-positivist realists implies that inquiry should not be directed toward acquiring the single "correct" account of the world. Rather, realist theory speaks in terms of better and worse accounts, and the pursuit of knowledge is understood as the pursuit of increasingly less distorted explanations of social phenomena. Objectivity, thus understood, signifies the theoretical standpoint that any given time and place provides the most coherent and least distorted explanation of a particular event.

That is, objectivity is not a single fixed account or theory, and different people will be more objective in different situations.

In this context, error is an inevitable part of learning as individuals work to refine their theories in light of circumstances that they cannot currently explain. Satya Mohanty writes

> An essential part of this conception of inquiry would be an understanding of fallibility which is developed and specified through our explanations of how different kinds and degrees of error arise. Precision and depth in understanding the sources and causes of error or mystification help us define the nature of objectivity, and central to this definition would be the possibility of its revision and improvement on the basis of new information. (215)

This conception of error strikes a difficult but productive balance between positivism and relativism. Error, like objectivity, is characterized as socially situated and context-dependent, yet this characterization does not imply a solipsistic universe. I may believe that a particular hand gesture conveys a certain meaning, say "victory"; however, if I use this same gesture in another culture in which it is understood to convey an obscene meaning, I cannot simply will that people in that culture accept my interpretation. I will be in error if I use this gesture in such a situation because I would fail to convey my intended meaning. Yet, because such knowledge is highly situated and culturally contingent, it would have been difficult, though not impossible, for me to acquire it ahead of time. Out of context, such knowledge would be abstract for a foreigner, and hence difficult to internalize. Knowledge only becomes concrete and immediately relevant when actively applied or misapplied, and this point suggests why Mohanty focuses on error in his theory of knowledge. The analysis of error provides the necessary means for refining knowledge in light of unfamiliar or alien circumstances. The horrified or angry reactions elicited by my gesture provide immediate and directly relevant feedback about my knowledge of the customs and language of the people with whom I interact, feedback that provides both the basis and impetus to revise my knowledge. By analyzing my gesture in that particular context—an analysis that will probably require a conversation with those whom I have offended—I have the opportunity to gain greater knowledge of what that particular gesture means in cultures outside of my own. Put in more theoretical terms, knowledge can often be understood as a set of codes or discourses with which individuals can gain greater fluency, and error can be understood as the beliefs or habits that, at any given moment, inhibit fluency.

To characterize knowledge in terms of fluency suggests that the acquisition of knowledge does not involve learning a set of abstract facts so much as engaging in a process of continual practice and self-correction. And to make such a process central to the classroom requires, before anything else, a certain humility on the part of instructors about what constitutes knowledge. Instructors and students alike tend to identify the instructor as the final arbiter of what constitutes truth. Yet, realist theory suggests that every individual possesses a unique body of knowledge that arises from their identity and experience. Courses inevitably value certain forms of knowledge over others, but such determinations should not simply dismiss the relevance of student knowledge and the values they might have. Values are always determined with respect to a particular reference group, Moya suggests, and instructors need to acknowledge that their own values are not necessarily universal (*Learning from Experience*, 161).³ Such knowledge is potentially lost to the student, his or her peers, and instructors if all expertise in the classroom is perceived to be vested with the instructor and error is cast as the disjunction between what the instructor and the student knows. The loss of such knowledge is damaging not only because it represents unique insights that the instructor may lack but also because it provides clues about the sources of student errors and how they might be corrected. For, as David Bartholomae argues, errors are not only examples of random accidents or behavior; they often indicate a kind of logical reasoning or "systematic, coherent, rule-governed behavior" (257; cf. Williams, 153). Keeping with the example of my "victory sign"/obscene gesture, my use of the gesture was not motivated by ignorance in any absolute sense; to the contrary, I used the gesture because I "knew" what it meant. I was operating with a set of rules in mind about what constitutes nonverbal communication. Acquiring knowledge, in this instance, does not require me to repudiate previous knowledge, but to situate it and to recognize that operating in other circumstances may require the acquisition of other models of rule-governed behavior from those that I currently possess. Thus, one of the central lessons of this experience is that different rule systems are applicable in different situations. This is to deny neither that there are definite bodies of knowledge or expertise that instructors possess nor that there are many situations in which questions have definitive answers that are independent of many facets of a person's identity. The United States declared independence from Great Britain in 1776 and parallel lines never meet regardless of my gender, ethnicity, or sexual orientation (though not regardless of my religious beliefs. Nor are the statements true if we are using a non-Georgian calendar or non-Euclidean geometry respectively). But frequently, the more sophisticated and complex questions in the various disciplines, such as how economic resources should be distributed in a society or whether the founding fathers should have outlawed slavery, can yield different answers depending on the social or cultural contexts in which such questions are discussed. Thus, the challenge for instructors in the classroom is to create an environment in which forms of student knowledge are not denigrated or disregarded while still maintaining the importance of the knowledge that the instructor can make available.

To meet this challenge requires the creation of a classroom evaluation system that encourages attention to error and how it can be used to revise and improve existing knowledge. The system of evaluation that operates within any particular class shapes the direction it takes, identifying for students what skills and materials to prioritize and what can be neglected. Traditional grading techniques, in which feedback is provided only after the final version of an assignment is submitted, discourage error analysis on the part of students. Students perceive very little direct and immediate benefit from error analysis in such classrooms, and hence errors become relevant only insofar as they indicate a failure to conform to the expectations of the instructor. Only if there are subsequent assignments that are fairly similar in form and if a student is enterprising and diligent enough to generalize comments from one assignment to the next will there be much likelihood of success.

Caroline Hau's reading of Mohanty suggests that a more fruitful emphasis in the classroom would be on what she calls the "*uses* of error" rather than on error itself (160, italics in original). Hau conceives of error analysis as having both descriptive and normative components, whereby individuals respectively examine how error arises and how similar errors can be identified in the future. This distinction is useful for present purposes because it emphasizes how the analysis of error can be seen as a recursive process of refining an individual's theories and the ways in which they influence his or her practices in light of new information. A similar impulse within composition studies has motivated scholars such as Brian Huot to emphasize the dialectical relationship between theory and practice as a useful model for thinking about pedagogy (165). Although error is less of a significant term in Huot's thinking, he shares with Hau a vision of a classroom environment in which error analysis is integrated into the daily work in which a class is engaged; this altered form for running classroom sessions would focus class time on the identification of student error, its possible sources and outcomes, and how to reduce if not overcome the same error on subsequent occasions.

The more encompassing notion of student evaluation proposed here needs to start from the first day of class with the presentation of clear course objectives and the establishment of explicit criteria for how all significant assignments will be assessed (recommendation one). By providing such criteria in written form, students will be better able to address errors that arise from implicit or poorly communicated expectations. More importantly, providing such criteria will help students to identify more precisely what their goals are in any given assignment, and this in turn encourages the idea that students can increasingly identify for themselves potential errors in their work. Explicit descriptions of how course assignments are evaluated are particularly important for more amorphous factors, such as class participation, for which students often have a very difficult time identifying what skills are being cultivated.

In my own evaluation sheets for class participation (see appendix 1), four broad categories are identified as factors considered when assessing student performance: the ability to ask questions, to offer personal interpretations, to support/critique other members' ideas (including my own, of course), and general citizenship (i.e., respect and generosity toward other members of the class). The assessment itself combines quantitative and qualitative components. Each grade is characterized by the frequency with which the various factors are employed by a student; each grade also has subtly different qualitative distinctions within a given factor. For example, questions of a thematic nature typically demonstrate a higher degree of intellectual engagement in classroom discussion than questions of factual nature and are more likely to sustain an ongoing conversation; as such, the higher grades require more frequent demonstration of such questions.

The presentation of these criteria can itself be a powerful learning opportunity for students, and should be made a part of classroom discussion (recommendation two). The various differences described in my class participation sheet, for example, help to define what constitutes a more sophisticated and complex contribution to class discussion. By introducing such criteria as an explicit matter for discussion in the classroom, students are encouraged to assess periodically their own performance according to these standards. In this way, for example, a student whose contributions occur on a daily basis but are limited to asking factual questions about readings would be able to identify why he or she would not receive the highest evaluation. Students can identify the errors in their performance by themselves or in consultation with the instructor, using the terminology provided by the sheet to discuss how participation skills might be developed and further refined.

Explicit discussions about the criteria by which students are evaluated are crucial to addressing the concerns raised by Lisa Delpit that instructors frequently fail to meet the needs of minority students. On Delpit's account, the "culture of power" operating within any given society has a set of rules and codes by which individuals are assessed, and families from minority populations in many cases lack access to these codes (286). As such, one of the central tasks of schooling from kindergarten through the university must be instruction in how to master such codes. "I suggest that students must be *taught* the codes needed to participate fully in the mainstream of American life," Delpit argues, "...they must be allowed the resource of the instructor's expert knowledge, while being helped to acknowledge their own 'expertness' as well" (296). Explicit descriptions of evaluation criteria make available the codes by which students are assessed in the classroom and in many areas of society more broadly. More so-called "process theory" pedagogies are often counterproductive in that they either tacitly assume student knowledge of these codes or attempt to ignore them altogether.[4] This focus becomes apparent, for example, in Christopher C. Weaver's conception of grading: "if we truly believe in our process pedagogies," he argues, "then we need to construct grading systems which foreground the complexities of the writing process and which minimize, defer, or possibly even ignore questions about the quality of the writing our students ultimately produce" (142). While the effort to challenge normative ideas of correctness is certainly laudable, students will be assessed on their products when they leave their schools and enter the workforce. Hence, we do students no favors by pretending that such systems do not exist outside the classroom or even within it.[5]

It is not only disingenuous but also detrimental to learning to pretend that there is a rigid separation between the classroom and the social environments in which students live. And this is why efforts by Macdonald, Sánchez-Casal, and the other contributors to this volume to mobilize identities in the classroom represent such a powerful pedagogical approach. The notion that the classroom can be a kind of laboratory for disinterested inquiry in which beliefs are checked in at the door ignores the fact that individuals are always already in the world, committed to a number of beliefs and goals. As Alcoff suggests, total disengagement from these beliefs is practically and conceptually impossible. Indeed, Alcoff's work in epistemology is particularly illuminating for classroom practice because she insists on the importance of accounting for the "real starting position" of individuals (*Real Knowing*, 221), and for thinking about how more accurate knowledge about the world can be achieved without postulating some idealized learner. Philosophy, for Alcoff, and pedagogy for present purposes, needs to recognize "the unavoidable fact of our prior commitments, our situatedness in specific contexts, and our general tendency from within a set of cognitively relevant practices to conserve that which we think we already know" (230).

This last point is particularly significant because it marks my most significant departure from positivism. The previous recommendations in this essay—for clear grading criteria and discussion of the various skills that a course seeks to cultivate—could well be central features of a positivist classroom. In contrast, this essay's third recommendation—to identify the methodology of evaluation and to discuss how the instructor's biases inform grading—is based on a theoretical principle that would be contrary to positivism. According to a rigidly positivistic model, such a discussion would be unnecessary or at least fairly cursory. For, as stated earlier, positivism encourages individuals to move beyond their theoretical biases, identities, and personal commitments in the pursuit of knowledge. The implication of this pursuit for student evaluation should be that a particular assignment (a critical essay, for example) should be evaluated by the same criteria and in the same manner by all instructors. Any variation in grading practices, according to this model, would presumably imply the failure of an instructor to be fully objective—that is, to evaluate without bias.

Such a notion, however, runs against the reality of grading practices in university classrooms across the country. As most university students recognize, instructors have different criteria, focus on different kinds of error, and reward different kinds of skills. The notion of a post-positivist objectivity on which I have been drawing provides a way of understanding this phenomenon without seeing it necessarily as a failure on the part of instructors. On this understanding, the context-specific nature of knowledge suggests that the varying foci of instructors in different courses represent not inconsistency but different modes of producing knowledge. Indeed, multiple modes of evaluation might be seen as the hallmark of a healthy university environment, one that provides students the opportunity to perceive the variety and diversity of human perspectives and the ways in which a particular kind of assignment might be approached.

Even viewed from a more pessimistic perspective, presenting to students the methodologies by which they will be evaluated minimally teaches students the importance of learning how to approach a problem in a way they might not normally yet which appeases an authority figure. The ability to approach a problem in a flexible manner according to the desires of an authority figure or "customer" is a very powerful skill in the marketplace, and one that most of students will have to cultivate in an ever increasingly service-oriented economy. Alcoff's philosophical defense for the importance of accounting for our "situatedness," like Delpit's conception of social interactions as "codes," challenges the false division in many classrooms between "pure" knowledge and "vocational" skills. The tendency to demean the importance of the latter again assumes a vision of the classroom as isolated from social environments in which students live and interact.

The appendix describing my own grading practices for critical essays thus is formatted in terms of the approach I take in the process of grading rather than in terms of the defining characteristics of each given grade, as was the case with the previous appendix discussed [appendix 2].[6] My discussion of how essays are evaluated still distinguishes among more and less sophisticated features in essay; note, for example, that theses are discussed in terms of four progressively more sophisticated forms. Such descriptions are absolutely crucial for encouraging students to think of themselves as participating in a cumulative process of acquiring knowledge by progressively eliminating more and more subtle errors. Yet, the overall description of evaluation is organized chronologically according to how I approach and assess a given essay.

Presentation of the methodology of assessment recognizes that different instructors are more attentive to certain kinds of errors than others, and helps students to perceive why a given instructor focuses on the errors that he/she does. In my own case, the fact that formatting and prose are listed as the very first criteria to be assessed indicates that inadequate prose can have such a significant impact on the final grade assigned to a student essay. The discussion of this fact also provides the opportunity to discuss with students why this is the case. Poor grammar and proofreading are so detrimental to an essay that otherwise presents a powerful thesis that is discussed in a logical and coherent manner, according to my analysis, because before I arrive even at the thesis, I have already been presented with the prose, format, and general appearance of the essay. And, I explain, because these are the first features I notice, they shape the ways in which I will value whatever occurs later. Students and other instructors may dispute this perception, but this is precisely the point. By providing very clear written guidelines about my own methodology, I can help students to appreciate the theoretical foci that inform my own work; or, more modestly, I can provide students the ability to foresee what kinds of errors need to be addressed most urgently within my courses.

My claim here can be seen as an application of Alcoff's notion of "positional perspective" ("Cultural Feminism," 349). For Alcoff, the beliefs and assumptions that individuals bring to bear in any given interaction are profoundly shaped by the locations and cultures with which they identify. And this "locus" of values will inform how individuals, including both instructors and students, make choices. That individuals lack a kind of existential freedom to make decisions based purely upon their own personal interpretation of their experiences is not seen as a negative situation, according to Alcoff. Rather, the idea of an individual free of all ideological biases is an unhelpful illusion because it encourages individuals to disregard their own backgrounds, to perceive them as a burden rather than a potential source of knowledge. The idealized existential subject is also an unhelpful heuristic in the classroom because it tends to encourage criticism of those who draw upon crucial features of identity including gender, ethnicity, and religious belief. By recognizing that all individuals have positional perspectives, Alcoff encourages focus on how such perspectives can usefully be brought to bear in our interactions with others and how they might help to produce accurate knowledge about features of our world. When instructors conceal or deny the existence of their own theoretical assumptions, they are not in fact promoting openness or a multicultural classroom. Quite the opposite, they produce an environment in which there exists a normative but unstated perspective that is nonetheless recognizable to many students. Such environments encourage students to focus on "what the instructor wants" rather than on what will cultivate their knowledge or provide them the skills to achieve their life goals. As Juan C. Guerra suggests, if an instructor's goal is to foster a classroom environment "in which we encourage critical thinking and all participants can challenge one another's beliefs, it follows that we cannot avoid letting our students know where we stand, what our positional perspective is" (259).

The reality that different courses and instructors will value skills and errors somewhat differently means that the notion of defining knowledge through the analysis of error proposed by realist theory is all the more important in the humanities classroom. If students are to provide increasingly sophisticated analyses, then they need to know very precisely the sort of errors they are committing. Are they employing the wrong codes, to reinvoke Delpit's term? Are they failing to utilize a particular code in an idiomatic way recognizable to and validated by the "culture of power"? Are they using techniques that undermine their fluency with particular codes of techniques or styles? After the type of error is identified, students need explicit guidance on how such errors might be avoided in subsequent assignments (recommendation four). Such guidance must be given carefully, and instructors must be careful to avoid creating the perception that there is a single process by which error can be overcome. The critique of this more traditional notion of learning has been usefully made by so-called "post-process" theorists in composition studies, who urge greater focus on challenging "received notions or entrenched understandings" in the classroom (see Olson, 8). To accomplish this requires a kind of ongoing dialogue that, in the ideal circumstances, would mean that instructors have individual conferences with students before all large assignments are due. Of course, the time commitment this would require is prohibitive for many instructors, and here ideal pedagogy runs against the realities of an instructor's limited resources. Many of the same benefits can be had through much less time-intensive measures. One of the most successful I have found is the use of student critiques (see appendix 3). While students will certainly be less capable of identifying errors than their instructors, they are often far more capable of identifying errors in the works of their colleagues than in their own. Given that all students would have explicit written guidelines for how assignments are evaluated, they can be encouraged to use these guidelines to focus their critiques. Such an assignment has the added benefit of encouraging students to think of themselves as a learning community whose members can benefit from the knowledge and expertise of each other.

This fourth recommendation also has implications for how instructors should formulate their responses to students' work. To use the example of the critical essay again, it is very difficult for students to identify what errors must be addressed most immediately on the basis of marginal comments alone. Nor do such comments typically help students to address the errors presented. To help address these problems, a typed summary note appended to assignments can be very useful (I emphasize typed because, from my experience, students who cannot read the handwriting of their instructors are unlikely to go to them for clarification). In my own practice (see appendix 4), I try to model the realist notion of acquiring knowledge. After identifying what was successful and what were the central errors of an essay, I explicitly identify one or two areas for each student to focus on in his or her next essay.[7] I also encourage students to consult me if they have any questions, and to bring their essay with them when they come to subsequent conferences on their next assignment. In this way, I try to encourage students to see themselves as engaged in a process of continually refining their work rather than being faced with a set of relatively unrelated assignments.

To cultivate further the sense that a course involves an ongoing process of refinement, it is crucial to develop assignments in clusters or series, and to reward demonstrated progress in achieving the course's learning objectives (recommendation five). One of the more obvious ways to accomplish this end is to assign, where possible, a particular kind of assignment multiple times and to weight disproportionately the later versions. Equally important, however, is to design a series of assignments that target a particular kind of error. In my own courses, for example, I have found that one of the biggest struggles students have is to distinguish between arguable theses and theses that restate the ideas presented in the texts students read. This particular error often leads to a number of other errors including a tendency toward plot summary, repetitious argumentation, and insufficient consideration of the implications of a thesis. To address this problem, I have created a set of assignments that begin from the first day of class. On the syllabus, there are listed what I call "pinpoint questions" and "discussion questions" for most class sessions (see appendix 5).[8] Pinpoint questions are factual questions that can be answered on the basis of careful reading; discussion questions, in contrast, are thematic questions that result from the answers to pinpoint questions. By asking students to consider both on a daily basis, I make the distinction between what the text says and what arguments can be made about it an explicit feature of classroom discussion. The second set of assignments help students to build on the distinction established by the pinpoint/discussion questions. In a series of pop quizzes, students are asked to compose paragraphs answering questions similar to the discussion questions. This assignment moves beyond the first in that it encourages students to formulate a basic argument in writing. Because all of the quiz questions are based off of classroom discussion, responses are simpler than what will be asked of them later in critical essays, where they will need to formulate an original argument. In the third step of this series, students take a midterm examination that asks them to perform not only quiz-type questions but also to compare and contrast the various texts they have read on a single question. This latter form of question moves students one step further toward constructing an original argument because it demands that students synthesize materials discussed from several texts. By formulating assignments in such a series that progressively requires more and more sophisticated reasoning, students are then more capable of accomplishing the most difficult task, composing a critical essay.

Finally, it is important to rearticulate the criteria of evaluation at periodic intervals and in varying formats throughout the term/semester (recommendation six). Realist theory's understanding of the importance of subjective bias provides a theoretical confirmation of the truism in developmental psychology and education that different individuals learn in very different ways. Thus, if sheets discussing grading criteria are distributed at the beginning of the semester, various kinds of review handouts can be distributed throughout. For example, after rough drafts for essays have been submitted, I distribute a handout discussing tips for revision strategies (see appendix 6). Unlike the appendix devoted to the evaluation of essays, which focuses heavily on the methodology, this later appendix is presented in a much more telegraphic format, noting key areas on which to focus, and providing a short description of each of these areas to remind students of issues and techniques discussed in class.[9]

All of the recommendations presented here were designed to require fairly minimal additional labor on the part of the instructor. The enormous time that must be devoted to teaching, service, and research means that whatever improvements are suggested must be guided by a basic principle of feasibility. More sophisticated methodologies of evaluation do no service to students if they require so much extra work that the instructor feels embittered toward his or her students. The process of implementing such a program can be done over time and be refined with each subsequent semester. For myself, I find that I am continually refining my own evaluation practices as students help me to discover my own errors. Currently, for example, I am thinking about how to increase student input into the design of course objectives. Ultimately, evaluation needs to be understood less in terms of a means of doling out rewards and punishments than an ongoing process of conversation among students and instructors, a conversation that will admittedly always be fraught with the dynamics of power and privilege that are an inescapable feature of the university grading system.

APPENDIX 1

Appendix 1

Description of Evaluation—Class Participation

In an effort to make the process of evaluation clear and fair to everyone, I am providing general guidelines for how I evaluate class participation. Class participation demands an active engagement of the mind with your peers. Hence, the first responsibility is attendance. Let me requote the syllabus:

> ATTENDANCE: As we are members of a community, with responsibilities to each other, regular attendance is required. Your classmates have the right of your mind, just as you have the right of theirs. You will be expected to attend every session, to have read thoroughly the reading for the day, and to participate regularly. In accordance with university policy, more than six absences will lower your grade.

Beyond attendance, evaluation of class participation takes into account four categories: the ability to ask questions, to offer personal interpretations, to support/critique other members' ideas (including my own, of course), and general citizenship (i.e., respect and generosity toward other members of the class).

Based on these categories, I generally break down grading as follows:

> A (excellent). Daily questions of factual and thematic nature; such questions periodically try to link the texts under discussion to other texts and to the central themes and issues of the course. Able to offer thoughtful personal interpretations of the material daily. Frequent attempts to develop and/or to challenge the ideas presented by classmates.
>
> AB (very good). Frequent questions concerning both factual and thematic issues; such questions periodically try to link the texts under discussion to other texts in the course or others the student has read elsewhere. Frequently offers personal interpretations of course material. Periodically supports and/or critiques the ideas of classmates.
>
> B (good). Frequent questions, periodically concerning thematic issues. Able to offer personal interpretations when elicited by me. Occasional efforts to support/critique classmates' ideas.
>
> BC (decent). Periodic questions relating to both factual and thematic issues. Some ability to offer personal interpretations. Fairly limited interaction with other members' ideas.

C (adequate). Consistent class attendance and note taking. Periodic questions, generally limited to factual issues. Inconsistently able to offer personal interpretations of reading material. Fairly limited interaction with other members' ideas. Respectful of other members. **Please note: this category represents the performance of a diligent student who always attends class, takes notes, but rarely if ever speaks.**

CD (inadequate). Inconsistent class attendance. Demonstrating a lack of preparation to discuss the material for the day.

D (unacceptable). Inconsistent class attendance and/or demonstrating disrespect to classmates.

I do reserve the right to modify these criteria under exceptional circumstances. However, in the interests of uniformity, I will avoid doing so if at all possible. If you ever have questions about how to improve your participation in the classroom, I am always happy to meet with you to plan strategies.

Appendix 2
Appendix 2
Description of Evaluation—Formal Written Essays

The process of evaluating a formal written essay occurs over the course of several readings and is a bit difficult to quantify. What I will do here is to describe the process by which I tend to evaluate an essay.

The basic task of all of the essays we write in this class is to present a coherent, original, and intellectually engaging argument that is developed over the course of the paper in a logical manner. Evaluation of an essay is based upon the degree to which the author succeeds in this task.

Format and presentation. In my first reading, I'm simply trying to work out what the argument is and how the author attempts to compel his/her audience. Hence, clear prose and correct format are essential. In the classroom, as in the workplace, a well-polished and professional looking essay is likely to impress readers and make them more likely to take the author seriously. Drawing upon industry standards, I permit one grammatical/stylistic/factual error per page without penalty. Such errors include, but are not limited to: errors in grammar, spelling, incorrect format, the absence of page numbers or staple/fastener, incorrect citations or quotations, erroneous information. Depending upon the frequency and severity of these errors, the final grade of the essay will be reduced 1–2 increments. Repeated grammatical errors are considered particularly unprofessional; as such, three or more instances of the same error are likely to result in a 2 increment grade reduction.

Thesis. The basis of any essay we write must be a clearly defined and original argument. There are many degrees of sophistication in thesis writing, so I'll begin from least sophisticated and move toward most. First, the thesis must be clearly identifiable and an essay should not have multiple theses, although a complex thesis might have several parts to it. Without a thesis, the reader will have a difficult time appreciating the intelligence and persuasiveness of your thoughts, and will be likely to dismiss or overlook what might be an otherwise compelling paper. Such an essay would probably receive a grade no higher than CD. Second, the thesis must be arguable. Remember: a thesis that no one would disagree with is not a thesis. The point of an essay is to convince a reader who might disagree with you to be persuaded to appreciate, if not agree with, your perception of the world. An essay without an arguable thesis will be unlikely to receive a final grade higher than C. Third, the thesis must be original. A thesis, in other words, must be the product of the author's own reading. A thesis that essentially restates an idea I or other members of the class presented does not represent an original idea. An essay with such a thesis would be unlikely to receive a final grade higher than BC. Fourth, a thesis must be an argument about the text, not a restatement of it. This is probably the most challenging aspect of thesis writing for university students. Sophisticated theses attempt to address **why** a text says what it does or explore the significance of an idea presented by a text. Adequate but less sophisticated university essays take ideas presented by a text as the thesis for the student's essay. When in doubt, ask yourself: does my thesis state an idea that is already presented by the text I'm writing about? Essays with such theses tend not to receive final grades higher than B.

Logic and argumentation. In my second and third readings, I'm attempting to assess how successful the essay is at developing its thesis and making a compelling case for it. As with the thesis, there are degrees of sophistication. The most basic level of mastery demands that every paragraph explicitly develop the essay's stated thesis. Papers in the C range often argue multiple theses or a thesis different from the one initially stated by the author. For papers of the length we are writing, every single paragraph must relate to and develop the thesis explicitly; otherwise, the author is likely to confuse the reader and make it difficult for him/her to appreciate what the most important points are. The second level of mastery comes from emphasizing the author's ideas and interpretations of the text rather than restating the plot of the text. Papers in the C range tend to spend too much time restating the plot; papers in the B range offer an interpretation of the text but still spend a majority of space describing plot; papers in the A range describe plot details only insofar as they assist the author's argument. Anything beyond this is extraneous. The third level of mastery comes from well-organized paragraphs. The paragraphs in the most sophisticated essays have a clear topic sentence which is developed in the body of the paragraph; less sophisticated essays often have multiple topic sentences or ideas in a single paragraph, which are likely to confuse the reader or to appear like unsupported assertions.

The fourth level of mastering logical argumentation is by far the most challenging. The most sophisticated essays develop their argument over the course of the paper in a logical manner that establishes clear causal relationships between paragraphs. Less rigorous essays (like the five paragraph model we've been taught in high school) have two major flaws in this regard: (1) they do not establish a relationship between paragraphs—each individual paragraph develops the thesis, but they tend not to develop earlier paragraphs; (2) they do not develop the thesis beyond its initial statement in the essay's introduction—most sophisticated theses require the author to explore what he/she means by them, rather than simply stating them. If you find yourself beginning paragraphs with phrases like "another reason that...," or "this also means...," you are probably not developing the relationships between your ideas. Essentially, the organization of your paragraphs is dictated by what you feel readers needs to know first, second, third, etc. in order to be compelled by your thesis. Essays with weak argumentative development (like the five paragraph essay) tend to fall in the C range, although well-polished versions can earn a B.

Now, I recognize that this list is formidable—it takes me minimally 30 minutes to read a single essay under this system. But this is the model for the most professional and rigorous essays, and it is a model that we can all achieve through hard work. It requires that we be willing to revise our work and to seek advice of peers and/or myself. To this day, I need to have minimally three readers critique my work.

If you have any questions in the process of writing, please see me.

APPENDIX 3

Appendix 3

[note: the student addressed here wrote an essay on Toni Morrison's *Beloved*]

Dear X:

You have written a good paper. You do a very fine job of setting up a problem and presenting an interesting theory; you also offer your readers numerous insights into the significance of scars as a metaphor for memory. To take this essay to the next level, you need to develop your assertions more fully, so that readers who are not already predisposed to agree with you will appreciate why your theory is so credible. I have identified key areas in the margins of your essay, and placed questions to guide you for revision.

For future essays, I recommend working on two areas. First, identify the topic sentence of every paragraph, and revise to ensure that each paragraph tells us something more about your thesis than we understood before. Particularly since you focus on several different characters, ask yourself why it is necessary to analyze Sethe after you have analyze Beloved. Second, identify what you consider to be your most key assertion in every paragraph. Force yourself to add 2–3 sentences to develop each of these assertions, explaining what exactly you mean and why your readers should believe you. You may find that these added sentences are simply repetitive, in which case you can delete some or all of them. But this practice will help to ensure that you have fully developed your concepts.
Grade: B

I'm looking forward to reading your next piece!
Best wishes,
John

APPENDIX 4

Appendix 4

Models of Heroism and the Depiction of Women from Homer to Milton

English 005, 1002 Fall 2002   *John Su*

Please note: come to class prepared to answer the pinpoint and discussion questions.

Syllabus:

M	8/26	Introduction: what is a hero?	
W	8/28	*Odyssey*	books 1–3

Pinpoint Question: why do human suffer, according to Zeus?
Discussion Question: how do the gods in Homer's cosmos differ from the God associated with Judaism and Christianity?

F	8/30	*Odyssey*	books 4–6

Pinpoint Question: who has foreknowledge of Odysseus' fate among the Phaeacians?
Discussion Question: why would a quest narrative reveal the outcome of events
in advance?

M	9/2	**Labor Day**	No Class
W	9/4	*Odyssey*	books 7–9

PQ: how does Odysseus outwit Polyphemus?
DQ: what do Odysseus' adventures tell us about his character?

F	9/6	*Odyssey*	books 10–12

PQ: how does the destruction of Odysseus' ship and men occur?
DQ: has Odysseus failed as a leader? What makes for a virtuous leader?

M	9/9	*Odyssey*	books 13–16

PQ: why is Odysseus told not to reveal himself in Ithaca?
DQ: Odysseus, Telemachus, and Penelope are all tested in the *Odyssey*: what is each character's test and why are they tested?

W	9/11	*Odyssey*	books 17–20

PQ:
DQ:

F	9/13	*Odyssey*	books 21–24

PQ: how do the Ithacans respond to the news of the suitors' deaths?
DQ: why does the story end as it does?

Appendix 5

Appendix 5

Description of Evaluation—Critiques

The ability to write coherent and thoughtful critiques of other people's writing is one of the most important skills to develop. Critiques help the author to see another perspective on his/her work; they also help the critique writer to understand problems in writing that may well affect his/her own work.

Let me cite the syllabus to remind us of the procedure for this assignment:

> WORKSHOPPING. This course relies heavily on workshopping our rough drafts. Bring 3 copies of your rough draft to class on the day the rough draft is due. The first copy is for me, so that I know you are making adequate progress (not to be graded or edited). Give the other copies to your peer editors (groups of peer editors will be determined in the second week of class). You will in turn read, edit, and write a two page critique for each of the other two members of your small group. Bring two copies of each of your critiques to the next class session: one copy for me and the second for

the writer. Please note: failure to turn in rough drafts on time will result in a half-grade reduction on the final paper.

Format: 1–2 pages typed, double spaced. I would recommend addressing the critique as a letter to the author of the essay. Bring two copies of each critique to class—one copy will be for the author and the other copy will be for myself.

Guidelines: critiques must address the three following areas: (1) what you liked about the essay; (2) questions you have about the essay; (3) areas for improving the essay. **I would strongly recommend consulting my appendix, "Description of Evaluation—Formal Written Essays," for areas on which to evaluate essays.**

Here are a few things I keep in mind when I write critiques:

1) What you liked about the essay: because most authors are uncertain about what their strengths or weaknesses are, it is important to begin your critique by addressing what the author did well.
2) Questions you have about the essay: in this section, address uncertainties you had with the argument.
3) Areas for improving the essay: it is vitally important not only to detail problems with the essay in this section but also to offer suggestions for improvement. Place yourself in the author's shoes and be as specific as possible. **Please note:** the emphasis of this section should be on thematic, argumentative, and thesis issues; grammatical problems should be mentioned only briefly and can be detailed on the copy of author's rough draft.

Evaluation: critique will receive one of the following marks: +, √, −critiques modify the final class participation percentage you will receive according to the following format:

+	+2%
√	+0%
−	−2%

Appendix 6

Appendix 6

John's "I Want to Receive an A Checklist"

1) introduction: establish an interesting and difficult problem for your readers. You must provide a context for readers to understand why your thesis is so significant.

2) thesis: provide your readers with a theory to explain or resolve the problem you have established.

3) topic sentences: make every paragraph begin with an assertion about why your theory explains the problem you are addressing better than other theories. Avoid plot summary.

4) relationship between topic sentences: each paragraph must tell your readers something about your theory they did not understand from previous paragraphs.

5) paragraphs: the body of each paragraph must explicitly clarify what you mean by your topic sentence and what it significance is.

6) quotations: to develop the significance of your claims, you need to draw evidence from the text, but the focus of your analysis must be on the significance of the quotations not on plot summary.

7) defining key terms: make sure that each of your key terms is clearly defined and define what the relationships between these terms are.

8) Proofreading: preferably, get at least one other person to help you with this task.

9) Correct formatting: see the MLA appendix I provided.

Notes

1. Although I am not unsympathetic to claims made by Sidney I. Dobrin and others that current social conditions prevent the possibility of an ideal translation of pedagogical theory to practice (147), it is crucial to develop theories that can guide pedagogy in the interim. Sidney I. Dobrin, "Paralogic Hermeneutic Theories, Power, and the Possibility for Liberating Pedagogies," in *Post-Process Theory: Beyond the Writing-Process Paradigm*, ed. Thomas Kent (Carbondale: Southern Illinois University Press, 1999), 132–148.

2. For a succinct description of post-positivist realism and its central tenets, see Moya's introduction to *Reclaiming Identity*. Paula M. L. Moya, "Introduction: Reclaiming Identity," in *Reclaiming Identity: Realist Theory and the Predicament of Postmodernism*, ed. Paula M. L. Moya and Michael Hames-García (Berkeley: University of California Press, 2000), 1–28.

3. Moya's assertion concerning the importance of acknowledging the situatedness of value represents one of eight recommendations she presents for creating learning environments in which the knowledge of all students is valued. Paula M. L. Moya, *Learning from Experience: Minority Identities, Multicultural Struggles* (Berkeley: University of California Press, 2002) 158–174.

4. Faigley's now classic article on process theory within composition studies provides an excellent summary of the various forms the theory has taken. Faigley identifies the three primary strands of process theory, which he refers to as the expressive, cognitive, and social views respectively. Lester Faigly, "Competing Theories of Process: A Critique and a Proposal," *College English*, 48.6 (1986): 527–541.

5. On this point, I am in agreement with Weaver, Straub, and others who caution against efforts to create purportedly "evaluation-free zones" in the classroom (Christopher C. Weaver, "Grading in a Process-based Writing Classroom," in *The Theory and Practice of Grading Writing: Problems and Possibilities*, eds. Frances Zak and Christopher C. Weaver [Albany: State University of New York Press, 1998], 143) or to distinguish between "directive" versus "facilitative" commentary on student assignments (Richard Straub, "The Concept of Control in Teacher Response: Defining the Varieties of "Directive" and "Facilitative" Commentary" in *College Composition and Communication*, 47.2 [1996]: 224). As these scholars suggest, all interactions with students, whether written or oral, inevitably involve a power hierarchy that cannot altogether be eliminated. To claim otherwise would be to mislead students.

6. For two examples of grading criteria for critical essays that do use the format of presenting the defining characteristics of each grade, see Xin Liu Gale, "Judgment Deferred: Reconsidering Institutional Authority in the Portfolio Writing Classroom," in *Grading in the Post-process Classroom: From Theory to Practice*, eds. Libby Allison, Lizbeth Bryant, and Maureen Hourigan (Portsmouth: Boynton/Cook, 1997), 91–93; and Kathleen and James Strickland, "Grades for Work: Giving Value for Value," in *Grading in the Post-process Classroom: From Theory to Practice*, eds. Libby Allison, Lizbeth Bryant, and Maureen Hourigan (Portsmouth: Boynton/Cook, 1997), 148–149.

7. Smith provides a very useful caution about becoming too formulaic in the comments instructors provide one student essays. For two excellent essays on improving written commentaries for students, see Summer Smith, "The Genre of the End Comment: Conventions in Teacher Responses to Student Writing," *College Composition and Communication*, 48.2 (1997): 249–268; and Nancy Sommers, "Responding to Student Writing," *College Composition and Communication* 33.2 (1982): 148–156.
8. I am drawing the terminology of "pinpoint" and "discussion questions" from my colleague Andrew Sofer.
9. Telegraphic handouts such as appendix 6 need to be balanced by more extensive commentaries or descriptions of criteria, such as my earlier and more extensive description of how I evaluate critical essays (appendix 2). Otherwise, they may encourage a "rubric" mentality among students, that requirements can be fulfilled in a fairly cursory manner.

REFERENCES

Alcoff, Linda Martín. "Cultural Feminism versus Post-Structuralism: The Identity Crisis in Feminist Theory," in *The Second Wave: A Reader in Feminist Theory*, ed. Linda Nicholson. New York: Routledge, 1997, 330–355.

———. *Real Knowing: New Versions of the Coherence Theory*. Ithaca, NY: Cornell University Press, 1996.

Allison, Libby, Lizbeth Bryant, and Maureen Hourigan, eds. *Grading in the Post-process Classroom: From Theory to Practice*. Portsmouth: Boynton/Cook, 1997.

Bartholomae, David. "The Study of Error." *College Composition and Communication*, 31.3 (1980): 253–269.

Delpit, Lisa D. "The Silenced Dialogue: Power and Pedagogy in Educating Other People's Children." *Harvard Educational Review*, 58.3 (1988): 280–298.

Dobrin, Sidney I. "Paralogic Hermeneutic Theories, Power, and the Possibility for Liberating Pedagogies," in *Post-Process Theory: Beyond the Writing-Process Paradigm*, ed. Thomas Kent. Carbondale: Southern Illinois University Press, 1999, 132–148.

Faigley, Lester. "Competing Theories of Process: A Critique and a Proposal." *College English*, 48.6 (1986): 527–541.

Gale, Xin Liu. "Judgment Deferred: Reconsidering Institutional Authority in the Portfolio Writing Classroom," in *Grading in the Post-process Classroom: From Theory to Practice*, eds. Libby Allison, Lizbeth Bryant, and Maureen Hourigan. Portsmouth: Boynton/Cook, 1997, 75–93.

Guerra, Juan C. "The Place of Intercultural Literacy in the Writing Classroom," in *Writing in Multicultural Settings*, eds. Carol Severino, Juan C. Guerra, and Johnnella E. Butler. New York: Modern Language Association of America, 1997, 248–260.

Huot, Brian. "Toward a New Discourse of Assessment for the College Writing Classroom." *College English*, 65.2 (2002): 163–180.

Macdonald, Amie A., and Susan Sánchez-Casal, eds. *Twenty-first-Century Feminist Classrooms: Pedagogies of Identity and Difference*. New York: Palgrave Macmillan, 2002.

Mohanty, Satya P. *Literary Theory and the Claims of History: Postmodernism, Objectivity, Multicultural Politics*. Ithaca, NY: Cornell University Press, 1997.

Moya, Paula M. L. "Introduction: Reclaiming Identity," in *Reclaiming Identity: Realist Theory and the Predicament of Postmodernism*, ed. Paula M. L. Moya and Michael R. Hames-García. Berkeley: University of California Press, 2000, 1–28.

———. *Learning from Experience: Minority Identities, Multicultural Struggles*. Berkeley: University of California Press, 2002.

Olson, Gary A. "Toward a Post-process Composition: Abandoning the Rhetoric of Assertion," in *Post-process Theory: Beyond the Writing-Process Paradigm*, ed. Thomas Kent. Carbondale: Southern Illinois University Press, 1999, 7–15.

Putnam, Hilary. *Renewing Philosophy*. Cambridge, MA: Harvard University Press, 1992.

Schiffman, Betty Garrison. "Grading Student Writing: The Dilemma from a Feminist Perspective," in *Grading in the Post-process Classroom: From Theory to Practice*, eds. Libby

Allison, Lizbeth Bryant, and Maureen Hourigan. Portsmouth: Boynton/Cook, 1997, 58–72.

Smith, Summer. "The Genre of the End Comment: Conventions in Teacher Responses to Student Writing." *College Composition and Communication*, 48.2 (1997): 249–268.

Sommers, Nancy. "Responding to Student Writing." *College Composition and Communication*, 33.2 (1982): 148–156.

Straub, Richard. "The Concept of Control in Teacher Response: Defining the Varieties of 'Directive' and 'Facilitative' Commentary." *College Composition and Communication*, 47.2 (1996): 223–251.

Strickland, Kathleen, and James Strickland. "Grades for Work: Giving Value for Value," in *Grading in the Post-process Classroom: From Theory to Practice*, eds. Libby Allison, Lizbeth Bryant and Maureen Hourigan. Portsmouth: Boynton/Cook, 1997, 141–156.

Weaver, Christopher C. "Grading in a Process-based Writing Classroom," in *The Theory and Practice of Grading Writing: Problems and Possibilities*, eds. Frances Zak and Christopher C. Weaver. Albany: State University of New York Press, 1998, 141–150.

Williams, Joseph M. "The Phenomenology of Error." *College Composition and Communication*, 32.2 (1981): 152–168.

Zak, Frances, and Christopher C. Weaver, eds. *The Theory and Practice of Grading Writing: Problems and Possibilities*. Albany: State University of New York Press, 1998.

Index

academic freedom, 163
access
   to educational resources, 2, 3, 11, 35, 37, 66
   to higher education, 1–4, 10–12, 196
   to social and economic resources, 2, 3, 196, 200
   see also critical access, epistemic access
accessibility
   and empathy, 86
additive approach, 15–16, 111, 142
   see also pluralist approach
affirmative action, 2, 10, 26, 103, 156, 225
Alcoff, Linda, 47, 49–50, 171, 244, 255, 259–61
   Visible Identities, 49–50
"American literature," 5–6, 103, 106–7
   selection of, 15, 107–10, 172
"amnesiac creolity," 137
Appiah, Kwame Anthony, 232, 238–9, 243, 246

Bell Curve, 47
bias
   experiential, 21
   in grading, 252–5, 260–1, 263
   racial, 2, 67
   theoretical, 20, 21, 260
Bible, 6, 212, 215–21
   and homosexuality, 219–20
   and slavery, 218–19
"bootstraps," 47, 162
Brown v. Board of Education, 10, 32

capitalism, 162, 163, 193
   racist effects of, 20, 22
   sexist structure of, 22

Census, 145, 148
   "Other" category in, 136–7
Christian, -ity, 6, 197–8, 211–12, 214–17, 219
Chronicle of Higher Education, 103
civil rights activism, 10, 29, 120, 131, 134–5, 143–5, 147, 152
Civil Rights Act of 1964, 10
colonization, 13, 111, 135, 230
   economic consequences of, 20
color-blind approach, 13, 63, 235–6
communities of conflict, 112
communities of meaning, 5, 6, 25–31, 38–9, 59–61, 64, 97, 154, 158, 181–2, 212–17, 220–1
   definition of, 25
conflict
   adjudicating, 60
   racial, 91, 118
consent of the governed, 18–19
Cornell University, 156–7, 169
"covering," 133–5
critical access, 2–6, 12–14, 18–20, 24, 27, 31, 66, 68, 89–90, 96, 151, 172, 182
   objectives, 2–3
   theory of, 2
Critical Race Pedagogy (CRP), 70, 72, 88–91, 93
Critical Race Theory (CRT), 70, 88–90, 93
critical theory of difference, 4, 88, 91, 93
cross-racial developmental relationships
   see under developmental relationships
cultural awareness, 165–6, 238
"cultural change," 144
"cultural competency," 165–6
cultural relativism, 5, 111, 113, 119
"cultural taxation," 67, 80–1, 96

Index

"culture of power," 14, 35, 259, 262
"cyborg feminism," 153

DaCosta, Kimberly McClain, 135, 137
democracy, 2, 3, 52–3, 57, 105, 110, 193, 206
   racial, 1, 2, 4, 9, 10, 11, 13, 14, 20, 27–9, 31, 33–4, 89, 97
democratization, 1, 4, 10, 11, 13, 15–20, 22–3, 27, 36, 39, 229
   of classroom, 3, 4, 6, 172, 205, 211, 251–2
denial and suppression, 71, 91
desegregation, 10
developmental relationships, 68–9, 79, 94–6
   cross-racial, 70–2, 88, 91, 93, 95
Díaz, Junot, 28–30
direct engagement, 71
diversity
   cultural, 5, 110, 151, 175, 226–8, 231
   "risk-free," 111, 142
Du Bois, W.E.B., 111, 114–15, 118, 119, 138

education
   democratize(d), *see* democratization
   multicultural, 9, 45, 53, 57, 63, 109–13, 142, 144, 165–7, 232, 237–8
   politicize(d), 103
   relationship to democracy, 206
Emerson, Ralph Waldo, 106–7, 119, 123, 126
empathy, 83, 86–91, 95, 160, 164, 205, 227, 238, 243, 246
English as a Foreign Language (EFL), 226–43
   theories of intercultural learning, 240–2
   tourist ideal of, 228
epistemic
   access, 16
   advantage, 17, 24
   authority, 27
   collectives, 27
   contact zones, 31
   function of social identity, 16
   resources, 4, 45, 49–52
   tools, subversive, 27
   see also related entries under identity
equality
   structural, 3
   see also inequality
equal rights, 20
error, 255–63
   analysis of, 257–9, 262
essentialism, 36, 37, 154, 181, 243, *see also* essentialist identity
ethnicity
   constructivist theories of, 154
   invention of, 193
Ethnic Studies Requirement (ESR), 5, 151, 155–67
ethnocentric, -ism, 119, 151, 156–8, 162–6, 225
Eurocentric curriculum(a), 2, 12, 14–15

fluency, 256, 262
"forced habitus," 234
Future of Minority Studies (FMS), 49

"gender tax," 80–1, *see also* "cultural taxation"
"glass ceiling," 196
grading, 6, 251–70
   realist methodology of, 252–3
   in the sciences, 253–5
Graff, Gerald, 112–13, 119, 124, 125
Grutter v. Bollinger, 10

Halfbreed, 191–2, 195–201, 205
Hames-García, Michael, 5, 15, 28, 52, 141–2, 172, 183, 203–4, 255
Harper, Michael, 119–20
Harper anthology, 115
Harvard, 4, 65–97
"hidden curriculum," 66, 68
"hidden service agenda," 67
Hughes, Langston, 140

identity, 3–4, 6, 16–20, 45–52, 54–6, 87–8, 137, 171–2, 192, 198, 212–14, 232–3, 243, 261
   "American," 104, 116, 119
   choosing, 199, 200–2, 205
   class, 196
      socioeconomic, 81–2
   in the classroom, 3, 16, 24, 27, 191–2

INDEX

conflictive, 228
contextual nature of, 48–9
contingencies, 45, 48–9, 52, 57, 65
deconstruction of, 152
defensive, 145
denaturalizing, 58–9, 237
dialectical concept of, 46, 49, 61
   ascriptive, 46–7, 53, 55
   subjective, 46–7
epistemic consequences of, 25, 31
epistemic significance of, 4, 16, 18, 20, 23, 24, 95, 181, 232, 235, 237
essentialist, 47, 167
gender, 39, 81, 212
idealist, 47
inessential, 194
inherited, 17
"invisible," 55
and knowledge, *see* knowledge
literary construction of, 182
and literature, 238–40
"mainstream," 51–2
minority, 6, 152, 171, 177, 182–3, 243, 244
"mixed race," 5, 63, 139, 145, 198
mobilizing, 4, 56–62, 243, 259
national, 122, 201
"negative," 195
oppressed, 25, 213
pedagogy and, 4, 16–20
political, 17, 194, 197, 201, 205
politics of, 5, 50, 108–9, 124, 125–6, 141–5, 191, 235
   neoconservative, 143
postcolonial, 171
poverty and, 196
privileged, 24, 213
racial, 4, 12, 16, 21–2, 24, 25, 31, 55, 79, 140, 143, 194, 200, 202, 203, 212
   positive, 24
   white, 5–6, 21–2
realist, 48–9, 65, 167, 181, 183
realist approach to, 5, 16, 36, 46, 65, 183, 191
religious, 6, 197–8, 211, 221
sexual, 29, 39, 155
social, 4, 16–19, 23, 24, 26, 36, 39, 50, 87, 192, 203

epistemic function of, 16, 18, 38, 181, 235–6, 239
suppressed, 191, 194, 205
undertheorized, 23
identity blindness, 5, 49, 153, 156, 162–3, 165, *see also* color-blind approach
inequality, 26, 79, 145, 162–3, 233–4, 237
   in access to education, 1, 10–11, 67
   in classroom, 18
   economic, 6, 237
   gender, 26
   in interaction, 13
   racial, 9, 11, 21, 26, 88, 237
   social, 1–4, 113
integrated classrooms, 52
integrationist policies
   limits of, 11–12, 13
intellectual cooperation, 51, 60
intercultural communication, 228–9, 236, 239, 240
intercultural communicative competence, 228–9, 244
intercultural competence, 227–8, 232, 237–8
interculturality, 234, 240
intercultural learning, 237, 240, 243
interracial
   adoption, 131, 146
   interaction, 23, 34, 236, 240
   subjects in the marketplace, 135
   union, 131, 135, 144
Invisible Man, 109–10, 113–14, 207
Iroquois Confederacy, 19

Jefferson, Thomas, 153
Jim Crow laws, 154

knowledge
   context-specific, 260
   empirical, 50
   experiential, 17, 20, 30, 31, 96, 154
   false, 22
   and identity, 23, 24, 25, 27, 29, 30, 51, 64, 172, 174, 178, 183, 191, 256
   ideological, 21

knowledge—*Continued*
  production of, 23, 25, 51, 52, 243, 260
    collective, 181, 192
    and identity, 23
    situated, 50, 256, 257
    subjugated, 30, 154
    theoretical, 30
    and values, 256–7

Levinson, Daniel, 68, 71, 92
literature
  intercultural potential of, 242
  minority, *see* minority texts
  mixed race, 5, 134, 138–41, 144
  multicultural, 172, 237
  resistance, 161
Loving v. Virginia, 144
lynching, 114, 116–18

Macdonald, Amie, 5–6, 154, 158, 172, 178, 181, 182–3, 194, 212–14, 251, 259
majority-minority, 12, 26
Manifest Destiny, 162
Martí, José, 15, 104–7, 110–12, 115–19
Martial India, 171, 174–7, 179–80, 182
mentoring, 4–5, 68–70, 92–7
  case studies in, 74–9
  "constellation," 94
  cross-racial, 67, 79–80, 91
mentors, 66–8
meritocracy
  American, 23, 26, 162, 172
  British Army, 177, 178, 183
Michaels, Walter Ben, 153–6, 167
microaggression(s)
  racial, 68, 71, 72, 89, 94, 95
microaggressive behaviors, 81
microaggressive situations, 89, 90, 94
Mills, Charles, 9
minimalist autonomy, 54
minoritized, 80, 86, 97
minority texts, 5, 12, 14, 161, 172, 192–3, 229
mixed race, 5, 131–45
  advocacy, 134, 137–9, 143, 145
  canonization of, 137–41
  education, 131–3, 137, 142, 144

  oppression of, 142–3
  -targeted marketing, 135–6
Mohanty, Satya, 16, 17, 28, 30, 36, 46, 119, 125, 171, 244, 255, 256, 258
Morrison, Toni, 134, 163, 233, 236
Moya, Paula, 4, 9, 14, 23, 65, 87, 90, 97, 111–13, 143–4, 171–2, 179, 192, 231, 232–3, 235, 237–9, 242, 255, 257
mulatto, 138, 143
multicultural approach, 243
  v. color-blind approach, 235–6
multicultural "doctrine," 225
multiculturalism, 110, 165, 171–5, 177–8, 182–3, 230–4, 240–3, 244
  additive approach to, 111
  martial mode of, 171, 173, 175, 183, 184
  v. intercultural learning, 234–7
multicultural theories, 227, 231, 235–8
multiple group membership, 203
"multiplicity," 28, 39, 203–4, 225

Native American history, 18–20
"native informants," 56, 81, 143
"not in my backyard" phenomenon, 61

objectivity, 18, 255–6
  positivist notion of, 255
  post-positivist notion of, 252, 260
Ondaatje, Michael
  The English Patient, 171–83
    postcolonial sensibility of, 173, 176–8
oppression, 17, 20, 39, 56, 68, 91, 97, 108, 110–14, 142, 154, 212, 213, 217
  effects of, commutative, 88
  hierarchies of, 89
  internalized, 24–5
  racial, 20, 70, 88–9, 120
Orientalism, 173
Other, 160, 164
Othering, 55–6, 80, 82–7, 97, 239

pedagogy, 112, 133, 211, 231, 258, 259
  anti-bias, 237
  ideal, 262

INDEX

multicultural,-ist, 113, 181
oppositional, 132
process theory, 259
radical, 113
realist, 4–5, 16–18, 20–1, 23–5, 28, 36, 37, 171–2, 182–3, 194, 251–3
transgressive, *see* oppositional *under this main level*
see also Critical Race Pedagogy (CRP)

perspectival difference, 49, 53
Plessy v. Ferguson, 10, 156
pluralist approach, 112, 113, 115
"positional perspective," 261
positivism, 252, 255, 256, 260
post-positivist
  objectivity, 252, 260
  realism, 141, 192, 221, 270
"post-process" theorists, 262
predominantly white institution(s), 4, 5, 13, 25, 26, 65, 66, 68, 69, 80, 88, 90, 93, 96, 97, 103, 151, 156, 159, 166
presentism, 138
"progressive" consciousness, 152, 168

race
  commercialization of, 140
  consciousness, 21, 153
  as a construct, 204
  race-based evaluations, 202
  race-based housing, 13, 35
  race-based society, 199
racial
  contract, 9, 14
  democracy, *see* democracy
  freedom, 31
  hierarchies, 13, 131, 163, 237
  identity, *see* identity
  integration, 10, 34
  oppression, 20, 70, 88, 89, 120
  particularism, 141
  privilege, 1, 9, 142, 194, 205
    white, 12–13, 15, 22, 23, 24, 200, 202
  social structure, 1, 11, 160
racially undemocratic, 3, 9, 11, 35
racism, 9–10, 14, 16–18, 22, 24, 26, 28–30, 117–18, 153–4, 163

  legacy of, 12, 13
  resistance to, 18
  structural, 1, 11
  white, 20–3
rates
  dropout, 11
  enrollment, 11
  graduation, 11
realism, 3, 17, 21, 38, 144, 152, 164, 167, 192, 204, 216, 221, 262
realist classroom, 16–17, 22–5, 27, 172
realist theory, 161, 171, 251, 255–6, 262–3
reductionism, 243
Regents of the University of California v. Bakke, 10
"restriction," 203–4
Riggs, Marlon, 29–30
Roth, Philip
  The Human Stain, 236, 239
Rousseau
  social contract theory, 18–19

Sánchez-Casal, Susan, 5–6, 37, 59, 154, 158, 172, 178, 181, 183, 194, 212–14, 251, 259
schools
  integrated, 52
  public v. private, 54
selfhood
  theories of, 154–6
self-racialization, 204
slavery, 10, 46, 104, 163, 218–19, 220
Social Darwinism, 162
social hierarchy, -ies, 26, 131, 153, 228
socialization, 55, 56
social location, 17, 20, 25–6, 28, 30, 39, 58, 64, 95, 152–6, 158–61, 181, 191, 203, 213, 214, 217, 237, 255
"soft skills," 93
standpoint epistemology, 212–14
Stanford University, 48, 74, 133
Steele, Claude, 45, 48, 63
stereotype threat, 48–9, 57, 68
Stowe, Harriet Beecher, 113–14
student-professor hierarchy, 94

subjectivity, 46–7
"survival" behaviors, 85

University of Michigan, 10
University of Wisconsin, 156

visual fetishism, 55–6
vouchers, school, 54

welfare
    racist politics surrounding, 21–2
"welfare queens," 21, 22, 24
whiteness, 193–200, 202, 204–5
Williams, William Carlos,
    109, 111

Yale University, 74, 133